# *Handbook of*
# BACTERIOLOGY

**Dr. Amrita Rohilla**

Oxford Book Company
Jaipur ▪ New Delhi

*Published By*

Oxford Book Company

**Regd. Off.** "Bony Residency", Gate No. 2
Opp. Tilak Public School, Vishweswariya Nagar,
Gopalpura Road, Jaipur - 302018, India
Ph: 0141 - 2761280, Telefax: 2761381
email - oxfordbookjpr@gmail.com
website - oxfordbookcompany.com

**Delhi Office:** 102, 1st Floor, Satyam House,
4327/3, Ansari Road, Darya Ganj,
New Delhi - 110002, India
Ph: 011-45652440

Published in India, 2025



ISBN : 9789380179476

Price : ₹ 2250/-



Typeset by Shivangi Computers, Jaipur

Printed and bound in India by Thomson Press Ltd.,
New Delhi

# Preface

Bacteriology, the branch of biology devoted to the study of bacteria. Bacteriologists are concerned with the structure and functions of bacteria, with the environmental conditions that affect bacterial growth, with the effects of bacteria on the other organisms, and with the uses for bacteria.

Bacteriology is important in medicine, public health and sanitation work, agriculture, food-processing, and industry. There are many branches of this science, each dealing with special problems that have arisen from the complex effects that bacteria have upon civilization.

The relation of bacteria to disease has given rise to the sciences of pathologic bacteriology (concerning treatment and causes of disease), immunology (concerning immunity to disease), antiseptic surgery, veterinary bacteriology, and sanitary bacteriology.

***Dr. Amrita Rohilla***

# Contents


| | | |
|---|---|---|
| | *Preface* | *v* |
| 1. | Introduction to Bacteriology | 1 |
| 2. | Structure | 4 |
| 3. | Classification | 22 |
| 4. | Bacterial Metabolism | 41 |
| 5. | Genetics | 68 |
| 6. | Normal Flora | 112 |
| 7. | Bacterial Pathogenesis | 122 |
| 8. | Specific Acquired Immunity | 143 |
| 9. | Epidemiology | 151 |
| 10. | Principles of Diagnosis | 164 |
| 11. | Antimicrobial Chemotherapy | 177 |
| 12. | Nutrition and Cultivation of Bacteria | 207 |
| 13. | Bacterial Identification | 218 |
| 14. | Staphylococcus | 224 |
| 15. | Streptococcus | 239 |
| 16. | Neisseria, Moraxella, Kingella and Eikenella | 261 |
| 17. | Bacillus | 281 |
| 18. | Miscellaneous Pathogenic Bacteria | 293 |
| | *Index* | 302 |

## Chapter 1

# Introduction to Bacteriology

Bacteria are single-celled microorganisms that lack a nuclear membrane, are metabolically active and divide by binary fission. Medically they are a major cause of disease. Superficially, bacteria appear to be relatively simple forms of life; in fact, they are sophisticated and highly adaptable. Many bacteria multiply at rapid rates, and different species can utilize an enormous variety of hydrocarbon substrates, including phenol, rubber, and petroleum. These organisms exist widely in both parasitic and free-living forms. Because they are ubiquitous and have a remarkable capacity to adapt to changing environments by selection of spontaneous mutants, the importance of bacteria in every field of medicine cannot be overstated.

The discipline of bacteriology evolved from the need of physicians to test and apply the germ theory of disease and from economic concerns relating to the spoilage of foods and wine. The initial advances in pathogenic bacteriology were derived from the identification and characterization of bacteria associated with specific diseases. During this period, great emphasis was placed on applying Koch's postulates to test proposed cause-and-effect relationships between bacteria and specific diseases. Today, most bacterial diseases of humans and their etiologic agents have been identified, although important variants continue to evolve and sometimes emerge, e.g., Legionnaire's Disease, tuberculosis and toxic shock syndrome.

Major advances in bacteriology over the last century resulted in the development of many effective vaccines (e.g., pneumococcal polysaccharide vaccine, diphtheria toxoid, and tetanus toxoid) as well as of other vaccines (e.g., cholera, typhoid, and plague vaccines) that are less effective or have side effects. Another major advance was the discovery of antibiotics. These antimicrobial substances have not eradicated bacterial diseases, but they are powerful therapeutic tools. Their efficacy is reduced by the emergence of antibiotic resistant bacteria (now an important medical management problem) In reality, improvements in sanitation and water purification have a greater effect on the incidence of bacterial infections in a community than does the availability of antibiotics or bacterial vaccines. Nevertheless, many and serious bacterial diseases remain.

Most diseases now known to have a bacteriologic etiology have been recognized for hundreds of years. Some were described as contagious in the writings of the ancient Chinese, centuries prior to the first descriptions of bacteria by Anton van Leeuwenhoek in 1677. There remain a few diseases (such as chronic ulcerative colitis) that are thought by some investigators to be caused by bacteria but for which no pathogen has been identified.

Occasionally, a previously unrecognized diseases is associated with a new group of bacteria. An example is Legionnaire's disease, an acute respiratory infection caused by the previously unrecognized genus, Legionella. Also, a newly recognized pathogen, *Helicobacter*, plays an important role in peptic disease. Another important example, in understanding the etiologies of venereal diseases, was the association of at least 50 per cent of the cases of urethritis in male patients with *Ureaplasma urealyticum* or *Chlamydia trachomatis*.

Recombinant bacteria produced by genetic engineering are enormously useful in bacteriologic research and are being employed to manufacture scarce biomolecules (e.g. interferons) needed for research and patient care.

The antibiotic resistance genes, while a problem to the physician, paradoxically are indispensable markers in performing genetic engineering. Genetic probes and the polymerase chain reaction (PCR) are useful in the rapid identification of microbial pathogens in patient specimens.

Genetic manipulation of pathogenic bacteria continues to be indispensable in defining virulence mechanisms. As more protective protein antigens are identified, cloned, and sequenced, recombinant bacterial vaccines will be constructed that should be much better than the ones presently available. In this regard, a recombinant-based and safer pertussis vaccine is already available in some European countries. Also, direct DNA vaccines hold considerable promise.

In developed countries, 90 per cent of documented infections in hospitalized patients are caused by bacteria. These cases probably reflect only a small percentage of the actual number of bacterial infections occurring in the general population, and usually represent the most severe cases.

In developing countries, a variety of bacterial infections often exert a devastating effect on the health of the inhabitants. Malnutrition, parasitic infections, and poor sanitation are a few of the factors contributing to the increased susceptibility of these individuals to bacterial pathogens. The

World Health Organization has estimated that each year, 3 million people die of tuberculosis, 0.5 million die of pertussis, and 25,000 die of typhoid. Diarrheal diseases, many of which are bacterial, are the second leading cause of death in the world (after cardiovascular diseases), killing 5 million people annually.

Many bacterial diseases can be viewed as a failure of the bacterium to adapt, since a well-adapted parasite ideally thrives in its host without causing significant damage. Relatively nonvirulent (i.e., well-adapted) microorganisms can cause disease under special conditions - for example, if they are present in unusually large numbers, if the host's defenses are impaired, (e.g., AIDS and chemotherapy) or if anaerobic conditions exist. Pathogenic bacteria constitute only a small proportion of bacterial species; many nonpathogenic bacteria are beneficial to humans (i.e. intestinal flora produce vitamin K) and participate in essential processes such as nitrogen fixation, waste breakdown, food production, drug preparation, and environmental bioremediation. This textbook emphasizes bacteria that have direct medical relevance.

In recent years, medical scientists have concentrated on the study of pathogenic mechanisms and host defenses. Understanding host-parasite relationships involving specific pathogens requires familiarity with the fundamental characteristics of the bacterium, the host, and their interactions.

Therefore, this section first presents with the basic concepts of the immune response, bacterial structure, taxonomy, metabolism, and genetics. Subsequent chapters emphasize normal relationships among bacteria on external surfaces; mechanisms by which microorganisms damage the host; host defense mechanisms; source and distribution of pathogens (epidemiology); principles of diagnosis; and mechanisms of action of antimicrobial drugs.

These chapters provide the basis for the next chapters devoted to specific bacterial pathogens and the diseases they cause. The bacteria in these chapters are grouped on the basis of physical, chemical, and biologic characteristics. These similarities do not necessarily indicate that their diseases are similar; widely divergent diseases may be caused by bacteria in the same group.

## Chapter 2

# Structure

All bacteria, both pathogenic and saprophytic, are unicellular organisms that reproduce by binary fission. Most bacteria are capable of independent metabolic existence and growth, but species of Chlamydia and Rickettsia are obligately intracellular organisms.

Bacterial cells are extremely small and are most conveniently measured in microns (10–6 m). They range in size from large cells such as Bacillus anthracis (1.0 to 1.3 μm × 3 to 10 μm) to very small cells such as Pasteurella tularensis (0.2 × 0.2 to 0.7 μm) Mycoplasmas (atypical pneumonia group) are even smaller, measuring 0.1 to 0.2 μm in diameter. Bacteria therefore have a surface-to-volume ratio that is very high: about 100,000.

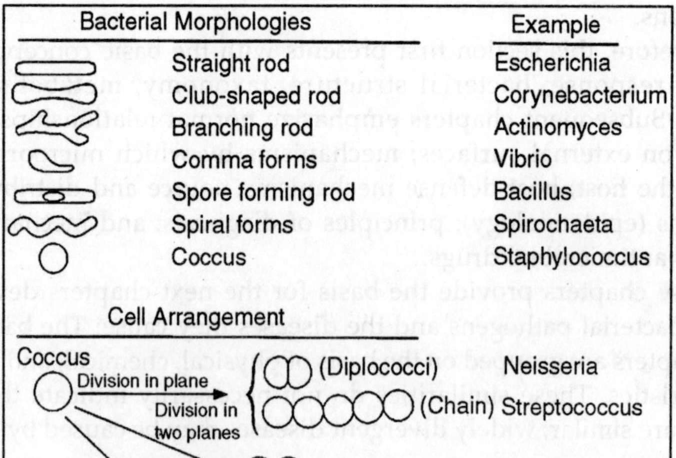


**Fig.** Typical Shapes and Arrangements of Bacterial Cells.

Bacteria have characteristic shapes. The common microscopic

morphologies are cocci (round or ellipsoidal cells, such as Staphylococcus aureus or Streptococcus respectively); rods, such as Bacillus and Clostridium species; long, filamentous branched cells, such as Actinomyces species; and comma-shaped and spiral cells, such as Vibrio cholerae and Treponema pallidum, respectively.

The arrangement of cells is also typical of various species or groups of bacteria. Some rods or cocci characteristically grow in chains; some, such as Staphylococcus aureus, form grapelike clusters of spherical cells; some round cocci form cubic packets. Bacterial cells of other species grow separately. The microscopic appearance is therefore valuable in classification and diagnosis.

## THE NUCLEOID

Prokaryotic and eukaryotic cells were initially distinguished on the basis of structure: the prokaryotic nucleoidthe equivalent of the eukaryotic nucleusis structurally simpler than the true eukaryotic nucleus, which has a complex mitotic apparatus and surrounding nuclear membrane. As the electron micrograph in Fig. shows, the bacterial nucleoid, which contains the DNA fibrils, lacks a limiting membrane. Under the light microscope, the nucleoid of the bacterial cell can be visualized with the aid of Feulgen staining, which stains DNA.

Gentle lysis can be used to isolate the nucleoid of most bacterial cells. The DNA is then seen to be a single, continuous, "giant" circular molecule with a molecular weight of approximately 3 × 109.

The unfolded nuclear DNA would be about 1 mm long (compared with an average length of 1 to 2 μm for bacterial cells). The bacterial nucleoid, then, is a structure containing a single chromosome. The number of copies of this chromosome in a cell depends on the stage of the cell cycle (chromosome replication, cell enlargement, chromosome segregation, etc). Although the mechanism of segregation of the two sister chromosomes following replication is not fully understood, all of the models proposed require that the chromosome be permanently attached to the cell membrane throughout the various stages of the cell cycle.

Bacterial chromatin does not contain basic histone proteins, but low-molecular-weight polyamines and magnesium ions may fulfill a function similar to that of eukaryotic histones. Despite the differences between prokaryotic and eukaryotic DNA, prokaryotic DNA from cells infected with bacteriophage g, when visualized by electron microscopy, has a beaded, condensed appearance not unlike that of eukaryotic chromatin.

## SURFACE APPENDAGES

Two types of surface appendage can be recognized on certain bacterial species: the flagella, which are organs of locomotion, and pili (Latin hairs), which are also known as fimbriae (Latin fringes). Flagella occur on both Gram-positive and Gram-negative bacteria, and their presence can be useful in identification. For example, they are found on many species of bacilli but rarely on cocci. In contrast, pili occur almost exclusively on Gram-negative bacteria and are found on only a few Gram-positive organisms (e.g., Corynebacterium renale).

Some bacteria have both flagella and pili. The electron micrograph in Fig. shows the characteristic wavy appearance of flagella and two types of pili on the surface of Escherichia coli.

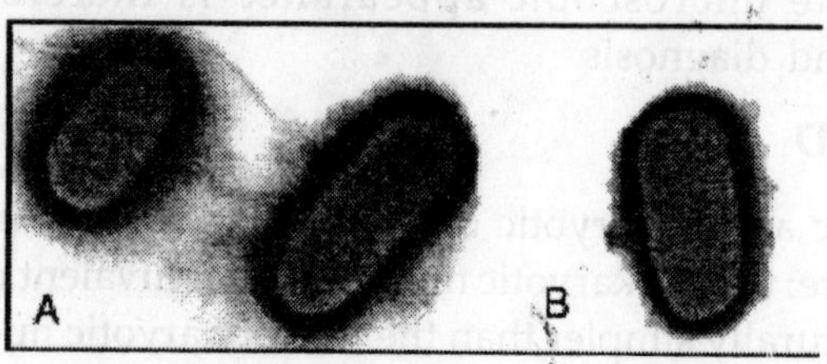


**Fig.** (A) Electron Micrograph of Negatively Stained E Coli Showing Wavy Flagella and Numerous Short, Thinner, and More Rigid Hairlike Structures, the Pili. (B) The Long Sex Pilus can be Distinguished from the Shorter Common Pili by Mixing E Coli Cells with a Male Bacteriophage that Binds Specifically to Sex Pili.

## FLAGELLA

Structurally, bacterial flagella are long (3 to 12 μm), filamentous surface appendages about 12 to 30 nm in diameter. The protein subunits of a flagellum are assembled to form a cylindrical structure with a hollow core. A flagellum consists of three parts: (1) the long filament, which lies external to the cell surface; (2) the hook structure at the end of the filament; and (3) the basal body, to which the hook is anchored and which imparts motion to the flagellum. The basal body traverses the outer wall and membrane structures.

It consists of a rod and one or two pairs of discs. The thrust that propels the bacterial cell is provided by counterclockwise rotation of the basal body, which causes the helically twisted filament to whirl. The movement of the basal body is driven by a proton motive force rather than by ATP directly.

The ability of bacteria to swim by means of the propeller-like action of the flagella provides them with the mechanical means to perform chemotaxis (movement in response to attractant and repellent substances in the environment).

Response to chemical stimuli involves a sophisticated sensory system of receptors that are located in the cell surface and/or periplasm and that transmit information to methyl-accepting chemotaxis proteins that control the flagellar motor. Genetic studies have revealed the existence of mutants with altered biochemical pathways for flagellar motility and chemotaxis. Chemically, flagella are constructed of a class of proteins called flagellins. The hook and basal-body structures consist of numerous proteins.

Mutations affecting any of these gene products may result in loss or impairment of motility. Flagellins are immunogenic and constitute a group of protein antigens called the H antigens, which are characteristic of a given species, strain, or variant of an organism. The species specificity of the flagellins reflects differences in the primary structures of the proteins. Antigenic changes of the flagella known as the phase variation of H1 and $H_2$ occurs in Salmonella typhimurium.

The number and distribution of flagella on the bacterial surface are characteristic for a given species and hence are useful in identifying and classifying bacteria. Figure illustrates typical arrangements of flagella on or around the bacterial surface. For example, V cholerae has a single flagellum at one pole of the cell (i.e., it is monotrichous), whereas Proteus vulgaris and E coli have many flagella distributed over the entire cell surface (i.e., they are peritrichous). The flagella of a peritrichous bacterium must aggregate as a posterior bundle to propel the cell in a forward direction. Flagella can be sheared from the cell surface without affecting the viability of the cell. The cell then becomes temporarily nonmotile. In time it synthesizes new flagella and regains motility. The protein synthesis inhibitor chloramphenicol, however, blocks regeneration of flagella.

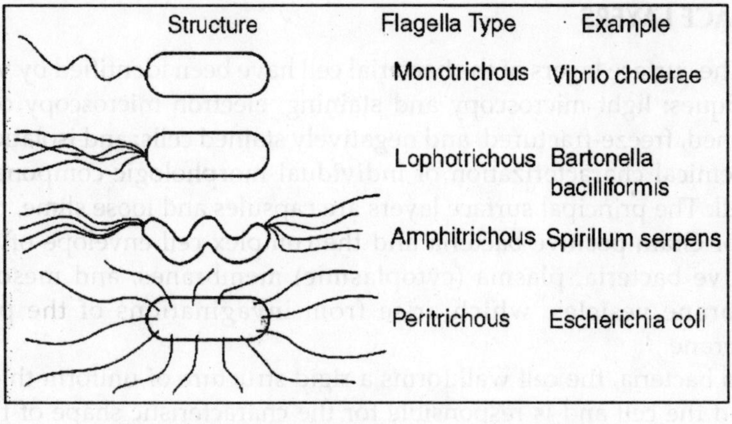


**Fig.** Typical Arrangements of Bacterial Flagella.

## PILI

The terms pili and fimbriae are usually used interchangeably to describe the thin, hairlike appendages on the surface of many Gram-negative bacteria and proteins of pili are referred to as pilins.

Pili are more rigid in appearance than flagella. In some organisms, such as Shigella species and E coli, pili are distributed profusely over the cell surface, with as many as 200 per cell.

As is easily recognized in strains of E coli, pili can come in two types: short, abundant common pili, and a small number (one to six) of very long pili known as sex pili. Sex pili can be distinguished by their ability to bind male-specific bacteriophages (the sex pilus acts as a specific receptor for these bacteriophages). The sex pili attach male to female bacteria during conjugation.

Pili in many enteric bacteria confer adhesive properties on the bacterial cells, enabling them to adhere to various epithelial surfaces, to red blood cells (causing hemagglutination), and to surfaces of yeast and fungal cells. These adhesive properties of piliated cells play an important role in bacterial colonization of epithelial surfaces and are therefore referred to as colonization factors.

The common pili found on E coli exhibit a sugar specificity analogous to that of phytohemagglutinins and lectins, in that adhesion and hemagglutinating capacities of the organism are inhibited specifically by mannose. Organisms possessing this type of hemagglutination are called mannose-sensitive organisms. Other piliated organisms, such as gonococci, are adhesive and hemagglutinating, but are insensitive to the inhibitory effects of mannose. Extensive antigenic variations in pilins of gonococci are well known.

## SURFACE LAYERS

The surface layers of the bacterial cell have been identified by various techniques: light microscopy and staining; electron microscopy of thin-sectioned, freeze-fractured, and negatively stained cells; and isolation and biochemical characterization of individual morphologic components of the cell. The principal surface layers are capsules and loose slime, the cell wall of Gram-positive bacteria and the complex cell envelope of Gram-negative bacteria, plasma (cytoplasmic) membranes, and mesosomal membrane vesicles, which arise from invaginations of the plasma membrane.

In bacteria, the cell wall forms a rigid structure of uniform thickness around the cell and is responsible for the characteristic shape of the cell (rod, coccus, or spiral). Inside the cell wall (or rigid peptidoglycan layer) is the plasma (cytoplasmic) membrane; this is usually closely apposed to the wall layer. The topographic relationships of the cell wall and envelope layers to the plasma membrane are indicated in the thin section of a Gram-positive organism (Micrococcus lysodeikticus) in Figure A and in the freeze-fractured cell of a Gram-negative organism (Bacteroides melaninogenicus) in Figure B. The latter shows the typical fracture planes seen in most Gram-negative bacteria, which are weak cleavage planes through the outer membrane of the envelope and extensive fracture planes through the bilayer region of the underlying plasma membrane.

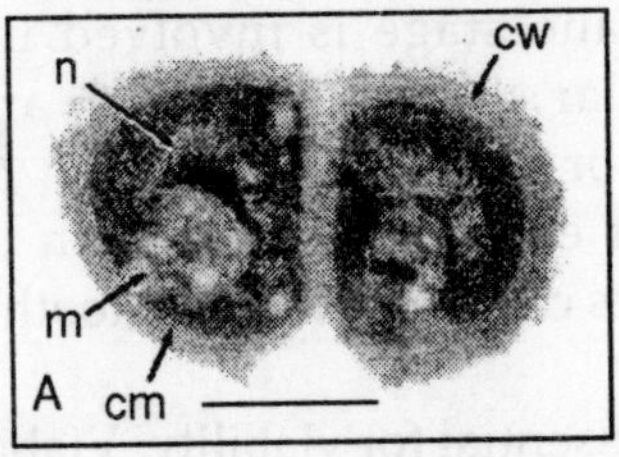


**Fig.** (A) Electron Micrograph of a Thin Section of the Gram-positive M Lysodeikticus Showing the Thick Peptidoglycan Cell Wall (cw), Underlying Cytoplasmic (Plasma) Membrane (cm), Mesome (m), and Nucleus (n).

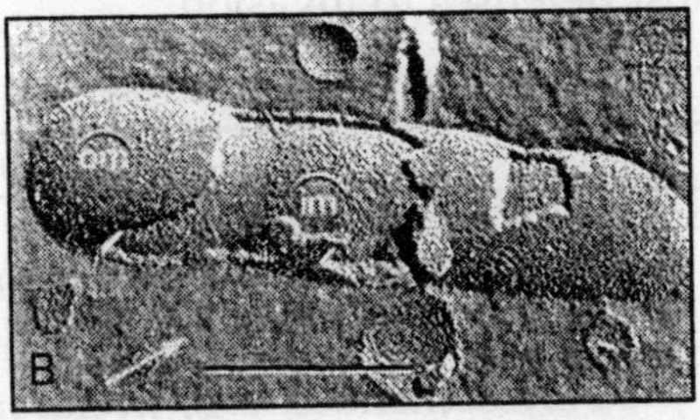


**FIg.**(B) Freeze-fractured Bacteriodes cell Showing Typical Major Convex Fracture Faces Through the Inner (im) and Outer (om) Membranes. Bars = 1 μm; Circled Arrow in Fig. B Indicates Direction of Shadowing.

## CAPSULES AND LOOSE SLIME

Some bacteria form capsules, which constitute the outermost layer of the bacterial cell and surround it with a relatively thick layer of viscous gel.

Capsules may be up to 10 μm thick. Some organisms lack a well-defined capsule but have loose, amorphous slime layers external to the cell wall or cell envelope.

The a hemolytic Streptococcus mutans, the primary organism found in dental plaque is able to synthesis a large extracellular mucoid glucans from sucrose. Not all bacterial species produce capsules; however, the capsules of encapsulated pathogens are often important determinants of virulence.

Encapsulated species are found among both Gram-positive and Gram-negative bacteria. In both groups, most capsules are composed of highmolecular-weight viscous polysaccharides that are retained as a thick gel outside the cell wall or envelope. The capsule of Bacillus anthracis (the causal agent of anthrax) is unusual in that it is composed of a g-glutamyl polypeptide.

Table presents the various capsular substances formed by a selection of Gram-positive and Gram-negative bacteria.

A plasma membrane stage is involved in the biosynthesis and assembly of the capsular substances, which are extruded or secreted through the outer wall or envelope structures.

Mutational loss of enzymes involved in the biosynthesis of the capsular polysaccharides can result in the smooth-to-rough variation seen in the pneumococci.

The capsule is not essential for viability. Viability is not affected when capsular polysaccharides are removed enzymatically from the cell surface. The exact functions of capsules are not fully understood, but they do confer resistance to phagocytosis and hence provide the bacterial cell with protection against host defenses to invasion.

**Table. Nature of Capsular Substances Formed by Various Bacteria**

| *Genus and Species* | *Capsular Substances* |
|---|---|
| Gram-positive bacteria | |
| S. *pneumoniae* | Polysaccharides: E.g., type 111, glucose, glucoronic |
| acid (cellobiuronic acid); | other types, various sugars and amino sugars. |
| Streptococcus spp | Polysaccharides: e.g., hyaluronic acid (group A), others containing amino sugars, uronic acids |
| B. *anthracis* | γ-Glutamyl polypeptide |
| Gram-negative bacteria | |
| H. *influenzae* | Polyribosephosphate |
| *Kiebsiella* spp | Polysaccharides: sugars such as hexoses, fucose, uronic acids |
| N. *meningitidis* | Polysaccharidos:N-acetylmannosamine phosphate polymer (group A): siatic acid polymers (groups B and C) |

## CELL WALL AND GRAM-NEGATIVE CELL ENVELOPE

The Gram stain broadly differentiates bacteria into Gram-positive and Gram-negative groups; a few organisms are consistently Gram-variable. Gram-positive and Gram-negative organisms differ drastically in the organization of the structures outside the plasma membrane but below the capsule: in Gram-negative organisms these structures constitute the cell envelope, whereas in Gram-positive organisms they are called a cell wall.

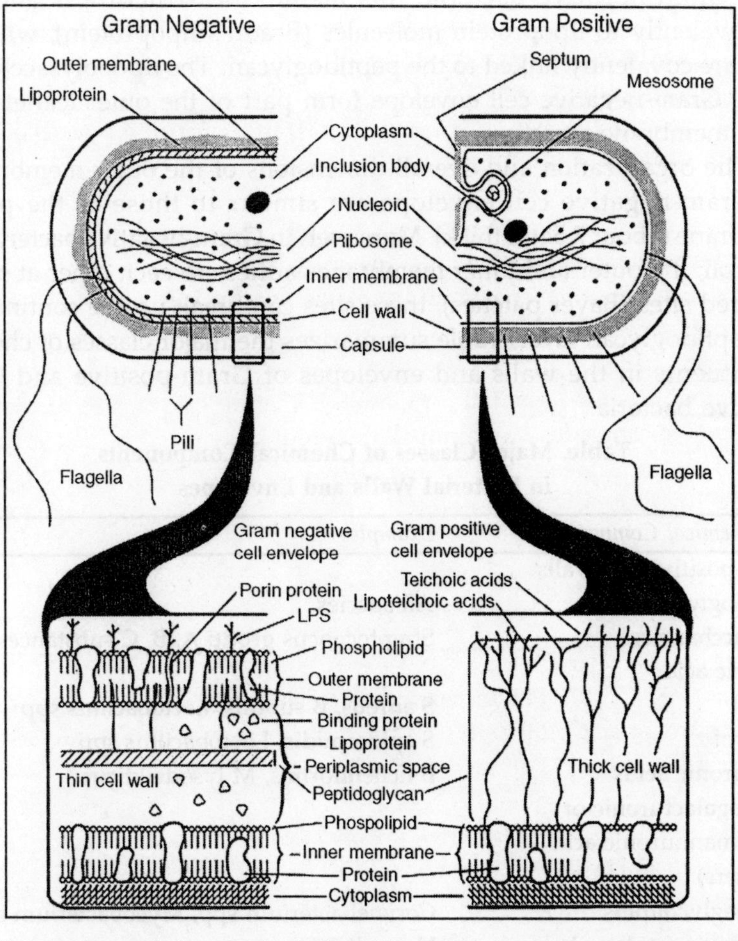


**Fig.** Comparison of the Thick Cell Wall of Gram-positive Bacteria with the Comparatively Thin Cell Wall of Gram-negative Bacteria.

Note the Complexity of the Gram-negative Cell Envelope (outer Membrane, its Hydrophobic Lipoprotein Anchor; Periplasmic Space).

Most Gram-positive bacteria have a relatively thick (about 20 to 80 nm), continuous cell wall (often called the sacculus), which is composed largely of peptidoglycan (also known as mucopeptide or murein). In thick cell walls, other cell wall polymers (such as the teichoic acids, polysaccharides, and peptidoglycolipids) are covalently attached to the peptidoglycan.

In contrast, the peptidoglycan layer in Gram-negative bacteria is thin (about 5 to 10 nm thick); in E coli, the peptidoglycan is probably only a monolayer thick. Outside the peptidoglycan layer in the Gram-negative envelope is an outer membrane structure (about 7.5 to 10 nm thick). In most Gram-negative bacteria, this membrane structure is anchored noncovalently to lipoprotein molecules (Braun's lipoprotein), which, in turn, are covalently linked to the peptidoglycan. The lipopolysaccharides of the Gram-negative cell envelope form part of the outer leaflet of the outer membrane structure.

The organization and overall dimensions of the outer membrane of the Gram-negative cell envelope are similar to those of the plasma membrane (about 7.5 nm thick). Moreover, in Gram-negative bacteria such as E coli, the outer and inner membranes adhere to each other at several hundred sites (Bayer patches); these sites can break up the continuity of the peptidoglycan layer. Table summarizes the major classes of chemical constituents in the walls and envelopes of Gram-positive and Gram-negative bacteria.

**Table. Major Classes of Chemical Components in Bacterial Walls and Envelopes**

| *Chemical Component* | *Examples* |
|---|---|
| Gram-positive cell walls | |
| Peplidogiycan | All species |
| Polysaccharides | Streptocoocus group A, B, C substances |
| Teichoic acid | |
| Ribitol | S aureus, B subtilis, Lactobacillus spp |
| Glycerol | S epidermidis, Lactobacillus spp |
| Teichuronic acids (aminogalacturonic or aminomannuronic acid polymers) | B licheniformis, M lysodeikticus |
| Peptideglycolipids (muramylpeptide-polysac charide-mycolates) | Corynebacterium spp, Mycobacterium spp. Nocardia spp |
| Glycolipids ('waxes') (polysacoharide-mycolates) | |
| Gram-negative envelopes | |
| LPS | All species |
| Lipoprotein | E coil and many enteric bacteria, Pseudomonas acruginosa |
| Porins (Major outer membrane proteins) | Eoil, Saimonella typhimurium |
| Phospholipids and proleins | All species |
| Peptidoglycan | Almost all species |

The basic differences in surface structures of Gram-positive and Gram-negative bacteria explain the results of Gram staining. Both Gram-positive and Gram-negative bacteria take up the same amounts of crystal violet (CV) and iodine (I).

The CV-I complex, however, is trapped inside the Gram-positive cell by the dehydration and reduced porosity of the thick cell wall as a result of the differential washing step with 95 per cent ethanol or other solvent mixture. In contrast, the thin peptidoglycan layer and probable discontinuities at the membrane adhesion sites do not impede solvent extraction of the CV-I complex from the Gram-negative cell.

The above mechanism of the Gram stain based on the structural differences between the two groups has been confirmed by sophisticated methods of electron microscopy. The sequence of steps in the Gram stain differentiation is illustrated diagrammatically in Figure.

Moreover, mechanical disruption of the cell wall of Gram-positive organisms or its enzymatic removal with lysozyme results in complete extraction of the CV-I complex and conversion to a Gram-negative reaction. Therefore, autolytic wall-degrading enzymes that cause cell wall breakage may account for Gram-negative or variable reactions in cultures of Gram-positive organisms (such as Staphylococcus aureus, Clostridium perfringens, Corynebacterium diphtheriae, and some Bacillus spp).

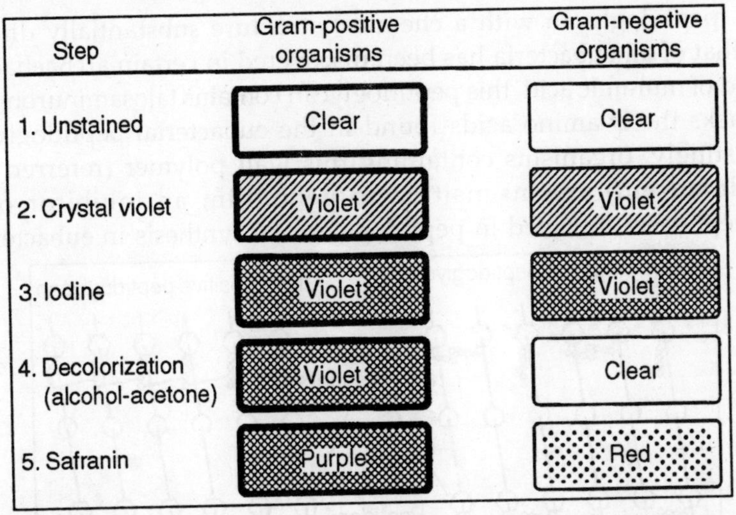


**Fig.** General Sequence of Steps in the Gram Stain Procedure and the Resultant Staining of Gram-positive and Gram-negative Bacteria.

## Peptidoglycan

Unique features of almost all prokaryotic cells (except for Halobacterium halobium and mycoplasmas) are cell wall peptidoglycan and the specific enzymes involved in its biosynthesis. These enzymes are target sites for inhibition of peptidoglycan synthesis by specific antibiotics. The primary chemical structures of peptidoglycans of both Gram-positive and Gram-negative bacteria have been established; they consist of a glycan backbone of repeating groups of b1, 4-linked disaccharides of b1,4-N-acetylmuramyl-N-acetylglucosamine.

Tetrapeptides of L-alanine-D-isoglutamic acid-L-lysine (or diaminopimelic acid)-n-alanine are linked through the carboxyl group by amide linkage of muramic acid residues of the glycan chains; the D-alanine residues are directly cross-linked to the e-amino group of lysine or diaminopimelic acid on a neighboring tetrapeptide, or they are linked by a peptide bridge. In S aureus peptidoglycan, a glycine pentapeptide bridge links the two adjacent peptide structures.

The extent of direct or peptide-bridge cross-linking varies from one peptidoglycan to another. The staphylococcal peptidoglycan is highly cross-linked, whereas that of E coli is much less so, and has a more open peptidoglycan mesh.

The diamino acid providing the e-amino group for cross-linking is lysine or diaminopimelic acid, the latter being uniformly present in Gram-negative peptidoglycans. The structure of the peptidoglycan is illustrated in Figure.

A peptidoglycan with a chemical structure substantially different from that of all eubacteria has been discovered in certain archaebacteria. Instead of muramic acid, this peptidoglycan contains talosaminuronic acid and lacks the D-amino acids found in the eubacterial peptidoglycans. Interestingly, organisms containing this wall polymer (referred to as pseudomurein) are insensitive to penicillin, an inhibitor of the transpeptidases involved in peptidoglycan biosynthesis in eubacteria.

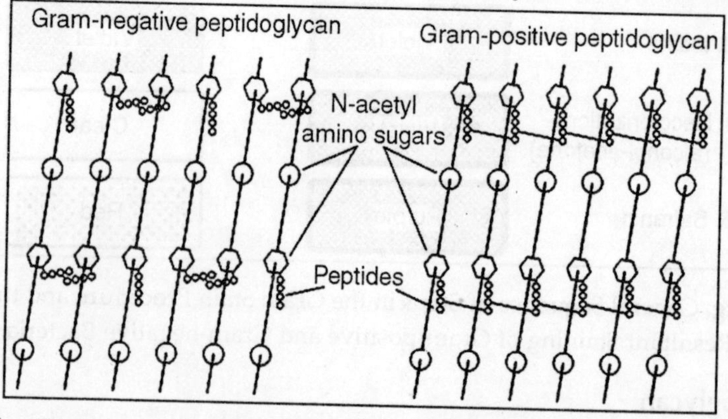


**Fig.** Diagrammatic Representation of Peptidoglycan Structures With Adjacent Glycan Strands Cross-linked Directly from the Carboxyterminal.D-alanine to the e-amino Group of an Adjacent Tetrapeptide or Through a Peptide Cross Bridge, N-acetylmuramic Acid; N-acetylglucosamine.

The β-1,4 glycosidic bond between N-acetylmuramic acid and N-acetylglucosamine is specifically cleaved by the bacteriolytic enzyme lysozyme. Widely distributed in nature, this enzyme is present in human tissues and secretions and can cause complete digestion of the peptidoglycan walls of sensitive organisms.

When lysozyme is allowed to digest the cell wall of Gram-positive bacteria suspended in an osmotic stabilizer (such as sucrose), protoplasts are formed. These protoplasts are able to survive and continue to grow on suitable media in the wall-less state.

Gram-negative bacteria treated similarly produce spheroplasts, which retain much of the outer membrane structure. The dependence of bacterial shape on the peptidoglycan is shown by the transformation of rod-shaped bacteria to spherical protoplasts (spheroplasts) after enzymatic breakdown of the peptidoglycan. The mechanical protection afforded by the wall peptidoglycan layer is evident in the osmotic fragility of both protoplasts and spheroplasts.

There are two groups of bacteria that lack the protective cell wall peptidoglycan structure, the Mycoplasma species, one of which causes atypical pneumonia and some genitourinary tract infections and the L-forms, which originate from Gram-positive or Gram-negative bacteria and are so designated because of their discovery and description at the Lister Institute, London.

The mycoplasmas and L-forms are all Gram-negative and insensitive to penicillin and are bounded by a surface membrane structure. L-forms arising "spontaneously" in cultures or isolated from infections are structurally related to protoplasts and spheroplasts; all three forms (protoplasts, spheroplasts, and L-forms) revert infrequently and only under special conditions.

### Teichoic Acids

Wall teichoic acids are found only in certain Gram-positive bacteria (such as staphylococci, streptococci, lactobacilli, and Bacillus spp); so far, they have not been found in gram- negative organisms. Teichoic acids are polyol phosphate polymers, with either ribitol or glycerol linked by phosphodiester bonds; their structures are illustrated in Figure. Substituent groups on the polyol chains can include D-alanine (ester linked), N-acetylglucosamine, N-acetylgalactosamine, and glucose; the substituent is characteristic for the teichoic acid from a particular bacterial species and can act as a specific antigenic determinant.

Teichoic acids are covalently linked to the peptidoglycan. These highly negatively charged polymers of the bacterial wall can serve as a cation-sequestering mechanism.

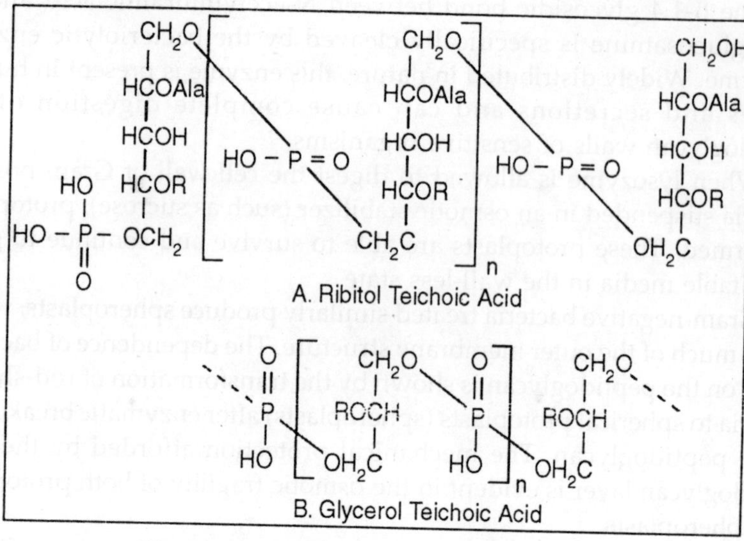

Fig. Structures of Cell Wall Teichoic Acids.

- Ribitol teichoic acid with repeating units of 1,5-phosphodiester linkages of D-ribitol and D-alanyl ester on position 2 and glycosyl substituents (R) on position 4. The glycosyl groups may abe N-acetylglucosaminyl (a or b) as in S aureus or a-glucosyl as in B subtilis W23.
- Glycerol teichoic acid with 1,3-phosphodiester linkages of glycerol repeating units (1,2-linkages in some species). In the glycerol teichoic acid structure shown, the polymer may be unsubstituted (R – H) or substituted (R – D-alanyl or glycosyl).

## Accessory Wall Polymers

In addition to the principal cell wall polymers, the walls of certain Gram-positive bacteria possess polysaccharide molecules linked to the peptidoglycan. For example, the C polysaccharide of streptococci confers group specificity. Acidic polysaccharides attached to the peptidoglycan are called teichuronic acids. Mycobacteria have peptidoglycolipids, glycolipids, and waxes associated with the cell wall.

## Lipopolysaccharides

A characteristic feature of Gram-negative bacteria is possession of various types of complex macromolecular lipopolysaccharide (LPS). So far, only one Gram-positive organism, Listeria monocytogenes, has been found to contain an authentic LPS. The LPS of this bacterium and those of all Gram-negative species are also called endotoxins, thereby distinguishing these cell-bound, heat-stable toxins from heat-labile, protein exotoxins secreted into culture media. Endotoxins possess an array of powerful biologic activities and play an important role in the pathogenesis of many Gram-negative bacterial infections.

In addition to causing endotoxic shock, LPS is pyrogenic, can activate macrophages and complement, is mitogenic for B lymphocytes, induces interferon production, causes tissue necrosis and tumor regression, and has adjuvant properties. The endotoxic properties of LPS reside largely in the lipid A components.

Usually, the LPS molecules have three regions: the lipid A structure required for insertion in the outer leaflet of the outer membrane bilayer; a covalently attached core composed of 2-keto-3deoxyoctonic acid (KDO), heptose, ethanolamine, N-acetylglucosamine, glucose, and galactose; and polysaccharide chains linked to the core. The polysaccharide chains constitute the O-antigens of the Gram-negative bacteria, and the individual monosaccharide constituents confer serologic specificity on these components.

Figure depicts the structure of LPS. Although it has been known that lipid A is composed of b1,6-linked D-glucosamine disaccharide substituted with phosphomonester groups at positions 4' and 1, uncertainties have existed about the attachment positions of the six fatty acid acyl and KDO groups on the disaccharide.

The demonstration of the structure of lipid A of LPS of a heptoseless mutant of Salmonella typhimurium has established that amide-linked hydroxymyristoyl and lauroxymyristoyl groups are attached to the nitrogen of the 2- and 2'-carbons, respectively, and that hydroxymyristoyl and myristoxymyristoyl groups are attached to the oxygen of the 3- and 3'-carbons of the disaccharide, respectively. Therefore, only position 6' is left for attachment of KDO units.

| Lipid A | Core | O Antigen |
|---|---|---|
| Glucosamine<br>b-hydroxymyristate<br>Fatty acids | Ketodeoxyoctonate<br>Phosphoethanolamine<br>Heptose<br>Glucose, galactose,<br>N-acetylglucosamine | Polysaccharide chains: repeating units of species-specific mano-saccharides, e.g., gal-actose, rhamnose, mannose and abequose in S typhimurium LPS |

Fig. The Three Major, Covalently Linked Regions that Form the Typical LPS.

LPS and phospholipids help confer asymmetry to the outer membrane of the Gram-negative bacteria, with the hydrophilic polysaccharide chains outermost. Each LPS is held in the outer membrane by relatively weak cohesive forces (ionic and hydrophobic interactions) and can be dissociated from the cell surface with surface-active agents.

As in peptidoglycan biosynthesis, LPS molecules are assembled at the plasma or inner membrane. These newly formed molecules are initially inserted into the outer-inner membrane adhesion sites.

### Outer Membrane of Gram-Negative Bacteria

In thin sections, the outer membranes of Gram-negative bacteria appear broadly similar to the plasma or inner membranes; however, they differ from the inner membranes and walls of Gram-positive bacteria in numerous respects. The lipid A of LPS is inserted with phospholipids to create the outer leaflet of the bilayer structure; the lipid portion of the lipoprotein and phospholipid form the inner leaflet of the outer membrane bilayer of most Gram-negative bacteria.

In addition to these components, the outer membrane possesses several major outer membrane proteins; the most abundant is called porin. The assembled subunits of porin form a channel that limits the passage of hydrophilic molecules across the outer membrane barrier to those having molecular weights that are usually less than 600 to 700. Evidence also suggests that hydrophobic pathways exist across the outer membrane and are partly responsible for the differential penetration and effectiveness of certain β-lactam antibiotics (ampicillin, cephalosporins) that are active against various Gram-negative bacteria.

Although the outer membranes act as a permeability barrier or molecular sieve, they do not appear to possess energy-transducing systems to drive active transport. Several outer membrane proteins, however, are involved in the specific uptake of metabolites (maltose, vitamin B12, nucleosides) and iron from the medium.

Thus, outer membranes of the Gram-negative bacteria provide a selective barrier to external molecules and thereby prevent the loss of metabolite-binding proteins and hydrolytic enzymes (nucleases, alkaline phosphatase) found in the periplasmic space. The periplasmic space is the region between the outer surface of the inner (plasma) membrane and the inner surface of the outer membrane.

Thus, Gram-negative bacteria have a cellular compartment that has no equivalent in Gram-positive organisms. In addition to the hydrolytic enzymes, the periplasmic space holds binding proteins (proteins that specifically bind sugars, amino acids, and inorganic ions) involved in membrane transport and chemotactic receptor activities. Moreover, plasmid-encoded β-lactamases and aminoglycoside-modifying enzymes (phosphorylation or adenylation) in the periplasmic space produce antibiotic resistance by degrading or modifying an antibiotic in transit to its target sites on the membrane (penicillin-binding proteins) or on the ribosomes (aminoglycosides). These periplasmic proteins can be released by subjecting the cells to osmotic shock and after treatment with the chelating agent ethylenediaminetetraacetic acid.

## INTRACELLULAR COMPONENTS

### Plasma (Cytoplasmic) Membranes

Bacterial plasma membranes, the functional equivalents of eukaryotic plasma membranes, are referred to variously as cytoplasmic, protoplast, or (in Gram-negative organisms) inner membranes. Similar in overall dimensions and appearance in thin sections to biomembranes from eukaryotic cells, they are composed primarily of proteins and lipids (principally phospholipids). Protein-to-lipid ratios of bacterial plasma membranes are approximately 3: 1, close to those for mitochondrial membranes. Unlike eukaryotic cell membranes, the bacterial membrane (except for Mycoplasma species and certain methylotrophic bacteria) has no sterols, and bacteria lack the enzymes required for sterol biosynthesis.

Although their composition is similar to that of inner membranes of Gram-negative species, cytoplasmic membranes from Gram-positive bacteria possess a class of macromolecules not present in the Gram-negative membranes. Many Gram-positive bacterial membranes contain membrane-bound lipoteichoic acid, and species lacking this component (such as Micrococcus and Sarcina spp) contain an analogous membrane-bound succinylated lipomannan.

Lipoteichoic acids are structurally similar to the cell wall glycerol teichoic acids in that they have basal polyglycerol phosphodiester 1-3 linked chains. These chains terminate with the phosphomonoester end of the polymer, which is linked covalently to either a glycolipid or a phosphatidyl glycolipid moiety. Thus, a hydrophobic tail is provided for anchoring in the membrane lipid layers. As in the cell wall glycerol teichoic acid, the lipoteichoic acids can have glycosidic and D-alanyl ester substituents on the C-2 position of the glycerol.

Both membrane-bound lipoteichoic acid and membrane-bound succinylated lipomannan can be detected as antigens on the cell surface, and the glycerol-phosphate and succinylated mannan chains appear to extend through the cell wall structure. This class of polymer has not yet been found in the cytoplasmic membranes of Gram-negative organisms. In both instances, the lipoteichoic acids and the lipomannans are negatively charged components and can sequester positivel charged substances. They have been implicated in adhesion to host cells, but their functions remain to be elucidated.

Multiple functions are performed by the plasma membranes of both Gram-positive and Gram-negative bacteria. Plasma membranes are the site of active transport, respiratory chain components, energy-transducing systems, the H+-ATPase of the proton pump, and membrane stages in the biosynthesis of phospholipids, peptidoglycan, LPS, and capsular polysaccharides. In essence, the bacterial cytoplasmic membrane is a multifunction structure that combines the mitochondrial transport and biosynthetic functions that are usually compartmentalized in discrete membranous organelles in eukaryotic cells. The plasma membrane is also the anchoring site for DNA and provides the cell with a mechanism (as yet unknown) for separation of sister chromosomes.

### Mesosomes

Thin sections of Gram-positive bacteria reveal the presence of vesicular or tubular-vesicular membrane structures called mesosomes, which are apparently formed by an invagination of the plasma membrane. These structures are much more prominent in Gram-positive than in Gram-negative organisms.

At one time, the mesosomal vesicles were thought to be equivalent to bacterial mitochondria; however, many other membrane functions have also been attributed to the mesosomes. At present, there is no satisfactory evidence to suggest that they have a unique biochemical or physiologic function. Indeed, electron-microscopic studies have suggested that the mesosomes, as usually seen in thin sections, may arise from membrane perturbation and fixation artifacts. No general agreement exists about this theory, however, and some evidence indicates that mesosomes may be related to events in the cell division cycle.

## OTHER INTRACELLULAR COMPONENTS

In addition to the nucleoid and cytoplasm (cytosol), the intracellular compartment of the bacterial cell is densely packed with ribosomes of the 70S type. These ribonucleoprotein particles, which have a diameter of 18 nm, are not arranged on a membranous rough endoplasmic reticulum as they are in eukaryotic cells. Other granular inclusions randomly distributed in the cytoplasm of various species include metabolic reserve particles such as poly-β-hydroxybutyrate (PHB), polysaccharide and glycogen-like granules, and polymetaphosphate or metachromatic granules.

Endospores are highly heat-resistant, dehydrated resting cells formed intracellularly in members of the genera Bacillus and Clostridium. Sporulation, the process of forming endospores, is an unusual property of certain bacteria.

The series of biochemical and morphologic changes that occur during sporulation represent true differentiation within the cycle of the bacterial cell. The process, which usually begins in the stationary phase of the vegetative cell cycle, is initiated by depletion of nutrients (usually readily utilizable sources of carbon or nitrogen, or both).

The cell then undergoes a highly complex, well-defined sequence of morphologic and biochemical events that ultimately lead to the formation of mature endospores. As many as seven distinct stages have been recognized by morphologic and biochemical studies of sporulating Bacillus species: stage 0, vegetative cells with two chromosomes at the end of exponential growth; stage I, formation of axial chromatin filament and excretion of exoenzymes, including proteases; stage II, forespore septum formation and segregation of nuclear material into two compartments; stage III, spore protoplast formation and elevation of tricarboxylic acid and glyoxylate cycle enzyme levels; stage IV, cortex formation and refractile appearance of spore; stage V, spore coat protein formation; stage VI, spore maturation, modification of cortical peptidoglycan, uptake of dipicolinic acid (a unique endospore product) and calcium, and development of resistance to heat and organic solvents; and stage VII, final maturation and liberation of endospores from mother cells (in some species).

When newly formed, endospores appear as round, highly refractile cells within the vegetative cell wall, or sporangium. Some strains produce autolysins that digest the walls and liberate free endospores. The spore protoplast, or core, contains a complete nucleus, ribosomes, and energy generating components that are enclosed within a modified cytoplasmic membrane.

The peptidoglycan spore wall surrounds the spore membrane; on germination, this wall becomes the vegetative cell wall. Surrounding the spore wall is a thick cortex that contains an unusual type of peptidoglycan, which is rapidly released on germination.

A spore coat of keratinlike protein encases the spore contained within a membrane (the exosporium).

During maturation, the spore protoplast dehydrates and the spore becomes refractile and resistant to heat, radiation, pressure, desiccation, and chemicals; these properties correlate with the cortical peptidoglycan and the presence of large amounts of calcium dipicolinate.

## Chapter 3

# Classification

Bacteria are classified and identified to distinguish one organism from another and to group similar organisms by criteria of interest to microbiologists or other scientists. The most important level of this type of classification is the species level.

A species name should mean the same thing to everyone. Within one species, strains and subgroups can differ by the disease they produce, their environmental habitat, and many other characteristics. Formerly, species were created on the basis of such criteria, which may be extremely important for clinical microbiologists and physicians but which are not a sufficient basis for establishing a species. Verification of existing species and creation of new species should involve biochemical and other phenotypic criteria as well as DNA relatedness. In numerical or phenetic approaches to classification, strains are grouped on the basis of a large number of phenotypic characteristics. DNA relatedness is used to group strains on the basis of overall genetic similarity.

Species are identified in the clinical laboratory by morphological traits and biochemical tests, some of which are supplemented by serologic assessments (e.g., identification of *Salmonella* and *Shigella* species). Because of differences in pathogenicity *(Escherichia coli)* or the necessity to characterize a disease outbreak (*Vibrio cholerae,* methicillin-resistant *Staphylococcus aureus),* strains of medical interest are often classified below the species level by serology or identification of toxins. Pathogenic or epidemic strains also can be classified by the presence of a specific plasmid, by their plasmid profile (the number and sizes of plasmids), or by bacteriophage susceptibility patterns (phage typing). Newer molecular biologic techniques have enabled scientists to identify some species and strains (without the use of biochemical tests) by identifying a specific gene or genetic sequence, sometimes directly from the clinical specimen.

Laboratories have no difficulty in identifying typical strains of common bacteria using commonly available test systems. Problems do arise, however, when atypical strains or rare or newly described species are not in the data base. Such difficulties are compounded when the strains are misidentified rather than unidentified, and so laboratory personnel and physicians (at least infectious diseases specialists) should be familiar with taxonomic reference texts and journals that publish papers on new species. Bacterial nomenclature at the genus and species level changes often, based primarily on the use of newer genetic techniques.

A species may acquire more than one name. In some cases the recognition of a new species results in a unique correlation with specific clinical problems. For example, recognition of *Porphyromonas gingivalis* as a unique species, separate from its previous inclusion within *Bacteroides melaninogenicus* (now known to be composed of several taxonomic groups of black-pigmenting anaerobic gram-negative bacilli), elucidated its role as a key pathogen in adult periodontitis. It is important to understand why these changes and synonyms exist in taxonomy.

The clinical laboratory is concerned with the rapid, sensitive, and accurate identification of microbes involved in producing disease. The number and types of tests done in such a laboratory depend on its size and the population it serves. Highly specialized or rarely performed tests should be done only by reference laboratories. Physicians, clinical laboratory personnel, and reference laboratory personnel must have a good working relationship if patients are to receive first-rate care.

In addition, the physician and the clinical laboratory personnel must know which diseases and isolates are reportable to public health laboratories and how to report them.

## APPROACHES TO TAXONOMY

### Numerical Approach

In their studies on members of the family Enterobacteriaceae, Edwards and Ewing established the following principles to characterize, classify, and identify organisms.

Classification and identification of an organism should be based on its overall morphologic and biochemical pattern. A single characteristic (pathogenicity, host range, or biochemical reaction), regardless of its importance, is not a sufficient basis for classifying or identifying an organism.

A large and diverse strain sample must be tested to determine accurately the biochemical characteristics used to distinguish a given species.

Atypical strains often are perfectly typical members of a given biogroup within an existing species, but sometimes they are typical members of an unrecognized new species.

In numerical taxonomy (also called computer or phenetic taxonomy) many (50 to 200) biochemical, morphological, and cultural characteristics, as well as susceptibilities to antibiotics and inorganic compounds, are used to determine the degree of similarity between organisms. In numerical studies, investigators often calculate the coefficient of similarity or percentage of similarity between strains (where strain indicates a single isolate from a specimen). A dendrogram or a similarity matrix is constructed that joins individual strains into groups and places one group with other groups on the basis of their percentage of similarity. In the dendrogram in Figure 3-1, group 1 represents three *Citrobacter freundii* strains that are about 95 per cent similar and join with a fourth *C freundii* strain at the level of 90 per cent similarity. Group 2 is composed of three *Citrobacter diversus* strains that are 95 per cent similar, and group 3 contains two *E coli* strains that are 95 per cent similar, as well as a third *E coli* strain to which they are 90 per cent similar. Similarity between groups 1 and 2 occurs at the 70 per cent level, and group 3 is about 50 per cent similar to groups 1 and 2.

In some cases, certain characteristics may be weighted more heavily; for example, the presence of spores in *Clostridium* might be weighted more heavily than the organism's ability to use a specific carbon source. A given level of similarity can be equated with relatedness at the genus, species, and, sometimes, subspecies levels. For instance, strains of a given species may cluster at a 90% similarity level, species within a given genus may cluster at the 70 per cent level, and different genera in the same family may cluster at the 50 per cent or lower level.

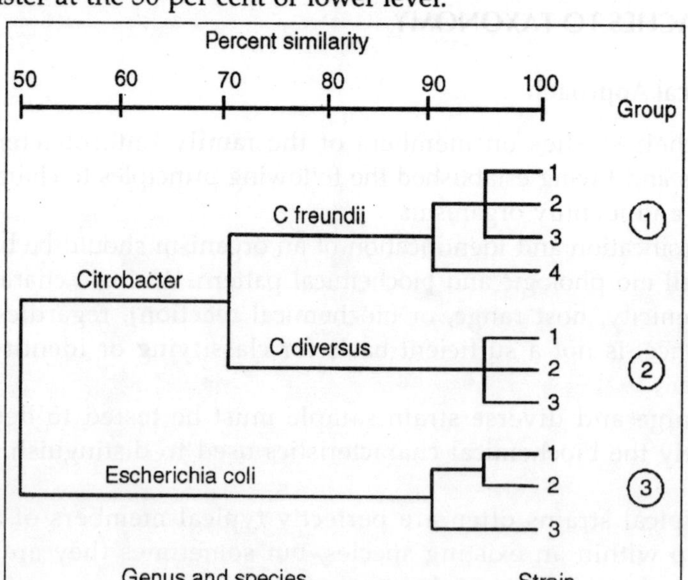


Fig. Example of Dendrogram.

When this approach is the only basis for defining a species, it is difficult to know how many and which tests should be chosen; whether and how the tests should be weighted; and what level of similarity should be chosen to reflect relatedness at the genus and species levels.

Most bacteria have enough DNA to specify some 1,500 to 6,000 average-sized genes. Therefore, even a battery of 300 tests would assay only 5 to 20 per cent of the genetic potential of a bacterium. Tests that are comparatively simple to conduct (such as those for carbohydrate utilization and for enzymes, presence of which can be assayed colorimetrically) are performed more often than tests for structural, reproductive, and regulatory genes, presence of which is difficult to assay. Thus, major differences may go undetected.

Other types of errors may occur when species are classified solely on the basis of phenotype. For example, different enzymes (specified by different genes) may catalyze the same reaction. Also, even if a metabolic gene is functional, negative reactions can occur because of the inability of the substrate to enter the cell, because of a mutation in a regulatory gene, or by production of an inactive protein. There is not necessarily a one-to-one correlation between a reaction and the number of genes needed to carry out that reaction. For instance, six enzymatic steps may be involved in a given pathway. If an assay for the end product is performed, a positive reaction indicates the presence of all six enzymes, whereas a negative reaction can mean the absence or nonfunction of one to six enzymes. Several other strain characteristics can affect phenotypic characterization; these include growth rate, incubation temperature, salt requirement, and pH. Plasmids that carry metabolic genes can enable strains to carry out reactions atypical for strains of that species.

The same set of "definitive" reactions cannot be used to classify all groups of organisms, and there is no standard number of specific reactions that allows identification of a species. Organisms are identified on the basis of phenotype, but, from the taxonomic standpoint, definition of species solely on this basis is subject to error.

## Phylogenetic Approach

The ideal means of identifying and classifying bacteria would be to compare each gene sequence in a given strain with the gene sequences for every known species. This cannot be done, but the total DNA of one organism can be compared with that of any other organism by a method called nucleic acid hybridization or DNA hybridization. This method can be used to measure the number of DNA sequences that any two organisms have in common and to estimate the percentage of divergence within DNA sequences that are related but not identical. DNA relatedness studies have been done for yeasts, viruses, bacteriophages, and many groups of bacteria.

Five factors can be used to determine DNA relatedness: genome size, guanine-plus-cytosine (G+C) content, DNA relatedness under conditions optimal for DNA reassociation, thermal stability of related DNA sequences, and DNA relatedness under conditions supraoptimal for DNA reassociation.

Because it is not practical to conduct these genotypic or phylogenetic evaluations in clinical laboratories, the results of simpler tests usually must be correlated with known phylogenetic data. For example, yellow strains of *Enterobacter cloacae* were shown, by DNA relatedness, to form a separate species, *Enterobacter sakazakii,* but were not designated as such until results of practical tests were correlated with the DNA data to allow routine laboratories to identify the new species.

### Genome Size

True bacterial DNAs have genome sizes (measured as molecular weight) between 1 × 109 and 8 × 109. Genome size determinations sometimes can distinguish between groups. They were used to distinguish *Legionella pneumophila* (the legionnaire's disease bacterium) from *Bartonella (Rickettsia) quintana,* the agent of trench fever. *L pneumophila* has a genome size of about 3 × 109; that of *B quintana* is about 1 × 109.

### Guanine-plus-Cytosine Content

The G+C content in bacterial DNA ranges from about 25 to 75 per cent. This percentage is specific, but not exclusive, for a species; two strains with a similar G+C content may or may not belong to the same species. If the G+C contents are very different, however, the strains cannot be members of the same species.

### DNA Relatedness under Conditions Optimal for DNA Reassociation

DNA relatedness is determined by allowing single-stranded DNA from one strain to reassociate with single-stranded DNA from a second strain, to form a double-stranded DNA molecule. This is a specific, temperature-dependent reaction. The optimal temperature for DNA reassociation is 25 to 30°C below the temperature at which native double-stranded DNA denatures into single strands. Many studies indicate that a bacterial species is composed of strains that are 70 to 100 per cent related. In contrast, relatedness between different species is 0 to about 65 per cent. It is important to emphasize that the term "related" does not mean "identical" or "homologous." Similar but nonidentical nucleic acid sequences can reassociate.

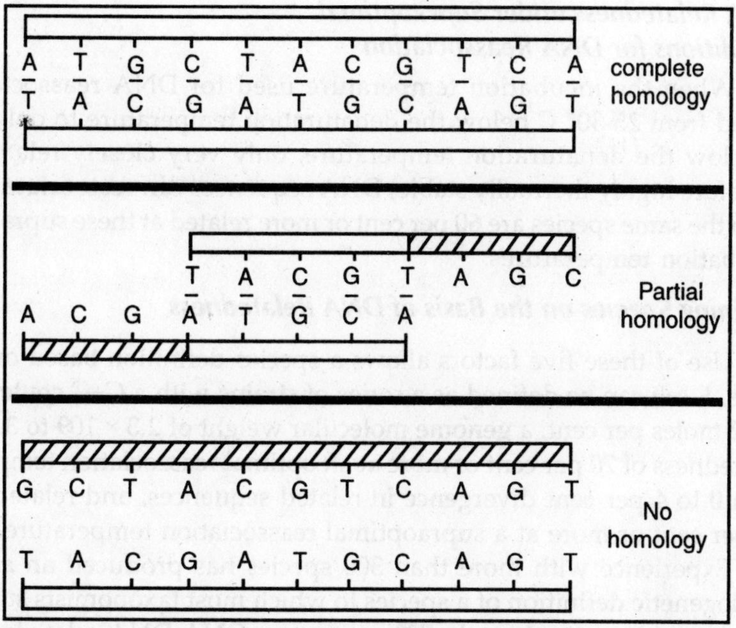


**Fig.** Diagram of DNA Reassociation.

DNA is composed of two purine nucleoside bases, adenine (A) and guanine (G), and two pyrimidine nucleoside bases, thymine (T) and cytosine (C).

Double-stranded DNA is formed through hydrogen bonds that can occur only between the complementary bases A and T or G and C.

(Top) Perfectly reassociated DNA base sequence in which all nucleosides are paired by hydrogen bonds. (Middle) Perfectly paired DNA base sequence in the centre with unpaired, single-strand ends on each strand.

(Bottom) None of the bases in the sequence (left to right) GCTACGTCAGT on the top strand are complementary to the sequence TACGATGCAGT in the bottom strand.

### *Thermal Stability of Related DNA Sequences*

Each 1 per cent of unpaired nucleotide bases in a double-stranded DNA sequence causes a 1 per cent decrease in the thermal stability of that DNA duplex.

Therefore, a comparison between the thermal stability of a control double-stranded molecule (in which both strands of DNA are from the same organism) and that of a heteroduplex (DNA strands from two different organisms) allows assessment of divergence between related nucleotide sequences.

### *DNA Relatedness under Supraoptimal Conditions for DNA Reassociation*

When the incubation temperature used for DNA reassociation is raised from 25-30° C below the denaturation temperature to only 10-15° C below the denaturation temperature, only very closely related (and therefore highly thermally stable) DNA sequences can reassociate. Strains from the same species are 60 per cent or more related at these supraoptimal incubation temperatures.

### *Defining Species on the Basis of DNA Relatedness*

Use of these five factors allows a species definition based on DNA. Thus, *E coli* can be defined as a series of strains with a G+C content of 49 to 52 moles per cent, a genome molecular weight of 2.3 × 109 to 3.0 × 109, relatedness of 70 per cent or more at an optimal reassociation temperature with 0 to 4 per cent divergence in related sequences, and relatedness of 60 per cent or more at a supraoptimal reassociation temperature.

Experience with more than 300 species has produced an arbitrary phylogenetic definition of a species to which most taxonomists subscribe: "strains with approximately 70% or greater DNA-DNA relatedness and with 5° C or less divergence in related sequences." When these two criteria are met, genome size and G+C content are always similar, and relatedness is almost always 60 per cent or more at supraoptimal incubation temperatures.

The 70 per cent species relatedness rule has been ignored occasionally when the existing nomenclature is deeply ingrained, as is that for *E coli* and the four *Shigella* species. Because these organisms are all 70 per cent or more related, DNA studies indicate that they should be grouped into a single species, instead of the present five species in two genera. This change has not been made because of the presumed confusion that would result.

DNA relatedness provides one species definition that can be applied equally to all organisms. Moreover, it cannot be affected by phenotypic variation, mutations, or the presence or absence of metabolic or other plasmids. It measures overall relatedness, and these factors affect only a very small percentage of the total DNA.

## POLYPHASIC APPROACH

In practice, the approach to bacterial taxonomy should be polyphasic. The first step is phenotypic grouping of strains by morphological, biochemical and any other characteristics of interest. The phenotypic groups are then tested for DNA relatedness to determine whether the observed phenotypic homogeneity (or heterogeneity) is reflected by phylogenetic homogeneity or heterogeneity. The third and most important step is reexamination of the biochemical characteristics of the DNA relatedness groups. This allows determination of the biochemical borders of each group and determination of reactions of diagnostic value for the group. For identification of a given organism, the importance of specific tests is weighted on the basis of correlation with DNA results. Occasionally, the reactions commonly used will not distinguish completely between two distinct DNA relatedness groups. In these cases, other biochemical tests of diagnostic value must be sought.

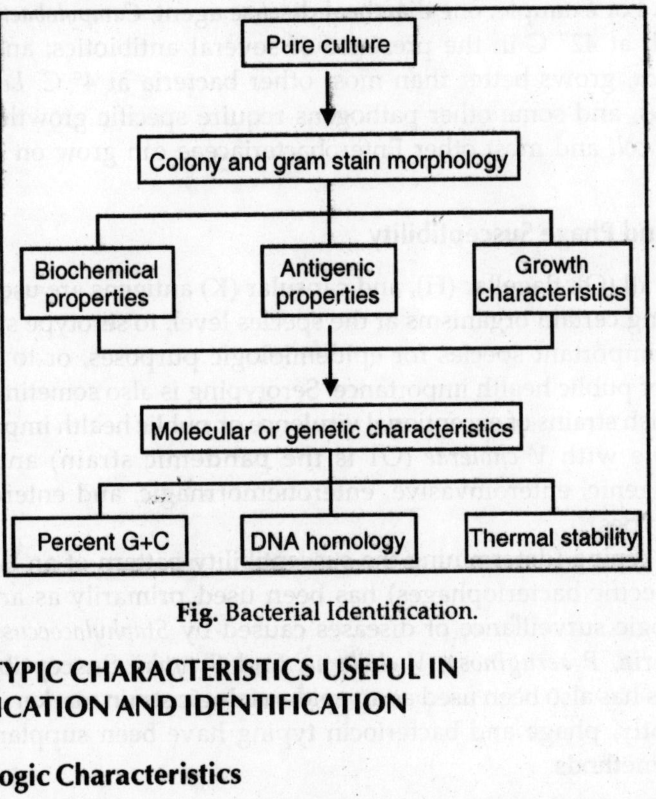


**Fig.** Bacterial Identification.

## PHENOTYPIC CHARACTERISTICS USEFUL IN CLASSIFICATION AND IDENTIFICATION

### Morphologic Characteristics

Both wet-mounted and properly stained bacterial cell suspensions can yield a great deal of information. These simple tests can indicate the Gram reaction of the organism; whether it is acid-fast; its motility; the arrangement of its flagella; the presence of spores, capsules, and inclusion bodies; and, of course, its shape. This information often can allow identification of an organism to the genus level, or can minimize the possibility that it belongs to one or another group. Colony characteristics and pigmentation are also quite helpful. For example, colonies of several *Porphyromonas* species autofluoresce under long-wavelength ultraviolet light, and *Proteus* species swarm on appropriate media.

## Growth Characteristics

A primary distinguishing characteristic is whether an organism grows aerobically, anaerobically, facultatively (i.e., in either the presence or absence of oxygen), or microaerobically (i.e., in the presence of a less than atmospheric partial pressure of oxygen).

The proper atmospheric conditions are essential for isolating and identifying bacteria. Other important growth assessments include the incubation temperature, pH, nutrients required, and resistance to antibiotics. For example, one diarrheal disease agent, *Campylobacter jejuni,* grows well at 42° C in the presence of several antibiotics; another, *Y enterocolitica,* grows better than most other bacteria at 4° C. *Legionella, Haemophilus,* and some other pathogens require specific growth factors, whereas *E coli* and most other Enterobacteriaceae can grow on minimal media.

## Antigens and Phage Susceptibility

Cell wall (O), flagellar (H), and capsular (K) antigens are used to aid in classifying certain organisms at the species level, to serotype strains of medically important species for epidemiologic purposes, or to identify serotypes of public health importance. Serotyping is also sometimes used to distinguish strains of exceptional virulence or public health importance, for example with *V cholerae* (O1 is the pandemic strain) and *E coli* (enterotoxigenic, enteroinvasive, enterohemorrhagic, and enteropathogenic serotypes).

Phage typing (determining the susceptibility pattern of an isolate to a set of specific bacteriophages) has been used primarily as an aid in epidemiologic surveillance of diseases caused by *Staphylococcus aureus,* mycobacteria, *P aeruginosa, V cholerae,* and *S typhi.* Susceptibility to bacteriocins has also been used as an epidemiologic strain marker. In most cases recently, phage and bacteriocin typing have been supplanted by molecular methods.

## Biochemical Characteristics

Most bacteria are identified and classified largely on the basis of their reactions in a series of biochemical tests. Some tests are used routinely for many groups of bacteria (oxidase, nitrate reduction, amino acid degrading enzymes, fermentation or utilization of carbohydrates); others are restricted to a single family, genus, or species (coagulase test for staphylococci, pyrrolidonyl arylamidase test for Gram-positive cocci).

Both the number of tests needed and the actual tests used for identification vary from one group of organisms to another. Therefore, the lengths to which a laboratory should go in detecting and identifying organisms must be decided in each laboratory on the basis of its function, the type of population it serves, and its resources. Clinical laboratories today base the extent of their work on the clinical relevance of an isolate to the particular patient from which it originated, the public health significance of complete identification, and the overall cost-benefit analysis of their procedures.

For example, the Centers for Disease Control and Prevention (CDC) reference laboratory uses at least 46 tests to identify members of the Enterobacteriaceae, whereas most clinical laboratories, using commercial identification kits or simple rapid tests, identify isolates with far fewer criteria.

## CLASSIFICATION BELOW AND ABOVE THE SPECIES LEVEL

### Below the Species Level

Particularly for epidemiological purposes, clinical microbiologists must distinguish strains with particular traits from other strains in the same species. For example, serotype O157:H7 *E coli* are identified in stool specimens because of their association with bloody diarrhea and subsequent hemolytic uremic syndrome.

Below the species level, strains are designated as groups or types on the basis of common serologic or biochemical reactions, phage or bacteriocin sensitivity, pathogenicity, or other characteristics. Many of these characteristics are already used and accepted: serotype, phage type, colicin type, biotype, bioserotype (a group of strains from the same species with common biochemical and serologic characteristics that set them apart from other members of the species), and pathotype (e.g., toxigenic *Clostridium difficile*, invasive *E coli*, and toxigenic *Corynebacterium diphtheriae*).

### Above the Species Level

In addition to species and subspecies designations, clinical microbiologists must be familiar with genera and families. A genus is a group of related species, and a family is a group of related genera.

An ideal genus would be composed of species with similar phenotypic and phylogenetic characteristics. Some phenotypically homogeneous genera approach this criterion *(Citrobacter, Yersinia,* and *Serratia)*. More often, however, the phenotypic similarity is present, but the genetic relatedness is not. *Bacillus, Clostridium,* and *Legionella* are examples of accepted phenotypic genera in which genetic relatedness between species is not 50 to 65 per cent, but 0 to 65 per cent. When phenotypic and genetic similarity are not both present, phenotypic similarity generally should be given priority in establishing genera. Identification practices are simplified by having the most phenotypically similar species in the same genus. The primary consideration for a genus is that it contain biochemically similar species that are convenient or important to consider as a group separate from other groups of organisms.

The sequencing of ribosomal RNA (rRNA) genes, which have been highly conserved through evolution, allows phylogenetic comparisons to be made between species whose total DNAs are essentially unrelated. It also allows phylogenetic classification at the genus, family, and higher taxonomic levels. The rRNA sequence data are usually not used to designate genera or families unless supported by similarities in phenotypic tests.

## DESIGNATION OF NEW SPECIES AND NOMENCLATURAL CHANGES

Species are named according to principles and rules of nomenclature set forth in the Bacteriological Code. Scientific names are taken from Latin or Greek. The correct name of a species or higher taxon is determined by three criteria: valid publication, legitimacy of the name with regard to the rules of nomenclature, and priority of publication (that is, it must be the first validly published name for the taxon).

To be published validly, a new species proposal must contain the species name, a description of the species, and the designation of a type strain for the species, and the name must be published in the *International Journal for Systematic Bacteriology (IJSB).* Once proposed, a name does not go through a formal process to be accepted officially; in fact, the opposite is truea validly published name is assumed to be correct unless and until it is challenged officially.

A challenge is initiated by publishing a request for an opinion (to the Judicial Commission of the International Association of Microbiological Societies) in the *IJSB.* This occurs only in cases in which the validity of a name is questioned with respect to compliance with the rules of the Bacteriological Code. A question of classification that is based on scientific data (for example, whether a species, on the basis of its biochemical or genetic characteristics, or both, should be placed in a new genus or an existing genus) is not settled by the Judicial Commission, but by the preference and usage of the scientific community.

This is why there are pairs of names such as *Providencia rettgeri/Proteus rettgeri, Moraxella catarrhalis/Branhamella catarrhalis,* and *Legionella micdadei/Tatlockia micdadei.* More than one name may thus exist for a single organism. This is not, however, restricted to bacterial nomenclature. Multiple names exist for many antibiotics and other drugs and enzymes.

A number of genera have been divided into additional genera and species have been moved to new or existing genera, such as *Arcobacter* (new genus for former members of *Campylobacter*) and *Burkholderia* species (formerly species of *Pseudomonas*). Two former *Campylobacter* species (*cinaedi* and *fennelliae*) have been moved to the existing genus *Helicobacter* in another example.

The best source of information for new species proposals and nomenclatural changes is the *IJSB*. In addition, the *Journal of Clinical Microbiology* often publishes descriptions of newly described microorganisms isolated from clinical sources. Information, including biochemical reactions and sources of isolation, about new organisms of clinical importance, disease outbreaks caused by newer species, and reviews of clinical significance of certain organisms may be found in the *Annals of Internal Medicine, Journal of Infectious Diseases, Clinical Microbiology Reviews,* and *Clinical Infectious Diseases*. The data provided in these publications supplement and update *Bergey's Manual of Systematic Bacteriology,* the definitive taxonomic reference text.

## ASSESSING NEWLY DESCRIBED BACTERIA

Since 1974, the number of genera in the family Enterobacteriaceae has increased from 12 to 28 and the number of species from 42 to more than 140, some of which have not yet been named. Similar explosions have occurred in other genera. In 1974, five species were listed in the genus *Vibrio* and four in *Campylobacter*; the genus *Legionella* was unknown. Today, there are at least 25 species in *Vibrio,* 12 *Campylobacter* species, and more than 40 species in *Legionella*. The total numbers of genera and species continue to increase dramatically.

The clinical significance of the agent of legionnaire's disease was well known long before it was isolated, characterized, and classified as *Legionella pneumophila*. In most cases, little is known about the clinical significance of a new species at the time it is first described. Assessments of clinical significance begin after clinical laboratories adopt the procedures needed to detect and identify the species and accumulate a body of data.

In fact, the detection and even the identification of uncultivatable microbes from different environments are now possible using standard molecular methods. The agents of cat scratch disease (*Bartonella henselae*) and Whipple's disease (*Tropheryma whippelii*) were elucidated in this manner. *Bartonella henselae* has since been cultured from several body sites from numerous patients; *T whippelii* remains uncultivated.

New species will continue to be described. Many will be able to infect humans and cause disease, especially in those individuals who are immunocompromised, burned, postsurgical, geriatric, and suffering from acquired immunodeficiency syndrome (AIDS). With today's severely immunocompromised patients, often the beneficiaries of advanced medical interventions, the concept of "pathogen" holds little meaning. Any organism is capable of causing disease in such patients under the appropriate conditions.

## ROLE OF THE CLINICAL LABORATORY

Clinical laboratory scientists should be able to isolate, identify, and determine the antimicrobial susceptibility pattern of the vast majority of human disease agents so that physicians can initiate appropriate treatment as soon as possible, and the source and means of transmission of outbreaks can be ascertained to control the disease and prevent its recurrence. The need to identify clinically relevant microorganisms both quickly and cost-effectively presents a considerable challenge.

To be effective, the professional clinical laboratory staff must interact with the infectious diseases staff. Laboratory scientists should attend infectious disease rounds. They must keep abreast of new technology, equipment, and classification and should communicate this information to their medical colleagues. They should interpret, qualify, or explain laboratory reports. If a bacterial name is changed or a new species reported, the laboratory should provide background information, including a reference.

The clinical laboratory must be efficient. A concerted effort must be made to eliminate or minimize inappropriate and contaminated specimens and the performance of procedures with little or no clinical relevance. Standards for the selection, collection, and transport of specimens should be developed for both laboratory and nursing procedure manuals and reviewed periodically by a committee composed of medical, nursing, and laboratory staff. Ongoing dialogues and continuous communication with other health care workers concerning topics such as specimen collection, test selection, results interpretation, and new technology are essential to maintaining high quality microbiological services.

### Biochemical and Susceptibility Testing

Most laboratories today use either commercially available miniaturized biochemical test systems or automated instruments for biochemical tests and for susceptibility testing.

The kits usually contain 10 to 20 tests. The test results are converted to numerical biochemical profiles that are identified by using a codebook or a computer. Carbon source utilization systems with up to 95 tests are also available. Most identification takes 4 to 24 hours. Biochemical and enzymatic test systems for which data bases have not been developed are used by some reference laboratories.

Automated instruments can be used to identify most Gram-negative fermenters, nonfermenters, and Gram-positive bacteria, but not for anaerobes. Antimicrobial susceptibility testing can be performed for some microorganisms with this equipment, with results expressed as approximate minimum inhibitory drug concentrations. Both tasks take 4 to 24 hours. If semiautomated instruments are used, some manipulation is done manually, and the cultures (in miniature cards or microdilution plates) are incubated outside of the instrument. The test containers are then read rapidly by the instrument, and the results are generated automatically. Instruments are also available for identification of bacteria by cell wall fatty acid profiles generated with gas-liquid chromatography (GLC), analysis of mycolic acids using high performance liquid chromatography (HPLC), and by protein-banding patterns generated by polyacrylamide gel electrophoresis (PAGE). Some other instruments designed to speed laboratory diagnosis of bacteria are those that detect (but do not identify) bacteria in blood cultures, usually faster than manual systems because of continuous monitoring. Also available are many rapid screening systems for detecting one or a series of specific bacteria, including certain streptococci, *N meningitidis*, salmonellae, *Chlamydia trachomatis*, and many others. These screening systems are based on fluorescent antibody, agglutination, or other rapid procedures.

It is important to inform physicians as soon as a presumptive identification of an etiologic agent is obtained so that appropriate therapy can be initiated as quickly as possible. Gram stain and colony morphology; acid-fast stains; and spot indole, oxidase, and other rapid enzymatic tests may allow presumptive identification of an isolate within minutes.

## Role of the Reference Laboratory

Despite recent advances, the armamentarium of the clinical laboratory is far from complete. Few laboratories can or should conduct the specialized tests that are often essential to distinguish virulent from avirulent strains. Serotyping is done only for a few species, and phage typing only rarely. Few pathogenicity tests are performed. Not many laboratories can conduct comprehensive biochemical tests on strains that cannot be identified readily by commercially available biochemical systems. Even fewer laboratories are equipped to perform plasmid profiles, gene probes, or DNA hybridization. These and other specialized tests for the serologic or biochemical identification of some exotic bacteria, yeasts, molds, protozoans, and viruses are best done in regional reference laboratories. It is not cost-effective for smaller laboratories to store and control the quality of reagents and media for tests that are seldom run or quite complex. In addition, it is impossible to maintain proficiency when tests are performed rarely. Sensitive methods for the epidemiologic subtyping of isolates from disease outbreaks, such as electrophoretic enzyme typing, rRNA fingerprinting, whole-cell protein electrophoretic patterns, and restriction endonuclease analysis of whole-cell or plasmid DNA, are used only in reference laboratories and a few large medical centers.

Specific genetic probes are now available commercially for identifying virulence factors and many bacteria and viruses. Genetic probes are among the most common methods used for identification of *Mycobacterium tuberculosis* and *M avium* complex in the U.S. today. Probes for *Neisseria gonorrhoeae* and *Chlamydia trachomatis* are now being used directly on clinical specimens with excellent sensitivity and almost universal specificity with same-day results. Mycobacterial probes are also being evaluated for direct specimen testing.

## Interfacing with Public Health

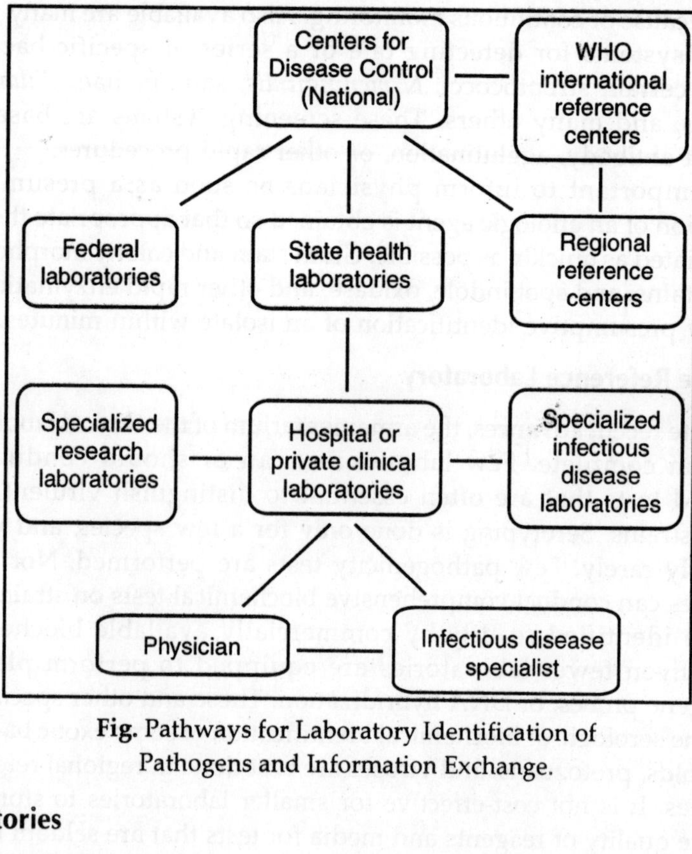


**Fig.** Pathways for Laboratory Identification of Pathogens and Information Exchange.

## Laboratories

Hospital and local clinical laboratories interact with district, state, and federal public health laboratories in several important ways. The clinical laboratories participate in quality control and proficiency testing programmes that are conducted by federally regulated agencies. The government reference laboratories supply cultures and often reagents for use in quality control, and they conduct training programs for clinical laboratory personnel.

All types of laboratories should interact closely to provide diagnostic services and epidemic surveillance. The primary concern of the clinical laboratory is identifying infectious disease agents and studying nosocomial and local outbreaks of disease.

When the situation warrants, the local laboratory may ask the state laboratory for help in identifying an unusual organism, discovering the cause or mode of transmission in a disease outbreak, or performing specialized tests not done routinely in clinical laboratories.

Cultures should be pure and should be sent on appropriate media following appropriate procedures for transport of biohazardous materials. Pertinent information, including the type of specimen; patient name (or number), date of birth, and sex; clinical diagnosis, associated illness, date of onset, and present condition; specific agent suspected, and any other organisms isolated; relevant epidemiologic and clinical data; treatment of patient; previous laboratory results (biochemical or serologic tests); and necessary information about the submitting party must accompany each request.

These data allow the state laboratory to test the specimen properly and quickly, and they provide information about occurrences within the state. For example, a food-borne outbreak might extend to many parts of the state (or beyond its boundaries). The state laboratory can alert local physicians to the possibility of such outbreaks.

Another necessary interaction between local and state laboratories is the reporting of notifiable diseases by the local laboratory. The state laboratory makes available to local laboratories summaries of the incidence of these diseases. The state laboratories also submit the summaries to the CDC weekly (or, for some diseases, yearly), and national summaries are published weekly in the *Morbidity and Mortality Weekly Report.*

Interaction between the CDC and state and federal laboratories is very similar to that between local and state laboratories. The CDC provides quality control cultures and reagents to state laboratories, and serves as a national reference laboratory for diagnostic services and epidemiologic surveillance. Local laboratories, however, must initially send specimens to the local or state public health laboratory, which, when necessary, forwards them to the CDC. The CDC reports its results back to the state laboratory, which then reports to the local laboratory.

## Hazards of Clinical Laboratory Work

Clinical laboratory personnel, including support and clerical employees, are subject to the risk of infection, chemical hazards, and, in some laboratories, radioactive contamination. Such risks can be prevented or minimized by a laboratory safety programme.

### *Radiation Hazards*

Personnel who work with radioactive materials should have taken a radioactivity safety course; they should wear radiation monitor badges and be aware of the methods for decontaminating hands, clothing, work surfaces, and equipment. They should wear gloves when working with radioactive compounds. When they work with high-level radiation, they should use a hood and stand behind a radiation shield. Preparative radioactive work should be done in a separate room with access only by personnel who are involved directly in the work.

### *Chemical Hazards*

Chemicals can harm laboratory personnel through inhalation or skin absorption of volatile compounds; bodily contact with carcinogens, acids, bases, and other harmful chemicals; or introduction of poisonous or skin-damaging liquids into the mouth. Good laboratory practices require that volatile compounds be handled only under a hood, that hazardous chemicals never be pipetted by mouth, and that anyone working with skin-damaging chemicals wear gloves, eye guards, and other personal protective equipment as necessary. Workers should be familiar with the materials safety data sheets (MSDS) posted in an accessible place in every laboratory. These forms contain information about chemical hazards and procedures for decontamination should an accident occur.

### *Biologic Hazards*

Microbiologic contamination is the greatest hazard in clinical microbiology laboratories. Laboratory infections are a danger not only to the clinical laboratory personnel but also to anyone else who enters the laboratory, including janitors, clerical and maintenance personnel, and visitors. The risk of infection is governed by the frequency and length of contact with the infectious agent, its virulence, the dose and route of administration, and the susceptibility of the host. The inherent hazard of any infectious agent is affected by factors such as the volume of infectious material used, handling of the material, effectiveness of safety containment equipment, and soundness of laboratory methods. Body fluids from patients, particularly those containing blood, are considered potentially infectious for blood-borne pathogens, and must be handled appropriately.

If possible, agents that are treated differently, such as viruses as opposed to bacteria, or *M tuberculosis* in contrast to *E coli*, should be handled in different laboratories or in different parts of the same laboratory. When the risk category of an agent is known, it should be handled in an area with appropriate containment. All specimens sent for microbiological studies and all organisms sent to the laboratory for identification should be assumed to be potentially infectious. A separate area should be set aside for the receipt of specimens. Personnel should be aware of the potential hazards of improperly packed, broken, or leaking packages and of the proper methods for their handling and decontamination.

To prevent infection, personnel should wear moisture-proof laboratory coats at all times, wash their hands before and after wearing gloves and at the conclusion of each potential exposure to etiologic agents, refrain from mouth pipetting, and not eat, drink, smoke, or apply cosmetics in the laboratory. Immunization may be appropriate for employees who are exposed often to certain infectious agents, including hepatitis B, yellow fever, rabies, polioviruses, meningococci, *Y pestis*, *S typhi*, and *Francisella tularensis*. Universal precautions, body substance isolation, and other mandated practices involve the use of personal protective equipment and engineering controls to minimize laboratory scientists' exposure to blood-borne pathogens, even when the risk of infection is unknown.

## Biosafety Levels

Infectious agents are assigned to a biosafety level from 1 to 4 on the basis of their virulence. The containment levels for organisms should correlate with the biosafety level assigned. Biosafety level 1 is for well-defined organisms not known to cause disease in healthy humans; it includes certain nonvirulent *E coli* strains (such as K-12) and *B subtilis*. Containment level 1 involves standard microbiologic practices, and safety equipment is not needed.

Biosafety level 2, the minimum level for clinical laboratories, is for moderate-risk agents associated with human disease. Containment level 2 includes limited access to the work area, decontamination of all infectious wastes, use of protective gloves, and a biologic safety cabinet for use in procedures that may create aerosols. Examples of biosafety level 2 agents include nematode, protozoan, trematode, and cestode human parasites; all human fungal pathogens except *Coccidioides immitis*; all members of the Enterobacteriaceae except *Y pestis*; *Bacillus anthracis*; *Clostridium tetani*; *Corynebacterium diphtheriae*; *Haemophilus* species; leptospires; legionellae; mycobacteria other than *M tuberculosis*; pathogenic *Neisseria* species;

staphylococci, streptococci, *Treponema pallidum; V cholerae;* and hepatitis and influenza viruses. Clinical specimens potentially containing some biosafety level 3 agents, such as *Brucella* spp., are usually handled using biosafety level 2 containment practices.

Biosafety level 3 is for agents that are associated with risk of serious or fatal aerosol infection. In containment level 3, laboratory access is controlled, special clothing is worn in the laboratory, and containment equipment is used for all work with the agent. *M tuberculosis, Coccidioides immitis, Coxiella burnetii,* and many of the arboviruses are biosafety 3 level agents. Containment level 3 usually is recommended for work with cultures of rickettsiae, brucellae, *Y pestis,* and a wide variety of viruses, including human immunodeficiency viruses.

Biosafety level 4 indicates dangerous and novel agents that cause diseases with high fatality rates. Maximum containment and decontamination procedures are used in containment level 4, which is found in only a few reference and research laboratories. Only a few viruses (including Lassa, Ebola, and Marburg viruses) are classified in biosafety level 4.

# Chapter 4

# Bacterial Metabolism

Metabolism refers to all the biochemical reactions that occur in a cell or organism. The study of bacterial metabolism focuses on the chemical diversity of substrate oxidations and dissimilation reactions (reactions by which substrate molecules are broken down), which normally function in bacteria to generate energy. Also within the scope of bacterial metabolism is the study of the uptake and utilization of the inorganic or organic compounds required for growth and maintenance of a cellular steady state (assimilation reactions). These respective exergonic (energy-yielding) and endergonic (energy-requiring) reactions are catalyzed within the living bacterial cell by integrated enzyme systems, the end result being self-replication of the cell. The capability of microbial cells to live, function, and replicate in an appropriate chemical milieu (such as a bacterial culture medium) and the chemical changes that result during this transformation constitute the scope of bacterial metabolism.

The bacterial cell is a highly specialized energy transformer. Chemical energy generated by substrate oxidations is conserved by formation of high-energy compounds such as adenosine diphosphate (ADP) and adenosine triphosphate (ATP) or compounds containing the thioester bond

$$(R-\overset{\overset{\displaystyle O}{||}}{C}\sim S-R)$$, such as acetyl ~ S – coenzyme A

(acetyl ~ SCoA) or succinyl ~ SCoA. ADP and ATP represent adenosine monophosphate (AMP) plus one and two high-energy phosphates (AMP ~ P and AMP ~ P~ P, respectively); the energy is stored in these compounds as high-energy phosphate bonds. In the presence of proper enzyme systems, these compounds can be used as energy sources to synthesize the new complex organic compounds needed by the cell. All living cells must maintain steady-state biochemical reactions for the formation and use of such high-energy compounds.

Kluyver and Donker (1924 to 1926) recognized that bacterial cells, regardless of species, were in many respects similar chemically to all other living cells. For example, these investigators recognized that hydrogen transfer is a common and fundamental feature of all metabolic processes. Bacteria, like mammalian and plant cells, use ATP or the high-energy phosphate bond (~ P) as the primary chemical energy source. Bacteria also require the B-complex vitamins as functional coenzymes for many oxidation-reduction reactions needed for growth and energy transformation.

An organism such as *Thiobacillus thiooxidans*, grown in a medium containing only sulfur and inorganic salts, synthesizes large amounts of thiamine, riboflavine, nicotinic acid, pantothenic acid, pyridoxine, and biotin.

Therefore, Kluyver proposed the unity theory of biochemistry (*Die Einheit in der Biochemie*), which states that all basic enzymatic reactions which support and maintain life processes within cells of organisms, had more similarities than differences.

This concept of biochemical unity stimulated many investigators to use bacteria as model systems for studying related eukaryotic, plant and animal biochemical reactions that are essentially "identical" at the molecular level.

From a nutritional, or metabolic, viewpoint, three major physiologic types of bacteria exist: the heterotrophs (or chemoorganotrophs), the autotrophs (or chemolithotrophs), and the photosynthetic bacteria (or phototrophs). These are discussed below.

## HETEROTROPHIC METABOLISM

Heterotrophic bacteria, which include all pathogens, obtain energy from oxidation of organic compounds.

Carbohydrates (particularly glucose), lipids, and protein are the most commonly oxidized compounds. Biologic oxidation of these organic compounds by bacteria results in synthesis of ATP as the chemical energy source. This process also permits generation of simpler organic compounds (precursor molecules) needed by the bacteria cell for biosynthetic or assimilatory reactions.

**Table. Nutritional Diversity Exhibited by Physiologically Different Bacteria.**

| Required Components for Bacterial Growth | | | | |
|---|---|---|---|---|
| Physiologic Type | Carbon Source | Nitrogen Source | Energy source | Hydrogen Source |
| Heterotrophic (chemoorganotrophic) | Organic | Organic or inorganic | Oxidation of organic compounds | —— |
| Autotrophic (chemolithotrophic) | $CO_2$ | Inorganic | Oxidation of inorganic compounds | —— |

| | | | | |
|---|---|---|---|---|
| Photosynthetic Photolithotrophic (Bacterial) | $CO_2$ | Inorganic | Sunlight | $H_2S$ or $H_2$ |
| Cyanobacterial | $CO_2$ | Inorganic | Sunlight | Photolysis of $H_2O$ |
| Photoorganotrophic | CO2 | Inorganic | Sunlight | Organic compounds |

- Common inorganic nitrogen sources are $NO_2$ or $NH_4$ ions: nitrogen fixers can use $N_2$.
- Many phototrophs and chemotrophs are nitrogen-fixing organisms.
- Results in $O_2$ evolution (or oxygenic photosynthesis) as commonly occurs in plants.
- Organic acids such as formate, acetate, and succinate can serve as hydrogen donors.

The Krebs cycle intermediate compounds serve as precursor molecules (building blocks) for the energy-requiring biosynthesis of complex organic compounds in bacteria. Degradation reactions that simultaneously produce energy and generate precursor molecules for the biosynthesis of new cellular constituents are called amphibolic.

All heterotrophic bacteria require preformed organic compounds. These carbon- and nitrogen-containing compounds are growth substrates, which are used aerobically or anaerobically to generate reducing equivalents (e.g., reduced nicotinamide adenine dinucleotide; NADH + $H^+$); these reducing equivalents in turn are chemical energy sources for all biologic oxidative and fermentative systems. Heterotrophs are the most commonly studied bacteria; they grow readily in media containing carbohydrates, proteins, or other complex nutrients such as blood. Also, growth media may be enriched by the addition of other naturally occurring compounds such as milk (to study lactic acid bacteria) or hydrocarbons (to study hydrocarbon-oxidizing organisms).

## RESPIRATION

Glucose is the most common substrate used for studying heterotrophic metabolism. Most aerobic organisms oxidize glucose completely by the following reaction equation:

$$C_6H_{12}O_6 + 6O_2 \longrightarrow 6CO_2 + 6H_2O + \text{energy}$$

This equation expresses the cellular oxidation process called respiration. Respiration occurs within the cells of plants and animals, normally generating 38 ATP molecules (as energy) from the oxidation of 1 molecule of glucose. This yields approximately 380,000 calories (cal) per mode of glucose (ATP ~ 10,000 cal/mole). Thermodynamically, the complete oxidation of one mole of glucose should yield approximately 688,000 cal; the energy that is not conserved biologically as chemical energy (or ATP formation) is liberated as heat (308,000 cal). Thus, the cellular respiratory process is at best about 55% efficient.

Glucose oxidation is the most commonly studied dissimilatory reaction leading to energy production or ATP synthesis. The complete oxidation of glucose may involve three fundamental biochemical pathways. The first is the glycolytic or Embden- Meyerhof-Parnas pathway, the second is the Krebs cycle (also called the citric acid cycle or tricarboxylic acid cycle), and the third is the series of membrane-bound electron transport oxidations coupled to oxidative phosphorylation.

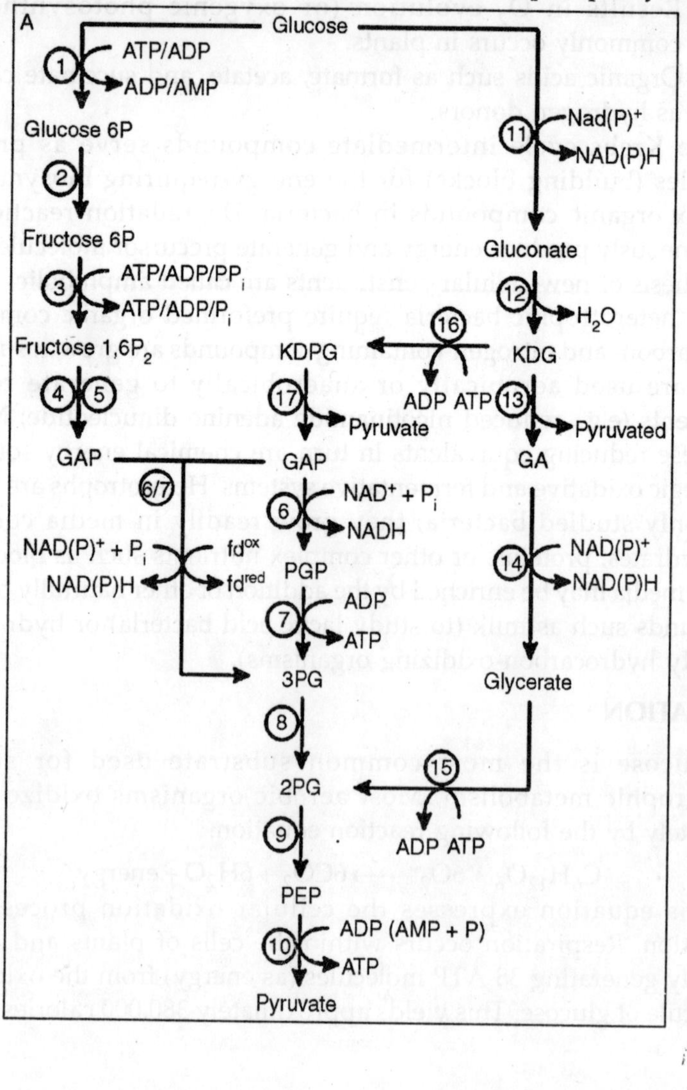

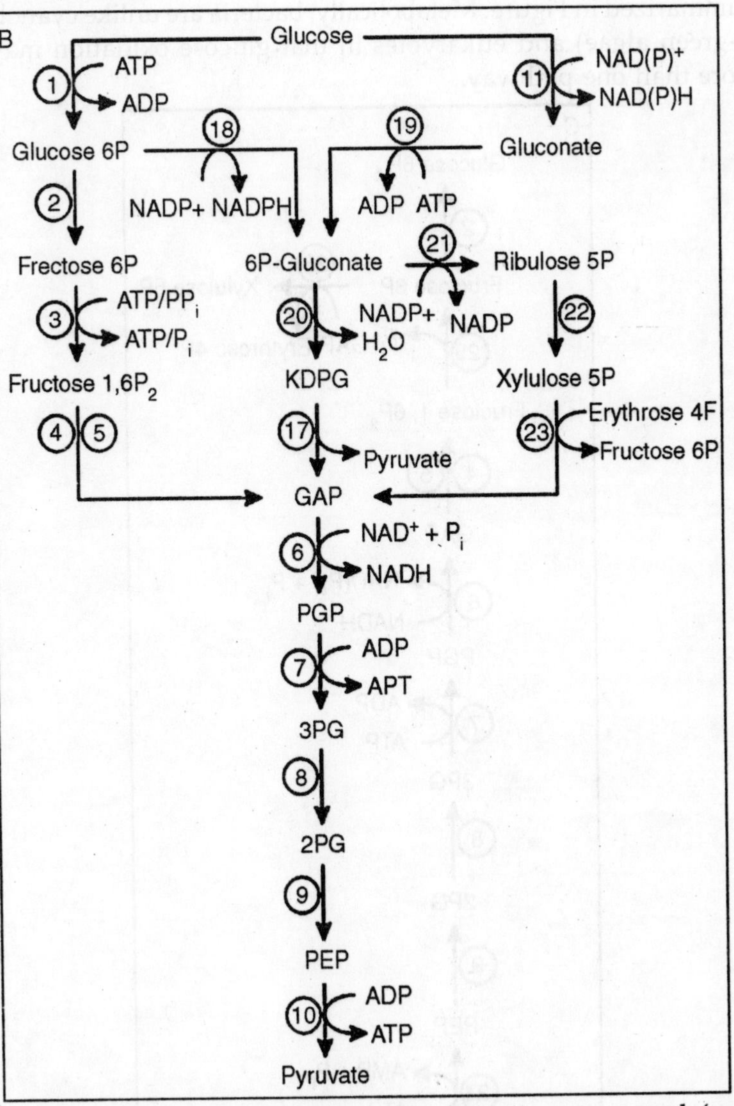


Respiration takes place when any organic compound (usually carbohydrate) is oxidized completely to $CO_2$ and $H_2O$. In aerobic respiration, molecular $O_2$ serves as the terminal acceptor of electrons.

For anaerobic respiration, $NO_3^-$, $SO_4^{2-}$, $CO_2$, or fumarate can serve as terminal electron acceptors (rather than $O_2$), depending on the bacterium studied. The end result of the respiratory process is the complete oxidation of the organic substrate molecule, and the end products formed are primarily $CO_2$ and $H_2O$. Ammonia is formed also if protein (or amino acid) is the substrate oxidized. The biochemical pathways normally involved in oxidation of various naturally occurring organic compounds are summarized in Figure. Metabolically, bacteria are unlike cyanobacteria (blue-green algae) and eukaryotes in that glucose oxidation may occur by more than one pathway.

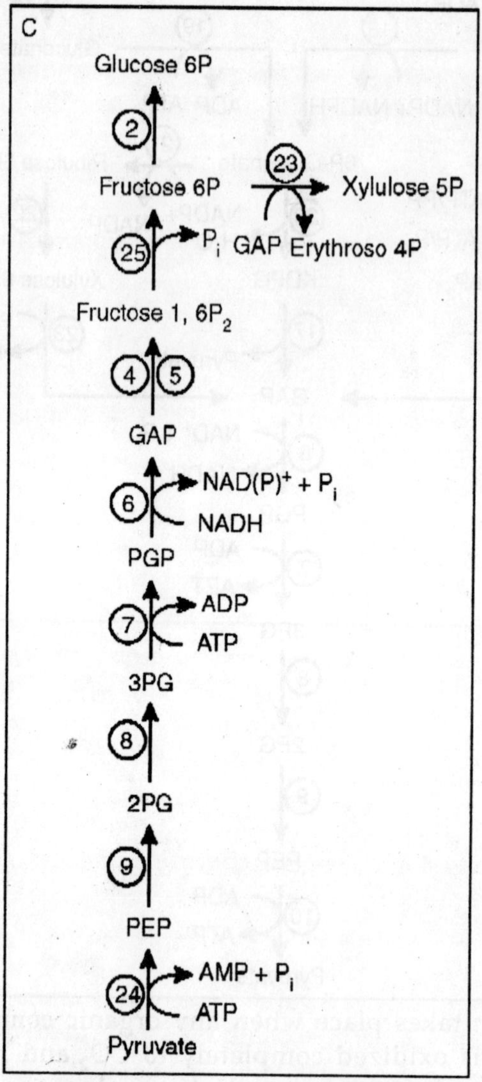


Fig. Glycolytic (EMP) Pathway.

In bacteria, glycolysis represents one of several pathways by which bacteria can catabolically attack glucose. The glycolytic pathway is most commonly associated with anaerobic or fermentative metabolism in bacteria and yeasts. In bacteria, other minor heterofermentative pathways, such as the phosphoketolase pathway, also exist.

In addition, two other glucose-catabolizing pathways are found in bacteria: the oxidative pentose phosphate pathway (hexose monophosphate shunt), and the Entner-Doudoroff pathway, which is almost exclusively found in obligate aerobic bacteria.

The highly oxidative *Azotobacter* and most *Pseudomonas* species, for example, utilize the Entner-Doudoroff pathway for glucose catabolism, because these organisms lack the enzyme phosphofructokinase and hence cannot synthesize fructose 1,6-diphosphate, a key intermediate compound in the glycolytic pathway. (Phospho-fructokinase is also sensitive to molecular $O_2$ and does not function in obligate aerobes).

Other bacteria, which lack aldolase (which splits fructose-1,6-diphosphate into two triose phosphate compounds), also cannot have a functional glycolytic pathway.

Although the Entner-Doudoroff pathway is usually associated with obligate aerobic bacteria, it is present in the facultative anaerobe *Zymomonas mobilis* (formerly *Pseudomonas lindneri*). This organism dissimilates glucose to ethanol and represents a major alcoholic fermentation reaction in a bacterium.

Glucose dissimilation also occurs by the hexose monophosphate shunt. This oxidative pathway was discovered in tissues that actively metabolize glucose in the presence of two glycolytic pathway inhibitors (iodoacetate and fluoride). Neither inhibitor had an effect on glucose dissimilation, and NADPH + $H^+$ generation occurred directly from the oxidation of glucose-6-phosphate (to 6-phosphoglucono-d-lactone) by glucose-6phosphate dehydrogenase. The pentose phosphate pathway subsequently permits the direct oxidative decarboxylation of glucose to pentoses. The capability of this oxidative metabolic system to bypass glycolysis explains the term shunt.

The biochemical reactions of the Entner-Doudoroff pathway are a modification of the hexose monophosphate shunt, except that pentose sugars are not directly formed. The two pathways are identical up to the formation of 6-phosphogluconate and then diverge.

In the Entner-Doudoroff pathway, no oxidative decarboxylation of 6-phosphogluconate occurs and no pentose compound is formed. For this pathway, a new 6 carbon compound intermediate (2-keto-3-deoxy6-phosphogluconate) is generated by the action of 6-phosphogluconate dehydratase (an $Fe^{2+}$- and glutathione-stimulated enzyme); this intermediate compound is then directly cleaved into the triose (pyruvate) and a triose-phosphate compound (glyceraldehyde-3-phosphate) by the 2-keto-3-deoxy6-phosphogluconate aldolase.

The glyceraldehyde-3-phosphate is further oxidized to another pyruvate molecule by the same enzyme systems that catalyze the terminal glycolytic pathway.

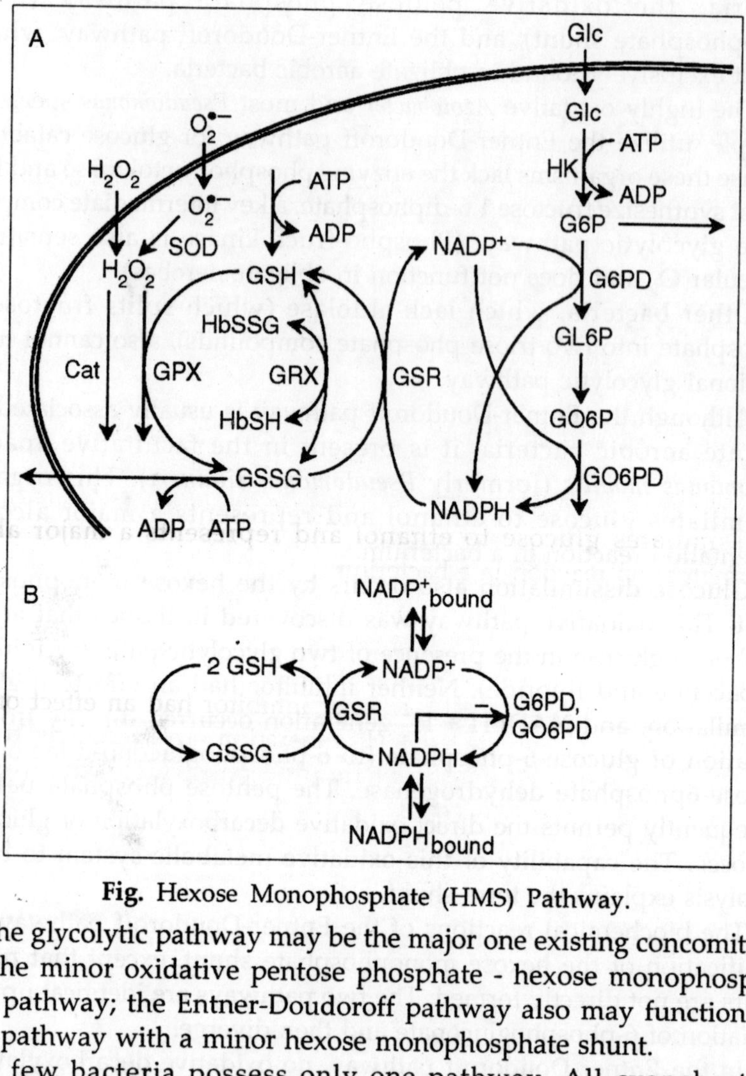


**Fig.** Hexose Monophosphate (HMS) Pathway.

The glycolytic pathway may be the major one existing concomitantly with the minor oxidative pentose phosphate - hexose monophosphate shunt pathway; the Entner-Doudoroff pathway also may function as a major pathway with a minor hexose monophosphate shunt.

A few bacteria possess only one pathway. All cyanobacteria, *Acetobacter suboxydans,* and *A xylinum* possess only the hexose monophosphate shunt pathway; *Pseudomonas saccharophilia* and *Z mobilis* possess solely the Entner-Doudoroff pathway. Thus, the end products of glucose dissimilatory pathways are as follows:

All major pathways of glucose or hexose catabolism have several metabolic features in common. First, there are the preparatory steps by which key intermediate compounds such as the triose-$PO_4$, glyceraldehyde-3-phosphate, and/or pyruvate are generated.

The latter two compounds are almost universally required for further assimilatory or dissimilatory reactions within the cell. Second, the major source of phosphate for all reactions involving phosphorylation of glucose or other hexoses is ATP, not inorganic phosphate (Pi).

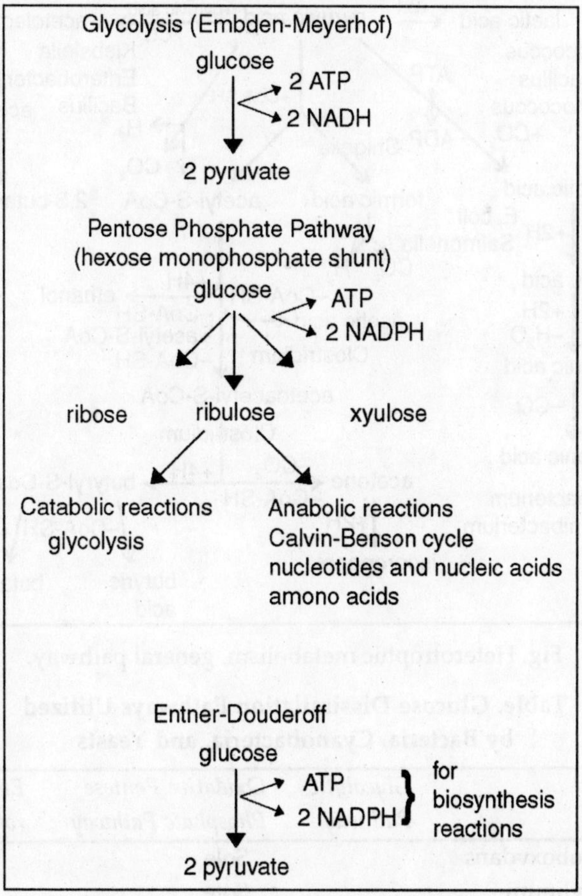


**Fig.** Entner-Doudoroff (ED) Pathway.

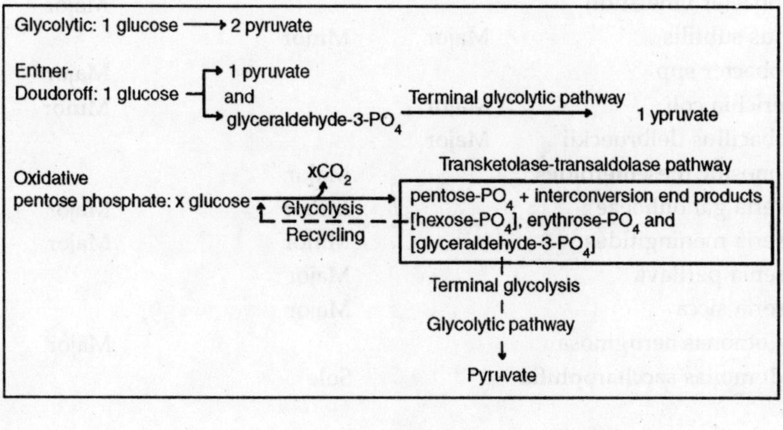

The glucose dissimilation pathways used by specific microorganisms are shown in Table.

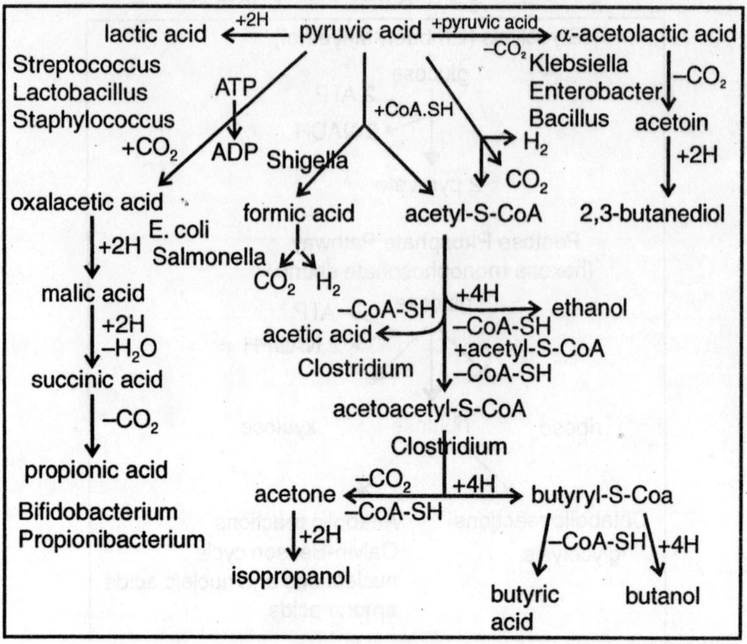


**Fig.** Heterotrophic metabolism, general pathway.

**Table. Glucose Dissimilation Pathways Utilized by Bacteria, Cyanobacteria, and Yeasts**

| *Bacteria* | *Glycolytic Pathway* | *Oxidative Pentose Phosphate Pathway* | *Entner-Doudoroff Pathway* |
|---|---|---|---|
| Acetobacter Suboxydans | | Sole | |
| Acetobacter xylinum | | Sote | |
| Agrobacterium spp | | | Major |
| Azotobacter vinelandii | | | Major |
| Bacillus subtilis | Major | Minor | |
| Caulobacter spp | | | Major |
| Escherichia coli | Major | | Minor |
| Lactobacillus delbrueckii | Major | | |
| Leuconostoc mesenteroides | | Major | |
| Neisseria ganonhoeae | | Minor | Major |
| Neisseria meningitides | | Minor | Major |
| Neissenia perllava | | Major | |
| Neisseria sicca | | Major | |
| Pseudomonas aeruginosa | | | Major |
| Pseudomonas saccharophifia | | Sole | |

| | | | |
|---|---|---|---|
| Rhizobium spp | | | Major |
| Sarcina jorea | Major | Minor | |
| Spirillum spp | | | Major |
| Streptococcus faecalis | | | (Major) |
| Streptomyces griseus | Major | Minor | |
| Zymomonas anaerobia | | | Sole |
| Zymomonas mobilis | | | Sole |
| All cyanobacteria | | | Sole |
| All yeasts | Major | Minor | |

Most species utilize the Enter. Doudoroff pathway as major pathway.
Induced by growth on gluconate.

Actually, chemical energy contained in ATP must be initially spent in the first step of glucose metabolism (via kinase-type enzymes) to generate glucose-6-phosphate, which initiates the reactions involving hexose catabolism. Third, NADH + $H^+$ or NADPH + $H^+$ is generated as reducing equivalents (potential energy) directly by one or more of the enzymatic reactions involved in each of these pathways.

## FERMENTATION

Fermentation, another example of heterotrophic metabolism, requires an organic compound as a terminal electron (or hydrogen) acceptor. In fermentations, simple organic end products are formed from the anaerobic dissimilation of glucose (or some other compound). Energy (ATP) is generated through the dehydrogenation reactions that occur as glucose is broken down enzymatically.

The simple organic end products formed from this incomplete biologic oxidation process also serve as final electron and hydrogen acceptors. On reduction, these organic end products are secreted into the medium as waste metabolites (usually alcohol or acid). The organic substrate compounds are incompletely oxidized by bacteria, yet yield sufficient energy for microbial growth. Glucose is the most common hexose used to study fermentation reactions.

In the late 1850s, Pasteur demonstrated that fermentation is a vital process associated with the growth of specific microorganisms, and that each type of fermentation can be defined by the principal organic end product formed (lactic acid, ethanol, acetic acid, or butyric acid). His studies on butyric acid fermentation led directly to the discovery of anaerobic microorganisms. Pasteur concluded that oxygen inhibited the microorganisms responsible for butyric acid fermentation because both bacterial mobility and butyric acid formation ceased when air was bubbled into the fermentation mixture. Pasteur also introduced the terms aerobic and anaerobic. His views on fermentation are made clear from his microbiologic studies on the production of beer:

In the experiments which we have described, fermentation by yeast is seen to be the direct consequence of the processes of nutrition, assimilation and life, when these are carried on without the agency of free oxygen. The heat required in the accomplishment of that work must necessarily have been borrowed from the decomposition of the fermentation matter.... Fermentation by yeast appears, therefore, to be essentially connected with the property possessed by this minute cellular plant of performing its respiratory functions, somehow or other, with the oxygen existing combined in sugar.

For most microbial fermentations, glucose dissimilation occurs through the glycolytic pathway. The simple organic compound most commonly generated is pyruvate, or a compound derived enzymatically from pyruvate, such as acetaldehyde, α-acetolactate, acetyl ~ SCoA, or lactyl ~ SCoA. Acetaldehyde can then be reduced by NADH + $H_+$ to ethanol, which is excreted by the cell.

The end product of lactic acid fermentation, which occurs in streptococci (e.g., *Streptococcus lactis*) and many lactobacilli (e.g., *Lactobacillus casei, L pentosus*), is a single organic acid, lactic acid. Organisms that produce only lactic acid from glucose fermentation are homofermenters.

Homofermentative lactic acid bacteria dissimilate glucose exclusively through the glycolytic pathway. Organisms that ferment glucose to multiple end products, such as acetic acid, ethanol, formic acid, and $CO_2$, are referred to as heterofermenters. Examples of heterofermentative bacteria include *Lactobacillus, Leuconostoc,* and *Microbacterium* species. Heterofermentative fermentations are more common among bacteria, as in the mixed-acid fermentations carried out by bacteria of the family Enterobacteriaceae (e.g., *Escherichia coli, Salmonella, Shigella,* and *Proteus* species).

Many of these glucose fermenters usually produce $CO_2$ and $H_2$ with different combinations of acid end products (formate, acetate, lactate, and succinate). Other bacteria such as *Enterobacter aerogenes, Aeromonas, Serratia, Erwinia,* and *Bacillus* species also form $CO_2$ and $H_2$ as well as other neutral end products (ethanol, acetylmethylcarbinol [acetoin], and 2,3-butylene glycol). Many obligately anaerobic clostridia (e.g., *Clostridium saccharobutyricum, C thermosaccharolyticum*) and *Butyribacterium* species ferment glucose with the production of butyrate, acetate, $CO_2$, and $H_2$, whereas other *Clostridum* species (*C acetobutylicum* and *C butyricum*) also form these fermentation end products plus others (butanol, acetone, isopropanol, formate, and ethanol).

Similarly, the anaerobic propionic acid bacteria (*Propionibacterium* species) and the related *Veillonella* species ferment glucose to form $CO_2$, propionate, acetate, and succinate. In these bacteria, propionate is formed by the partial reversal of the Krebs cycle reactions and involves a $CO_2$ fixation by pyruvate (the Wood-Werkman reaction) that forms oxaloacetate (a four-carbon intermediate).

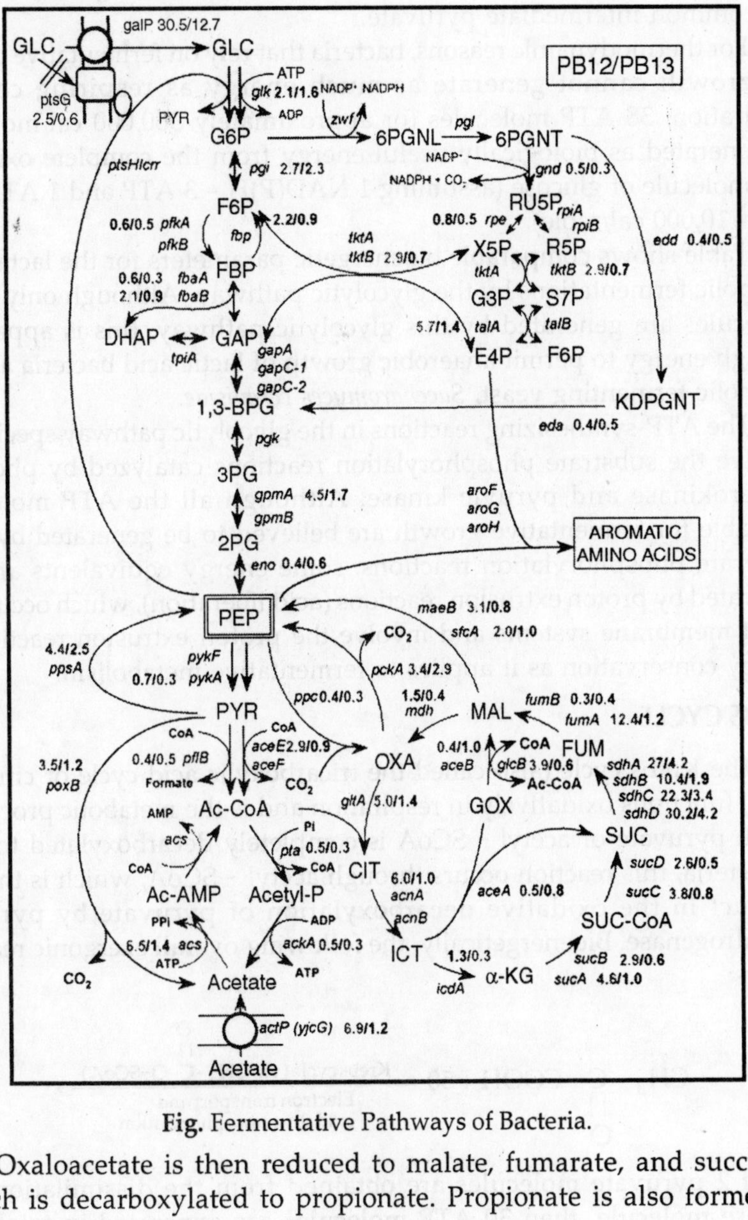


Fig. Fermentative Pathways of Bacteria.

Oxaloacetate is then reduced to malate, fumarate, and succinate, which is decarboxylated to propionate. Propionate is also formed by another three-carbon pathway in C *propionicum, Bacteroides ruminicola,* and *Peptostreptococcus* species, involving a lactyl ~ SCoA intermediate. The obligately aerobic acetic acid bacteria (*Acetobacter* and the related *Gluconobacter* species) can also ferment glucose, producing acetate and gluconate. Figure summarizes the pathways by which the various major fermentation end products form from the dissimilation of glucose through the common intermediate pyruvate.

For thermodynamic reasons, bacteria that rely on fermentative process for growth cannot generate as much energy as respiring cells. In respiration, 38 ATP molecules (or approximately 380,000 cal/mole) can be generated as biologically useful energy from the complete oxidation of 1 molecule of glucose (assuming 1 NAD(P)H = 3 ATP and 1 ATP ADP + Pi = 10,000 cal/mole).

Table shows comparable bioenergetic parameters for the lactate and ethanolic fermentations by the glycolytic pathway. Although only 2 ATP molecules are generated by this glycolytic pathway, this is apparently enough energy to permit anaerobic growth of lactic acid bacteria and the ethanolic fermenting yeast, *Saccharomyces cerevisiae.*

The ATP-synthesizing reactions in the glycolytic pathway specifically involve the substrate phosphorylation reactions catalyzed by phosphoglycerokinase and pyruvic kinase. Although all the ATP molecules available for fermentative growth are believed to be generated by these substrate phosphorylation reactions, some energy equivalents are also generated by proton extrusion reactions (acid liberation), which occur with intact membrane systems and involve the proton extrusion reactions of energy conservation as it applies to fermentative metabolism.

## KREBS CYCLE

The Krebs cycle (also called the tricarboxylic acid cycle or citic acid cycle) functions oxidatively in respiration and is the metabolic process by which pyruvate or acetyl ~ SCoA is completely decarboxylated to $CO_2$. In bacteria, this reaction occurs through acetyl ~ SCoA, which is the first product in the oxidative decarboxylation of pyruvate by pyruvate dehydrogenase. Bioenergetically, the following overall exergonic reaction occurs:

$$CH_3-\underset{\underset{O}{||}}{C}-COOH+50 \xrightarrow[\text{Electron transport and oxidative phosphorylation}]{\text{Krebs cycle (via } CH_3-\overset{\overset{O}{||}}{C}-O\sim SCoA)}$$

If 2 pyruvate molecules are obtained from the dissimilation of 1 glucose molecule, then 30 ATP molecules are generated in total. The decarboxylation of pyruvate, isocitrate, and a-ketoglutarate accounts for all $CO_2$ molecules generated during the respiratory process. Figure shows, the enzymatic reactions in the Krebs cycle. The chemical energy conserved by the Krebs cycle is contained in the reduced compounds is not available as ATP until the final step of respiration (electron transport and oxidative phosphorylation) occurs.

**Table. Energy Obtained from Bacterial Fermentations by Substrate Phosphorylations**

| *Fermentation* | *Actual Energy (cal/mole)* | *Theoretical Energy (cal/mole)* | *Efficiency (%)* |
|---|---|---|---|
| Homolactic | | | |
| $\underset{\text{(Glucose)}}{C_4H_{12}O_6} \xrightarrow{\text{Glycolysis}} 2CH_3-\underset{\text{OH}}{\overset{\text{H}}{\text{C}}}-COOH+$ (Lactic acid) | –20,000 | 57,000 | 35 |
| Alcoholic | | | |
| $\underset{\text{(Glucose)}}{C_4H_{12}O_6} \xrightarrow{\text{Glycolysis}} 2CH_3-\underset{\text{H}}{\overset{\text{H}}{\text{C}}}-OH+2CO_2+$ (Elhanol) | –20,000 | 58,000 | 34 |

## GLYOXYLATE CYCLE

In general, the Krebs cycle functions similarly in bacteria and eukaryotic systems, but major differences are found among bacteria. One difference is that in obligate aerobes, L-malate may be oxidized directly by molecular $O_2$ via an electron transport chain. In other bacteria, only some Krebs cycle intermediate reactions occur because α-ketoglutarate dehydrogenase is missing.

A modification of the Krebs cycle, commonly called the glyoxylate cycle, or shunt, which exists in some bacteria. This shunt functions similarly to the Krebs cycle but lacks many of the Krebs cycle enzyme reactions. The glyoxylate cycle is primarily an oxidative pathway in which acetyl~SCoA is generated from the oxidation, of acetate, which usually is derived from the oxidation of fatty acids. The oxidation of fatty acids to acetyl~SCoA is carried out by the β-oxidation pathway. Pyruvate oxidation is not directly involved in the glyoxylate shunt, yet this shunt yields sufficient succinate and malate, which are required for energy production.

The glyoxylate cycle also generates other precursor compounds needed for biosynthesis.

The glyoxylate cycle was discovered as an unusual metabolic pathway during an attempt to learn how lipid (or acetate) oxidation in bacteria and plant seeds could lead to the direct biosynthesis of carbohydrates.

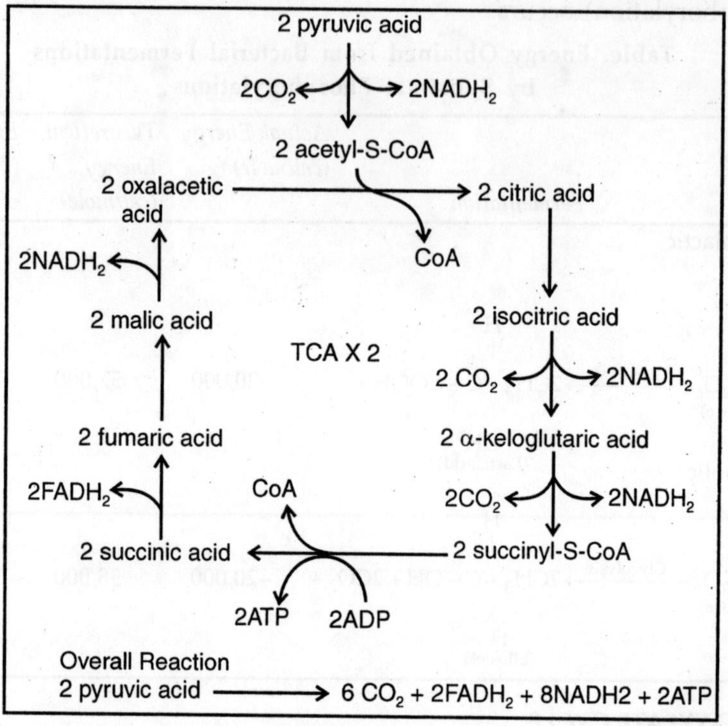


**Fig.** Krebs Cycle (also Tricarboxylic Acid or Citric Acid Cycle).

The glyoxylate cycle converts oxaloacetate either to pyruvate and $CO_2$ (catalyzed by pyruvate carboxylase) or to phosphoenolpyruvate and $CO_2$ (catalyzed by the inosine triphosphate [ITP]-dependent phosphoenolpyruvate carboxylase kinase).

Either triose compound can then be converted to glucose by reversal of the glycolytic pathway. The glyoxylate cycle is found in many bacteria, including *Azotobacter vinelandii* and particularly in organisms that grow well in media in which acetate and other Krebs cycle dicarboxylic acid intermediates are the sole carbon growth source.

One primary function of the glyoxylate cycle is to replenish the tricarboxylic and dicarboxylic acid intermediates that are normally provided by the Krebs cycle. A pathway whose primary purpose is to replenish such intermediate compounds is called anaplerotic.

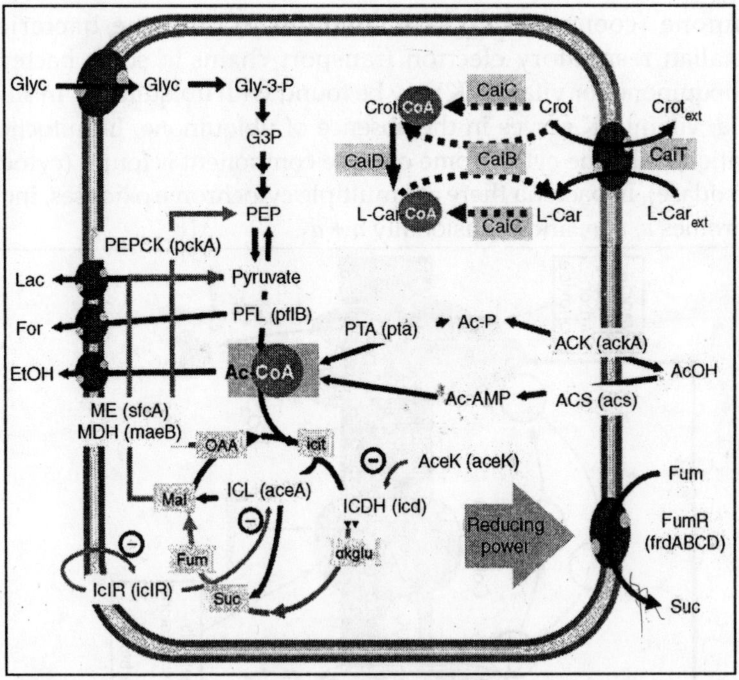


**Fig.** Glyoxylate Shunt.

## ELECTRON TRANSPORT AND OXIDATIVE PHOSPHORYLATION

The final stage of respiration occurs through a series of oxidation-reduction electron transfer reactions that yield the energy to drive oxidative phosphorylation; this in turn produces ATP. The enzymes involved in electron transport and oxidative phosphorylation reside on the bacterial inner (cytoplasmic) membrane. This membrane is invaginated to form structures called respiratory vesicles, lamellar vesicles, or mesosomes, which function as the bacterial equivalent of the eukaryotic mitochondrial membrane.

Respiratory electron transport chains vary greatly among bacteria, and in some organisms are absent. The respiratory electron transport chain of eukaryotic mitochondria oxidizes NADH + $H_+$, NADPH + H+, and succinate (as well as the coacylated fatty acids such as acetyl~SCoA). The bacterial electron transport chain also oxidizes these compounds, but it can also directly oxidize, via non-pyridine nucleotide-dependent pathways, a larger variety of reduced substrates such as lactate, malate, formate, α-glycerophosphate, $H_2$, and glutamate. The respiratory electron carriers in bacterial electron transport systems are more varied than in eukaryotes, and the chain is usually branched at the site(s) reacting with molecular $O_2$. Some electron carriers, such as nonheme iron centers and ubiquinone (coenzyme Q), are common to both the bacterial and mammalian respiratory electron transport chains.In some bacteria, the naphthoquinones or vitamin K may be found with ubiquinone. In still other bacteria, vitamin K serves in the absence of ubiquinone. In mitochondrial respiration, only one cytochrome oxidase component is found (cytochrome $a + a_3$ oxidase). In bacteria there are multiple cytochrome oxidases, including cytochromes *a, d, o,* and occasionally $a + a_3$.

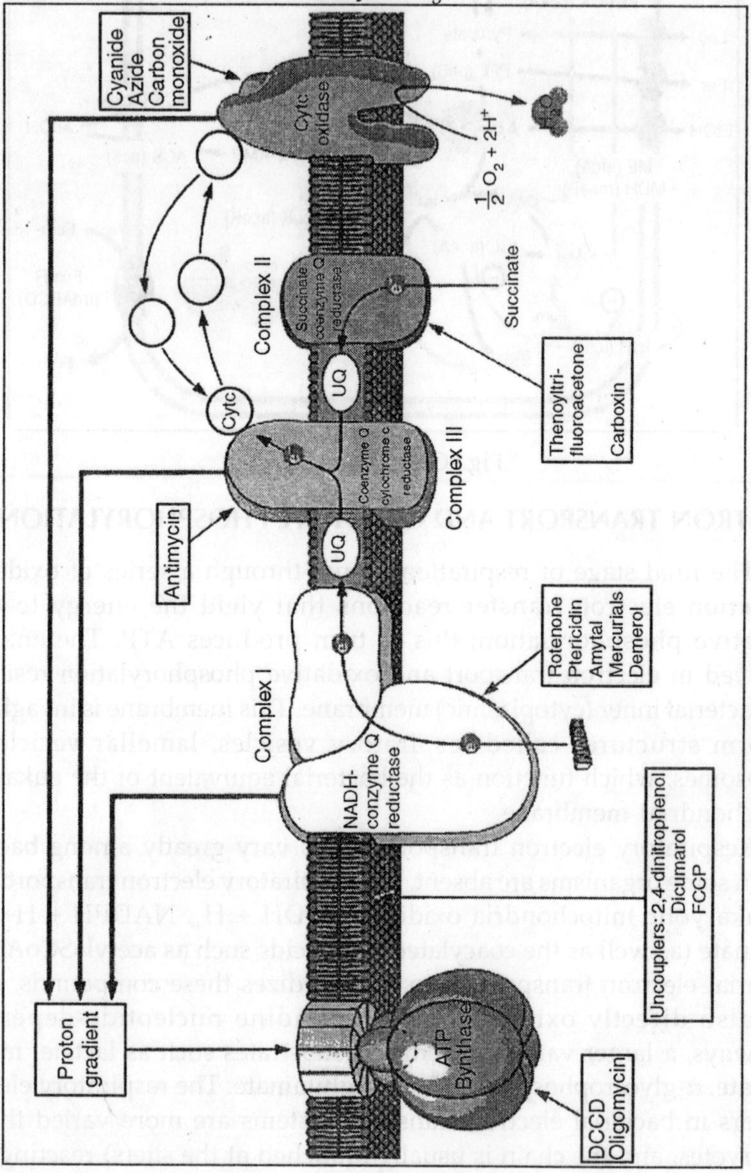


Fig. Respiratory Electron Transport Chains.

In bacteria cytochrome oxidases usually occur as combinations of *a1: d: o* and $a + a_3$: *o*. Bacteria also possess mixed-function oxidases such as cytochromes P-450 and P-420 and cytochromes *c′* and *c′c′*, which also react with carbon monoxide. These diverse types of oxygen-reactive cytochromes undoubtedly have evolutionary significance. Bacteria were present before $O_2$ was formed; when $O_2$ became available as a metabolite, bacteria evolved to use it in different ways; this probably accounts for the diversity in bacterial oxygen-reactive hemoproteins.

Cytochrome oxidases in many pathogenic bacteria are studied by the bacterial oxidase reaction, which subdivides Gram-negative organisms into two major groups, oxidase positive and oxidase negative. This oxidase reaction is assayed for by using N,N,N′, N′-tetramethyl-*p*-phenylenediamine oxidation (to Wurster's blue) or by using indophenol blue synthesis (with dimethyl-*p*-phenylenediamine and α-naphthol).

Oxidase-positive bacteria contain integrated (cytochrome *c* type:oxidase) complexes, the oxidase component most frequently encountered is cytochrome *o*, and occasionally $a + a_3$. The cytochrome oxidase responsible for the indophenol oxidase reaction complex was isolated from membranes of *Azotobacter vinelandii*, a bacterium with the highest respiratory rate of any known cell. The cytochrome oxidase was found to be an integrated cytochrome $c_4$*:o complex,* which was shown to be present in *Bacillus* species. These *Bacillus* strains are also highly oxidase positive, and most are found in morphologic group II.

Both bacterial and mammalian electron transfer systems can carry out electron transfer (oxidation) reactions with $NADH + H^+$, $NADPH + H^+$, and succinate.

Energy generated from such membrane oxidations is conserved within the membrane and then transferred in a coupled manner to drive the formation of ATP. The electron transfer sequence is accomplished entirely by membrane-bound enzyme systems. As the electrons are transferred by a specific sequence of electron carriers, ATP is synthesized from ADP + inorganic phosphate (Pi) or orthophosphoric acid ($H_3PO_4$).

In respiration, the electron transfer reaction is the primary mode of generating energy; electrons ($2e^-$) from a low-redox-potential compound such as $NADH + H_+$ are sequentially transferred to a specific flavoprotein dehydrogenase or oxidoreductase (flavin mononucleotide [FMN] type for NADH or flavin adenine dinucleotide [FAD] type for succinate); this electron pair is then transferred to a nonheme iron centre (FeS) and finally to a specific ubiquinone or a naphthoquinone derivative. This transfer of electrons causes a differential chemical redox potential change so that within the membrane enough chemical energy is conserved to be transferred by a coupling mechanism to a high-energy compound (e.g.,

ADP + Pi Æ ATP). ATP molecules represent the final stable high-energy intermediate compound formed.

A similar series of redox changes also occurs between ubiquinone and cytochrome c, but with a greater differential in the oxidation-reduction potential level, which allows for another ATP synthesis step. The final electron transfer reaction occurs at the cytochrome oxidase level between reduced cyotchrome c and molecular $O_2$; this reaction is the terminal ATP synthesis step.

## MITCHELL OR PROTON EXTRUSION HYPOTHESIS

A highly complex but attractive theory to explain energy conservation in biologic systems is the chemiosmotic coupling of oxidative and photosynthetic phosphorylations, commonly called the Mitchell hypothesis. This theory attempts to explain the conservation of free energy in this process on the basis of an osmotic potential caused by a proton concentration differential (or proton gradient) across a proton-impermeable membrane.

Energy is generated by a proton extrusion reaction during membrane-bound electron transport, which in essence serve as a proton pump; energy conservation and coupling follow. This represents an obligatory "intact" membrane phenomenon.

The energy thus conserved (again within the confines of the membrane and is coupled to ATP synthesis. This would occur in all biologic cells, even in the lactic acid bacteria that lack a cytochrome-dependent electron transport chain but still possesses a cytoplasmic membrane.

In this hypothesis, the membrane allows for charge separation, thus forming a proton gradient that drives all bioenergization reactions. By such means, electromotive forces can be generated by oxidation-reduction reactions that can be directly coupled to ion translocations, as in the separation of $H^+$ and OH ions in electrochemical systems. Thus, an enzyme or an electron transfer carrier on a membrane that undergoes an oxidation-reduction reaction serves as a specific conductor for OH (or $O_2$), and "hydrodehydration" provides electromotive power, as it does in electrochemical cells.

The concept underlying Mitchell's hypothesis is complex, and many modifications have been proposed, but the theory's most attractive feature is that it unifies all bioenergetic conservation principles into a single concept requiring an intact membrane vesicle to function properly. Figure shows how the Mitchell hypothesis might be used to explain energy generation, conservation, and transfer by a coupling process. The least satisfying aspect of the chemiosmotic hypothesis is the lack of understanding of how chemical energy is actually conserved within the membrane and how it is transmitted by coupling for ATP synthesis.

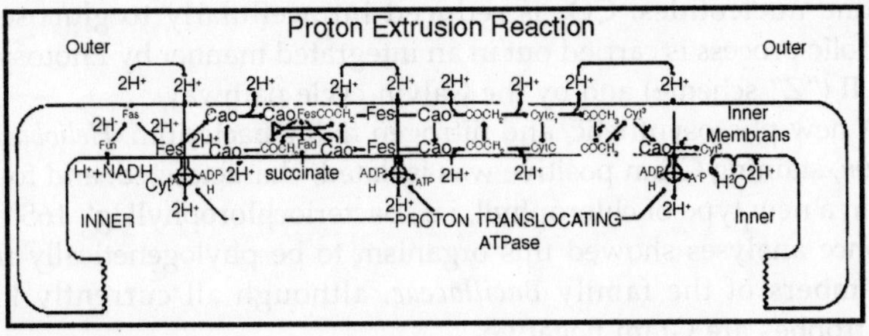


**Fig.** Mitchell Hypotheses, a Chemiosmotic Model of Energy Transduction.

## BACTERIAL PHOTOSYNTHESIS

Many prokaryotes (bacteria and cyanobacteria) possess phototrophic modes of metabolism. The types of photosynthesis in the two groups of prokaryotes differ mainly in the type of compound that serves as the hydrogen donor in the reduction of $CO_2$ to glucose. Phototrophic organisms differ from heterotrophic organisms in that they utilize the glucose synthesized intracellularly for biosynthetic purposes (as in starch synthesis) or for energy production, which usually occurs through cellular respiration.

Unlike phototrophs, heterotrophs require glucose (or some other preformed organic compound) that is directly supplied as a substrate from an exogenous source. Heterotrophs cannot synthesize large concentrations of glucose from $CO_2$ by specifically using $H_2O$ or ($H_2S$) as a hydrogen source and sunlight as energy. Plant metabolism is a classic example of photolithotrophic metabolism: plants need $CO_2$ and sunlight; $H_2O$ must be provided as a hydrogen source and usually $NO_3$ is the nitrogen source for protein synthesis. Organic nitrogen, supplied as fertilizer, is converted to $NO_3$ in all soils by bacteria via the process of ammonification and nitrification.

Although plant cells are phototrophic, they also exhibit a heterotrophic mode of metabolism in that they respire. For example, plants use classic respiration to catabolize glucose that is generated photosynthetically. Mitochondria as well as the soluble enzymes of the glycolytic pathway are required for glucose dissimilation, and these enzymes are also found in all plant cells. The soluble Calvin cycle enzymes, which are required for glucose synthesis during photosynthesis, are also found in plant cells. It is not possible to feed a plant by pouring a glucose solution on it, but water supplied to a plant will be "photolysed" by chloroplasts in the presence of light; the hydrogen(s) generated from $H_2O$

is used by Photosystems I and II (PSI and PSII) to reduce NADP+ to NADPH + H+. With the ATP generated by PSI and PSII, these reduced pyridine nucleotides, $CO_2$ is reduced intracellularly to glucose. This metabolic process is carried out in an integrated manner by Photosystems I and II ("Z" scheme) and by the Calvin cycle pathway.

A new photosynthetic, and nitrogen fixing bacterium, *Heliobacterium chlorum*, staining Gram positive was isolated, characterized, and found to contain a new type of chlorophyll, i.e., bacteriochlorophyll 'g'. 16S r-RNA sequence analyses showed this organism to be phylogenetically related to members of the family *Bacillaceae*, although all currently known phototrophes are Gram negative.

A few *Heliobacteriium* strains did show the presence of endospores. Another unusual phototrophe is the Gram negative *Halobacterium halobium* (now named *Halobacterium salinarium*), an archaebacterium growing best at 30°C in 4.0-5.0 M (or 25%, w/v) NaCl. This bacterium is a facultative phototrophe having a respiratory mode; it also possesses a purple membrane within which bacteriorhodopsin serves as the active photosynthetic pigment. This purple membranae possesses a light driven proton translocation pump which mediates photosynthetic ATP synthesis via a proton extrusion reaction. Table summarizes the characteristics of known photosynthetic bacteria.

**Table. Characteristics Commonly Exhibited by Phototrophic Bactetia**

| *Photosynthetic Type* | *Characteristics and Genera* | *Representative Families* |
|---|---|---|
| Purple bacteria Sulfur-type (formerly Thiorho daceae) Photalithotrophic bacteria | Obligate phototrophs Strict anaerobes $H_2S$(or $H_2$)serve as H source.Prossess S granules when $H_2S$ used.Contain bacteri-ochlorophyll a or b | Chromatiaceae(Chronatium) Thiospinillum ,Thiosarcina Thiocapsa |
| Non-sulfur-type(formerly (Rhodopseudo-Athior-hodaceae)-photo-organotrophic bacteria | Faculative phototrophs (have res-piratory mecha-nism and will grow helerolrephically) Oxygen-lolerant anaerobes.Most re quire one or more B vitamins Simple organic compounds serve as H source | Rhodospirillaceae monas Rhodosponillum, Rhodomirobium |

| | | |
|---|---|---|
| Green bacteria Photolithotraphic bacteria | Contain bacteriochlorphy a or b Obligate phototrophs Strict anaerobes Contains chlorobium chiorophy,which is currently referred to as bacteriochlorphyil type c and d.Many require vitamin $B_{12}$ $S_2$ deposited extracellularly. | Chlorcbiaceae (Chlorobium, Chloropseudomonas |

All are Gram negative; if motile, they exhibit polar flagellation. Most species are anaerobic, although some purple nonsulfur bacteria (family Athiorhodaceae) are facultative phototrophs and can grow as heterotrophs by using the anaerobic respiratory mode of metabolism; they are therefore oxygen tolerant. For further information, see Bergey's Manual of Determinative Bacteriology, 8th ed, part 1.

## AUTOTROPHY

Bacteria that grow solely at the expense of inorganic compounds (mineral ions), without using sunlight as an energy source, are called autotrophs, chemotrophs, chemoautotrophs, or chemolithotrophs. Like photosynthetic organisms, all autotrophs use $CO_2$ as a carbon source for growth; their nitrogen comes from inorganic compounds such as $NH_3$, $NO_3^-$, or $N_2$. Interestingly, the energy source for such organisms is the oxidation of specific inorganic compounds. Which inorganic compound is oxidized depends on the bacteria in question. Many autotrophs will not grow on media that contain organic matter, even agar.

**Table. Inorganic Oxidation Reactions Used by Autotrophic Bacteria as Energy Sources**

| *Chemosynthetic Type* | *Inorganic Compounds Oxidized as Energy(–E)Source* | *Representative Families, Genera, and Species"* | *Nitrogen Cycle Reaction* |
|---|---|---|---|
| $NH_3$ oxidizers | $NH_3 \rightarrow NO_2$ | Nitrobacteriaceae | |
| Nitrification (aerobic) | (–E) | (Nitrosomonas, | |
| Nitrification | (~E) = chemical energy | Nitrosococcus, | |
| Nitrification | or ATP produced | Nitrosospira) | |
| Nitrification | | | |
| $NO_2$ oxidizers | $NO_2 \rightarrow NO_4$ | Nitrobacteriaceae | |

| | | | |
|---|---|---|---|
| Nitrification (aerobic) | (–E) | | (Nitrobacter, |
| Nitrification | | | Nitrococcus) |
| Nitrification | | | |
| Sulfur (aerobic) | $S_2 \rightarrow SO_4$ (–E) | use both.... | Thiobacillus thicoxidans, |
| Iron oxidizers (aerobic) | $Fe^2 \rightarrow Fe^{3+}$ (–E) | reactions | Thiobacillus, ierrocxidans, Ferrobacillus, Leptothrix |
| Sulfur-compound oxidizers Denitrification (anaerobic) | $S_2O_3$ oxidized; $NO_3$ reduced | | Thiobacillus denitrificans |

All are Gram-negative species Strict autotrophic modes of metabolosm are not present in sulfur and sulfur compound-oxidizing bacteria, For example, heterotrophic sulfur compound oxidizers are known. the aerobic species being able to oxidize $H_2S\,H_2S \xrightarrow[(-E)]{} S_2$ (e.g., Beggiatoa and Thiothrix species).

Also found among the autotrophic microorganisms are the sulfur-oxidizing or sulfur-compound-oxidizing bacteria, which seldom exhibit a strictly autotrophic mode of metabolism like the obligate nitrifying bacteria. The representative sulfur compounds oxidized by such bacteria are $H_2S$, $S_2$, and $S_2O_3$.

Among the sulfur bacteria are two very interesting organisms; *Thiobacillus ferrooxidans*, which gets its energy for autotrophic growth by oxidizing elemental sulfur or ferrous iron, and *T denitrificans*, which gets its energy by oxidizing $S_2O_3$ anaerobically, using $NO_3^-$ as the sole terminal electron acceptor.

T denitrificans reduces $NO_3$ to molecular $N_2$, which is liberated as a gas; this biologic process is called denitrification.

All autotrophic bacteria must assimilate $CO_2$, which is reduced to glucose from which organic cellular matter is synthesized. The energy for this biosynthetic process is derived from the oxidation of inorganic compounds discussed in the previous paragraph.

Note that all autotrophic and phototrophic bacteria possess essentially the same organic cellular constituents found in heterotrophic bacteria; from a nutritional viewpoint, however, the autotrophic mode of metabolism is unique, occurring only in bacteria.

## ANEROBIC RESPIRATION

Some bacteria exhibit a unique mode of respiration called anaerobic respiration. These heterotrophic bacteria that will not grow anaerobically unless a specific chemical component, which serves as a terminal electron acceptor, is added to the medium. Among these electron acceptors are $NO_3$, $SO_4^2$, the organic compound fumarate, and $CO_2$. Bacteria requiring one of these compounds for anaerobic growth are said to be anaerobic respirers.

A large group of anaerobic respirers are the nitrate reducers. The nitrate reducers are predominantly heterotrophic bacteria that possess a complex electron transport system(s) allowing the $NO_3$ ion to serve anaerobically as a terminal acceptor of electrons ($NO_3$ $2e^-$ $NO_2$; $NO_3$ $5e^-$ $N_2$; or $NO_3$ $8e^-$ $NH_3$). The organic compounds that serve as specific electron donors for these three known nitrate reduction processes are shown in Table. The nitrate reductase activity is common in bacteria and is routinely used in the simple nitrate reductase test to identify bacteria.

$$4AH_2 + HNO_3 \xrightarrow{\text{Nitrate reduction}} 4A + NH_3 + 3H_2O + \text{energy}$$

($AH_2$ = organic substrate, which serves as electron donor)

A second group of anaerobic respirers, the sulfate reducers, utilizes $SO_4^{2-}$ ion in similar fashion

($SO_4^{2-}$ 8c: $H_2S$):

$$4AH_2 + H_2SO_4 \xrightarrow{\text{Suliatereduction}} 4A + H_2S + 4H_2O + \text{energy}$$

The third group, the fumarate respirers, are anaerobic bacteria that require exogenous HOOC – CH = CH – COOH for growth. Fumarate is reduced to succinate (HOOC – $CH_2$ – $CH_2$ – COOH), which is secreted as a by-product.

$$AH_2 + HOOC = \overset{\overset{\displaystyle H}{||}}{C} - \overset{\overset{\displaystyle H}{||}}{C}OOH \quad \underline{\text{Fumarate reduction}}$$

$$A + HOOC - CH_2 - CH_2 - COOH + \text{energy}$$

Organisms of still another specialized group of anaerobic respirers, the methanogens, produce methane gas $CO_2$ & $CH_4$ as a metabolic end product of microbial growth. $H_2$ gas is the growth substrate; $CO_2$ is the terminal electron acceptor.

$$4H_2 + CO_2 \xrightarrow{CO_2 \text{ reduction}} CH_4 + 2H_2O + \text{energy}$$

The methanogens are among the most anaerobic bacteria known, being very sensitive to small concentrations of molecular $O_2$. They are also archaebacteria, which typically live in unusual and deleterious environments.

All of the above anaerobic respirers obtain chemical energy for growth by using these anaerobic energy-yielding oxidation reactions.

**Table. Nitrate Reducers**

| *Physiologic Types of Nitrate Reductases* | *Electron Donor(s)* | *Representative Organisms* |
|---|---|---|
| Respiratory | Formate | Escherichia coli |
| $(NO_3^- \rightarrow N_2)$ | NADH | Klebsieffa acrogenes |
| Denitrifying | NADH | Pseudomonas aeroginosa |
| $(NO_3^- \rightarrow N_2)$ | Pyruvate | Clostridium perfingens |
| | NADH, Succinate | Paracoccus denitrificans |
| Assimilatory | Lactate | Staphylococcus aureus |
| $(NO_3^- \rightarrow NH_2)$ | $H_{2+}$ formate | Vibrio succinogenes |
| | NADH succinate | Bacillus stearothermophilus |
| | NADH | Enterobacter aerogenes |
| | NADH, lactale, glycerol phosphate | Escherichla coli |

## THE NITROGEN CYCLE

Nowhere can the total metabolic potential of bacteria and their diverse chemical-transforming capabilities be more fully appreciated than in the geochemical cycling of the element nitrogen. All the basic chemical elements (S, O, P, C, and H) required to sustain living organisms have geochemical cycles similar to the nitrogen cycle. The nitrogen cycle is an ideal demonstration of the ecologic interdependence of bacteria, plants, and animals. Nitrogen is recycled when organisms use one form of nitrogen for growth and excrete another nitrogenous compound as a waste product. This waste product is in turn utilized by another type of organism as a growth or energy substrate. Figure shows the nitrogen cycle.

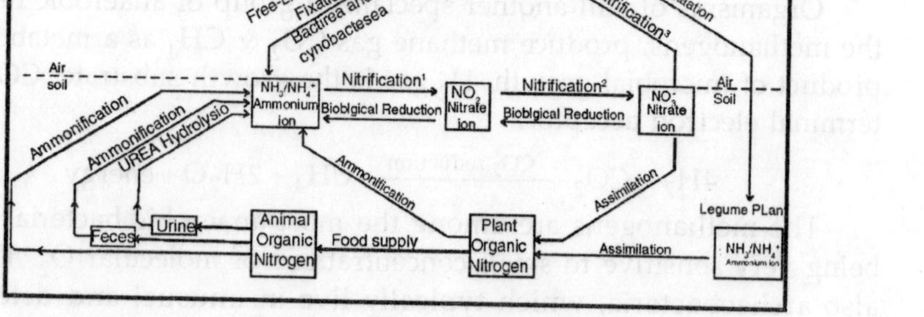


**Fig.** The Nitrogen Cycle.

When the specific breakdown of organic nitrogenous compounds occurs, that is, when proteins are degraded to amino acids (proteolysis) and then to inorganic $NH_3$, by heterotrophic bacteria, the process is called ammonification. This is an essential step in the nitrogen cycle. At death, the organic constituents of the tissues and cells decompose biologically to inorganic constituents by a process called mineralization; these inorganic end products can then serve as nutrients for other life forms. The $NH_3$ liberated in turn serves as a utilizable nitrogen source for many other bacteria. The breakdown of feces and urine also occurs by ammonification.

The other important biologic processes in the nitrogen cycle include nitrification (the conversion of $NH_3$ to $NO_3$ by autotrophes in the soil; denitrification (the anaerobic conversion of $NO_3$ to $N_2$ gas) carried out by many heterotrophs); and nitrogen fixation ($N_2$ to $NH_3$, and cell protein). The latter is a very specialized prokaryotic process called diazotrophy, carried out by both free-living bacteria (such as *Azotobacter, Derxia, Beijeringeia,* and *Azomona* species) and symbionts (such as *Rhizobium* species) in conjunction with legume plants (such as soybeans, peas, clover, and bluebonnets). All plant life relies heavily on $NO_3^-$ as a nitrogen source, and most animal life relies on plant life for nutrients.

# Chapter 5

# Genetics

## GENERAL CONCEPTS

### Genetic Information in Microbes

Genetic information in bacteria and many viruses is encoded in DNA, but some viruses use RNA. Replication of the genome is essential for inheritance of genetically determined traits. Gene expression usually involves transcription of DNA into messenger RNA and translation of mRNA into protein

### Genome Organization

The bacterial chromosome is a circular molecule of DNA that functions as a self-replicating genetic element (replicon). Extrachromosomal genetic elements such as plasmids and bacteriophages are nonessential replicons which often determine resistance to antimicrobial agents, production of virulence factors, or other functions. The chromosome replicates semiconservatively; each DNA strand serves as template for synthesis of its complementary strand.

### Mutation and Selection

The complete set of genetic determinants of an organism constitutes its genotype, and the observable characteristics constitute its phenotype. Mutations are heritable changes in genotype that can occur spontaneously or be induced by chemical or physical treatments. Organisms selected as reference strains are called wild type, and their progeny with mutations are called mutants. Selective media distinguish between wild type and mutant strains based on growth; differential media distinguish between them based on other phenotypic properties.

### Exchange of Genetic Information

Genetic exchanges among bacteria occur by several mechanisms. In transformation, the recipient bacterium takes up extracellular donor DNA.

In transduction, donor DNA packaged in a bacteriophage infects the recipient bacterium. In conjugation, the donor bacterium transfers DNA to the recipient by mating. Recombination is the rearrangement of donor and recipient genomes to form new, hybrid genomes. Transposons are mobile DNA segments that move from place to place within or between genomes.

### Recombinant DNA and Gene Cloning

Gene cloning is the incorporation of a foreign gene into a vector to produce a recombinant DNA molecule that replicates and expresses the foreign gene in a recipient cell. Cloned genes are detected by the phenotypes they determine or by specific nucleotide sequences that they contain. Recombinant DNA and gene cloning are essential tools for research in molecular microbiology and medicine. They have many medical applications, including development of new vaccines, biologics, diagnostic tests, and therapeutic methods.

### Regulation of Gene Expression

Expression of genes in microbes is often regulated by intracellular or environmental conditions. Regulation can affect any step in gene expression, including transcription initiation or termination, translation, or activity of gene products. An operon is a set of genes that is transcribed as a single unit and expressed coordinately. Specific regulation induces or represses a particular gene or operon. Global regulation affects a set of operons, which constitute a regulon. All operons in the regulon are coordinately controlled by the same regulatory mechanism.

## GENETIC INFORMATION IN MICROBES

The genetic material of bacteria and plasmids is DNA. Bacterial viruses (bacteriophages or phages) have DNA or RNA as genetic material. The two essential functions of genetic material are replication and expression. Genetic material must replicate accurately so that progeny inherit all of the specific genetic determinants (the genotype) of the parental organism.

Expression of specific genetic material under a particular set of growth conditions determines the observable characteristics (phenotype) of the organism.

Bacteria have few structural or developmental features that can be observed easily, but they have a vast array of biochemical capabilities and patterns of susceptibility to antimicrobial agents or bacteriophages. These latter characteristics are often selected as the inherited traits to be analyzed in studies of bacterial genetics.

## NUCLEIC ACID STRUCTURE

Nucleic acids are large polymers consisting of repeating nucleotide units. Each nucleotide contains one phosphate group, one pentose or deoxypentose sugar, and one purine or pyrimidine base. In DNA the sugar is D-2-deoxyribose; in RNA the sugar is D-ribose. In DNA the purine bases are adenine (A) and guanine (G), and the pyrimidine bases are thymine (T) and cytosine (C).

In RNA, uracil (U) replaces thymine. Chemically modified purine and pyrimidine bases are found in some bacteria and bacteriophages. The repeating structure of polynucleotides involves alternating sugar and phosphate residues, with phosphodiester bonds linking the 3'-hydroxyl group of one nucleotide sugar to the 5'-hydroxyl group of the adjacent nucleotide sugar.

These asymmetric phosphodiester linkages define the polarity of the polynucleotide chain. A purine or pyrimidine base is linked at the 1'-carbon atom of each sugar residue and projects from the repeating sugar-phosphate backbone. Double-stranded DNA is helical, and the two strands in the helix are antiparallel.

The double helix is stabilized by hydrogen bonds between purine and pyrimidine bases on the opposite strands. At each position, A on one strand pairs by two hydrogen bonds with T on the opposite strand, or G pairs by three hydrogen bonds with C. The two strands of double-helical DNA are, therefore, complementary.

Because of complementarity, double-stranded DNA contains equimolar amounts of purines (A + G) and pyrimidines (T + C), with A equal to T and G equal to C, but the mole fraction of G + C in DNA varies widely among different bacteria. Information in nucleic acids is encoded by the ordered sequence of nucleotides along the polynucleotide chain, and in double-stranded DNA the sequence of each strand determines what the sequence of the complementary strand must be. The extent of sequence homology between DNAs from different microorganisms is the most stringent criterion for determining how closely they are related.

The diagram shows the structure of DNA represented as a helical ladder. The backbone of each polynucleotide strand (represented as a ribbon) consists of alternating phosphate and deoxyribose residues linked by phosphodiester bonds, and the strands have opposite polarities (arrows).

The purine or pyrimidine base of each nucleotide on one strand projects toward the complementary base of the corresponding nucleotide from the other strand and is linked to it by hydrogen bonds. The double helix has a diameter of 2 nm. Each full turn of the double helix contains 10 nucleotide pairs and is 3.4 nm in length.

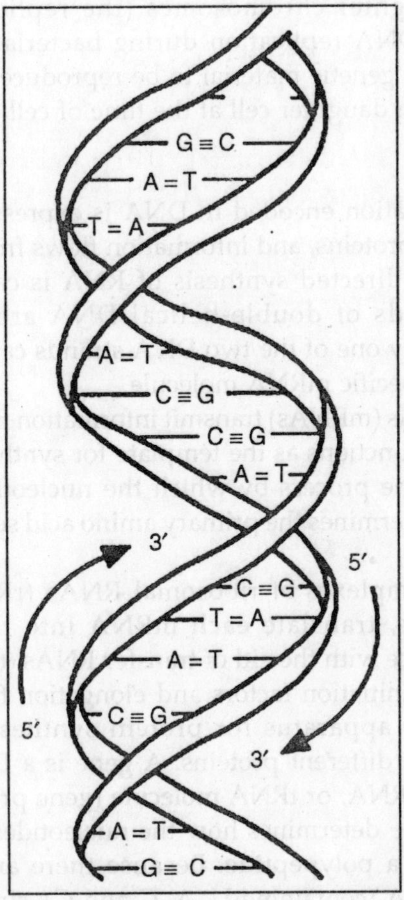


**Fig.** Double Helical Structure of DNA.

## DNA REPLICATION

During replication of the bacterial genome, each strand in double-helical DNA serves as a template for synthesis of a new complementary strand. Each daughter double-stranded DNA molecule thus contains one old polynucleotide strand and one newly synthesized strand. This type of DNA replication is called semiconservative.

Replication of chromosomal DNA in bacteria starts at a specific chromosomal site called the origin and proceeds bidirectionally until the process is completed. When bacteria divide by binary fission after completing DNA replication, the replicated chromosomes are partitioned into each of the daughter cells. The origin regions specifically and transiently associate with the celi membrane after DNA replication has been intitiated, leading to a model whereby membrane attachment directs separation of daughter chromosomes (the replicon model). These characteristics of DNA replication during bacterial growth fulfill the requirements of the genetic material to be reproduced accurately and to be inherited by each daughter cell at the time of cell division.

## GENE EXPRESSION

Genetic information encoded in DNA is expressed by synthesis of specific RNAs and proteins, and information flows from DNA to RNA to protein. The DNA-directed synthesis of RNA is called transcription. Because the strands of double-helical DNA are antiparallel and complementary, only one of the two DNA strands can serve as template for synthesis of a specific mRNA molecule.

Messenger RNAs (mRNAs) transmit information from DNA, and each mRNA in bacteria functions as the template for synthesis of one or more specific proteins. The process by which the nucleotide sequence of an mRNA molecule determines the primary amino acid sequence of a protein is called translation.

Ribosomes, complexes of ribosomal RNAs (rRNAs) and several ribosomal proteins, translate each mRNA into the corresponding polypeptide sequence with the aid of transfer RNAs (tRNAs), amino-acyl tRNA synthesases, initiation factors and elongation factors. All of these components of the apparatus for protein synthesis function in the production of many different proteins. A gene is a DNA sequence that encodes a protein, rRNA, or tRNA molecule (gene product).

The genetic code determines how the nucleotides in mRNA specify the amino acids in a polypeptide. Because there are only 4 different nucleotides in mRNA (containing U, A, C and G), single nucleotides do not contain enough information to specify uniquely all 20 of the amino acids. In dinucleotides 16 ($4 \times 4$) arrangements of the four nucleotides are possible, and in trinucleotides 64 ($4 \times 4 \times 4$) arrangements are possible. Thus, a minimum of three nucleotides is required to provide at least one unique sequence corresponding to each of the 20 amino acids.

The "universal" genetic code employed by most organisms is a triplet code in which 61 of the 64 possible trinucleotides (codons) encode specific amino acids, and any of the three remaining codons (UAG, UAA or UGA) results in termination of translation. The chain-terminating codons are also called nonsense codons because they do not specify any amino acids. The genetic code is described as degenerate, because several codons may be used for a single amino acid, and as nonoverlapping, because adjacent codons do not share any common nucleotides.

Exceptions to the "universal" code include the use of UGA as a tryptophan codon in some species of Mycoplasma and in mitochondrial

DNA, and a few additional codon differences in mitochondrial DNAs from yeasts, Drosophila, and mammals. Translation of mRNA is usually initiated at an AUG codon for methionine, and adjacent codons are translated sequentially as the mRNA is read in the 5' to 3' direction. The corresponding polypeptide chain is assembled beginning at its amino terminus and proceeding toward its carboxy terminus. The sequence of amino acids in the polypeptide is, therefore, colinear with the sequence of nucleotides in the mRNA and the corresponding gene. Specific enzymatic reactions involved in DNA, RNA, and protein synthesis are beyond the scope of this chapter. Expression of genetic determinants in bacteria involves the unidirectional flow of information from DNA to RNA to protein. In bacteriophages, either DNA or RNA can serve as genetic material. During infection of bacteria by RNA bacteriophages, RNA molecules serve as templates for RNA replication and as mRNAs.

Studies with the retrovirus group of animal viruses reveal that DNA molecules can be synthesized from RNA templates by enzymes designated as RNA-dependent DNA polymerases (reverse transcriptases). This reversal of the usual direction for flow of genetic information, from RNA to DNA instead of from DNA to RNA, is an important mechanism for enabling information from retroviruses to be encoded in DNA and to become incorporated into the genomes of animal cells.

**Table. The Genetic Code**

| First Nucleotide of Codon | Second Nucleotide of Codon | | | | Third Nucleotide of Codon |
|---|---|---|---|---|---|
| | U | C | A | G | |
| U | Phe | Ser | Tyr | Cys | U |
| | Phe | Ser | Tyr | Cys | C |
| | Leu | Ser | Termination | Termination | A |
| | Leu | Ser | Termination | TRp | G |
| C | Leu | Pro | His | Arg | U |
| | Leu | Pro | His | Arg | C |
| | Leu | Pro | His | Arg | A |
| | Leu | Pro | His | Arg | G |
| A | Ile | Thr | Asn | Ser | U |
| | Ile | Thr | Asn | Ser | C |
| | Ile | Thr | Lys | Arg | A |
| | Met | Thr | Lys | Arg | G |
| G | Val | Ala | Asp | Gly | U |
| | Val | Ala | Asp | Gly | C |
| | Val | Ala | Glu | Gly | A |
| | Val | Ala | Glu | Gly | C |

Abbreviations: Ala, alanine; Arg, arginine; Asn, asparagine; Asp, aspartic acid; Cys, cysteine; Gln, glutamine; Glu, glutamic acid; Gly, glycine; His, histidine; Ile, isoleucine; Leu, leucine; Lys, lysine; Met, methionine; Phe, Phenylalanine; Pro, proline; Ser, serine; Thr, threonine; Try, tyrosine; Trp, tryptophan; Val, valine.

## GENOME ORGANIZATION

DNA molecules that replicate as discrete genetic units in bacteria are called replicons. In some Escherichia coli strains, the chromosome is the only replicon present in the cell. Other bacterial strains have additional replicons, such as plasmids and bacteriophages.

## CHROMOSOMAL DNA

Bacterial genomes vary in size from about 0.4 × 109 to 8.6 × 109 daltons (Da), some of the smallest being obligate parasites (Mycoplasma) and the largest belonging to bacteria capable of complex differentiation such as Myxococcus. The amount of DNA in the genome determines the maximum amount of information that it can encode. Most bacteria have a haploid genome, a single chromosome consisting of a circular, double stranded DNA molecule.

However linear chromosomes have been found in Gram-positive Borrelia and Streptomyces spp., and one linear and one circular chromosome is present in the Gram-negative bacterium Agrobacterium tumefaciens.

The single chromosome of the common intestinal bacterium E coli is 3 × 109 Da (4,500 kilobase pairs [kbp]) in size, accounting for about 2 to 3 per cent of the dry weight of the cell. The E coli genome is only about 0.1% as large as the human genome, but it is sufficient to code for several thousand polypeptides of average size (40 kDa or 360 amino acids).

The chromosome of E coli has a contour length of approximately 1.35 mm, several hundred times longer than the bacterial cell, but the DNA is supercoiled and tightly packaged in the bacterial nucleoid. The time required for replication of the entire chromosome is about 40 minutes, which is approximately twice the shortest division time for this bacterium.

DNA replication must be initiated as often as the cells divide, so in rapidly growing bacteria a new round of chromosomal replication begins before an earlier round is completed. At rapid growth rates there may be four chromosomes replicating to form eight at the time of cell division, which is coupled with completion of a round of chromosomal replication. Thus, the chromosome in rapidly growing bacteria is replicating at more than one point. The replication of chromosomal DNA in bacteria is complex and involves many different proteins.

## PLASMIDS

Plasmids are replicons that are maintained as discrete, extrachromosomal genetic elements in bacteria. They are usually much smaller than the bacterial chromosome, varying from less than 5 to more than several hundred kbp, though plasmids as large as 2 Mbp occur in some bacteria. Plasmids usually encode traits that are not essential for bacterial viability, and replicate independently of the chromosome. Most plasmids are supercoiled, circular, double-stranded DNA molecules, but linear plasmids have also been demonstrated in Borrelia and Streptomyces. Closely related or identical plasmids demonstrate incompatibility; they cannot be stably maintained in the same bacterial host.

Classification of plasmids is based on incompatibility or on use of specific DNA probes in hybridization tests to identify nucleotide sequences that are characteristic of specific plasmid replicons. Some hybrid plasmids contain more than one replicon.Conjugative plasmids code for functions that promote transfer of the plasmid from the donor bacterium to other recipient bacteria, but nonconjugative plasmids do not. Conjugative plasmids that also promote transfer of the bacterial chromosome from the donor bacterium to other recipient bacteria are called fertility plasmids, and are discussed below. The average number of molecules of a given plasmid per bacterial chromosome is called its copy number.

Large plasmids (>40 kilobase pairs) are often conjugative, have small copy numbers (1 to several per chromosome), code for all functions required for their replication, and partition themselves among daughter cells during cell division in a manner similar to the bacterial chromosome. Plasmids smaller than 7.5 kilobase pairs usually are nonconjugative, have high copy numbers (typically 10-20 per chromosome), rely on their bacterial host to provide some functions required for replication, and are distributed randomly between daughter cells at division. Many plasmids control medically important properties of pathogenic bacteria, including resistance to one or several antibiotics, production of toxins, and synthesis of cell surface structures required for adherence or colonization. Plasmids that determine resistance to antibiotics are often called R plasmids (or R factors).

Representative toxins encoded by plasmids include heat-labile and heat-stable enterotoxins of E coli, exfoliative toxin of Staphylococcus aureus, and tetanus toxin of Clostridium tetani. Some plasmids are cryptic and have no recognizable effects on the bacterial cells that harbor them. Comparing plasmid profiles is a useful method for assessing possible relatedness of individual clinical isolates of a particular bacterial species for epidemiological studies. The role of plasmids in the evolution of resistance to antibiotics is discussed below.

## BACTERIOPHAGES

Bacteriophages (bacterial viruses, phages) are infectious agents that replicate as obligate intracellular parasites in bacteria. Extracellular phage particles are metabolically inert and consist principally of proteins plus nucleic acid (DNA or RNA, but not both). The proteins of the phage particle form a protective shell (capsid) surrounding the tightly packaged nucleic acid genome.

Phage genomes vary in size from approximately 2 to 200 kilobases per strand of nucleic acid and consist of double-stranded DNA, single-stranded DNA, or RNA. Phage genomes, like plasmids, encode functions required for replication in bacteria, but unlike plasmids they also encode capsid proteins and nonstructural proteins required for phage assembly. Several morphologically distinct types of phage have been described, including polyhedral, filamentous, and complex. Complex phages have polyhedral heads to which tails and sometimes other appendages (tail plates, tail fibers, etc.) are attached.

A single cycle of phage growth is shown in Fig. Infection is initiated by adsorption of phage to specific receptors on the surface of susceptible host bacteria.

The capsids remain at the cell surface, and the DNA or RNA genomes enter the target cells (penetration). Because infectivity of genomic DNA or RNA is much less than that of mature virus, there is a time immediately after infection called the eclipse period during which intracellular infectious phage cannot be detected. The infecting phage RNA or DNA is replicated to produce many new copies of the phage genome, and phage-specific proteins are produced.

For most phages assembly of progeny occurs in the cytoplasm, and release of the progeny occurs by cell lysis. In contrast, filamentous phages are formed at the cell envelope and released without killing the host cells. The eclipse period ends when intracellular infectious progeny appear. The latent period is the interval from infection until extracellular progeny appear, and the rise period is the interval from the end of the latent period until all phage are extracellular. The average number of phage particles produced by each infected cell, called the burst size, is characteristic for each virus and often ranges between 50 and several hundred. For discussions of structure, multiplication, and classification of animal viruses.

A culture of susceptible bacteria is synchronously infected with bacteriophage added at time 0 at low multiplicity of infection. Unabsorbed phage is inactivated shortly thereafter by addition of anti-phage antiserum, and the culture is then diluted to prevent further activity of the antiserum. Samples are taken at intervals for phage assays. Total phage (intracellular plus extracellular) is determined by testing the sample after treating it to disrupt infected bacteria, and extracellular phage is determined by testing supernatant after removal of bacteria by centrifugation or ultrafiltration. Phage titers are as the ratio of phage per infected bacterial cell.

Phages are classified into two major groups: virulent and temperate. Growth of virulent phages in susceptible bacteria destroys the host cells. Infection of susceptible bacteria by temperate phages can have either of two outcomes: lytic growth or lysogeny. Lytic growth of temperate and virulent bacteriophages is similar, leading to production of phage progeny and death of the host bacteria.

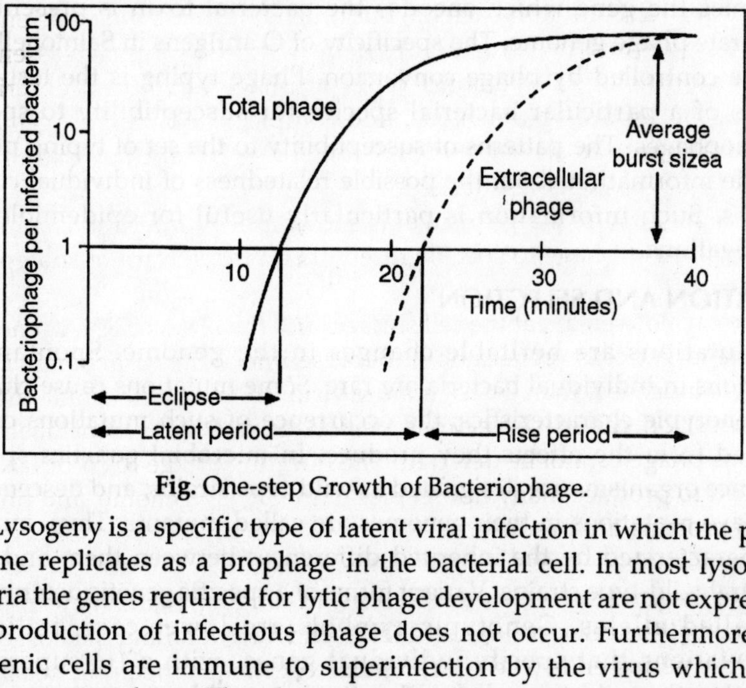


**Fig.** One-step Growth of Bacteriophage.

Lysogeny is a specific type of latent viral infection in which the phage genome replicates as a prophage in the bacterial cell. In most lysogenic bacteria the genes required for lytic phage development are not expressed, and production of infectious phage does not occur. Furthermore, the lysogenic cells are immune to superinfection by the virus which they harbor as a prophage. The physical state of the prophage is not identical for all temperate viruses. For example, the prophage of bacteriophage l in E coli is integrated into the bacterial chromosome at a specific site and replicates as part of the bacterial chromosome, whereas the prophage of bacteriophage P1 in E coli replicates as an extrachromosomal plasmid.

Lytic phage growth occurs spontaneously in a small fraction of lysogenic cells, and a few extracellular phages are present in cultures of lysogenic bacteria. For some lysogenic bacteria, synchronous induction of lytic phage development occurs in the entire population of lysogenic bacteria when they are treated with agents that damage DNA, such as ultraviolet light or mitomycin C. The loss of prophage from a lysogenic bacterium, converting it to the nonlysogenic state and restoring susceptibility to infection by the phage that was originally present as prophage, is called curing.

Some temperate phages contain genes for bacterial characteristics that are unrelated to lytic phage development or the lysogenic state, and expression of such genes is called phage conversion (or lysogenic conversion). Examples of phage conversion that are important for microbial virulence include production of diphtheria toxin by Corynebacterium diphtheriae, erythrogenic toxin by Streptococcus pyogenes (group A β-hemolytic streptococci), botulinum toxin by Clostridium botulinum, and Shiga-like toxins by E coli. In each of these examples the gene which encodes the bacterial toxin is present in a temperate phage genome. The specificity of O antigens in Salmonella can also be controlled by phage conversion. Phage typing is the testing of strains of a particular bacterial species for susceptibility to specific bacteriophages. The patterns of susceptibility to the set of typing phages provide information about the possible relatedness of individual clinical isolates. Such information is particularly useful for epidemiological investigations.

## MUTATION AND SELECTION

Mutations are heritable changes in the genome. Spontaneous mutations in individual bacteria are rare. Some mutations cause changes in phenotypic characteristics; the occurrence of such mutations can be inferred from the effects they produce. In microbial genetics specific reference organisms are designated as wild-type strains, and descendants that have mutations in their genomes are called mutants. Thus, mutants are characterized by the inherited differences between them and their ancestral wild-type strains. Variant forms of a specific genetic determinant are called alleles. Genotypic symbols are lower case, italicized abbreviations that specify individual genes, with a (+) superscript indicating the wild type allele. Phenotypic symbols are capitalized and not italicized, to distinguish them from genotypic symbols. For example, the genotypic symbol for the ability to produce β-galactosidase, required to ferment lactose, is lacZ+, and mutants that cannot produce β-galactosidase are lacZ. The lactose-fermenting phenotype is designated Lac+, and inability to ferment lactose is Lac-.

### Detection of Mutant Phenotypes

Selective and differential media are helpful for isolating bacterial mutants. Some selective media permit particular mutants to grow, but do not allow the wild-type strains to grow. Rare mutants can be isolated by using such selective media. Differential media permit wild-type and mutant bacteria to grow and form colonies that differ in appearance. Detection of rare mutants on differential media is limited by the total number of colonies that can be observed. Consider a wild-type strain of E coli that is susceptible to the antibiotic streptomycin (phenotype Strs) and can utilize lactose as the sole source of carbon (phenotype Lac+). Spontaneously occurring Strr mutants are rare and are usually found at frequencies of less than one per 109 bacteria in cultures of wild-type E coli.

Nevertheless, Strr mutants can be isolated easily by using selective media containing streptomycin, because the wild-type Strs bacteria are killed. Isolation of lactose-negative (phenotype Lac-) mutants of E coli poses a different problem. On minimal media with lactose as the sole source of carbon, Lac+ wild-type strains will grow, but Lac- mutants cannot grow. On differential media such as MacConkey-lactose agar or eosin-methylene blue-lactose agar, Lac+ wild-type and Lac- mutant strains of E coli can be distinguished by their colour, but spontaneous Lac-mutants are too rare to be isolated easily. Selective media for Lac- mutants of E coli can be made by incorporating chemical analogs of lactose that are converted into toxic metabolites by Lac+ bacteria but not by Lac-mutants. The Lac- mutants can then grow on such media, but the Lac+ wild-type bacteria are killed.

Mutations that inactivate essential genes in haploid organisms are usually lethal, but such potentially lethal mutations can often be studied if their expression is controlled by manipulation of experimental conditions. For example, a mutation that increases the thermolability of an essential gene product may prevent bacterial growth at 42°C, although the mutant bacterium can still grow at 25°C. Conversely, cold-sensitive mutants express the mutant phenotype at low temperature, but not at high temperature. Temperature-sensitive and cold-sensitive mutations are examples of conditional mutations, as are suppressible mutations described later in this chapter. A conditional lethal phenotype indicates that the mutant gene is essential for viability.

## Spontaneous and Induced Mutations

The mutation rate in bacteria is determined by the accuracy of DNA replication, the occurrence of damage to DNA, and the effectiveness of mechanisms for repair of damaged DNA.

For a particular bacterial strain under defined growth conditions, the mutation rate for any specific gene is constant and is expressed as the probability of mutation per cell division. In a population of bacteria grown from a small inoculum, the proportion of mutants usually increases progressively as the size of the bacterial population increases.

Mutations in bacteria can occur spontaneously and independently of the experimental methods used to detect them. This principle was first demonstrated by the fluctuation test. The numbers of phage-resistant mutants of E coli in replicate cultures grown from small inocula were measured and compared with those in multiple samples taken from a single culture.

If mutations to phage resistance occurred only after exposure to phage, the variability in numbers of mutants between cultures should be similar under both sets of conditions.

In contrast, if phage-resistant mutants occurred spontaneously before exposure of the bacteria to phage, the numbers of mutants should be more variable in the independently grown cultures, because differences in the size of the bacterial population when the first mutant appeared would contribute to the observed variability. The data indicated that the mutations to phage resistance in E coli occurred spontaneously with constant probability per cell division.

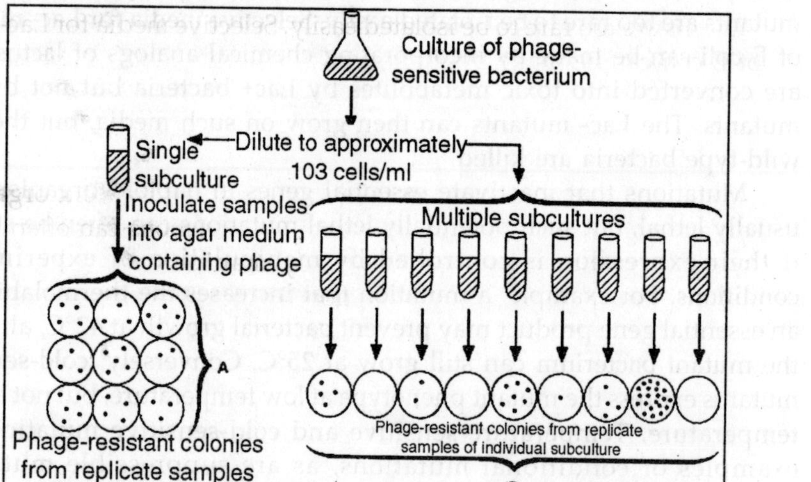


Fig. The Fluctuation Test.

Differences in numbers of colonies of phage-resistant mutants in replicate samples from single subculture were small and reflected only expected fluctuations due to sampling errors. In contrast, numbers of phage-resistant colonies in samples from individual subcultures were more variable and reflected both sampling errors and the independent origins of mutants in individual subcultures. Sizes of clonal populations of mutants in each culture reflected numbers of generations of growth between times that mutations occurred and time of sampling.

Replica plating confirmed that mutations in bacteria can occur spontaneously, without exposure of bacteria to selective agents. For replica plating, a flat, sterile, velveteen surface is used to pick up an inoculum from the surface of an agar master plate and transfer samples to other agar plates.

In this manner, samples of the bacterial population from the master plate are transferred to the replica plates without distorting their spatial arrangement. If the replica plates contain selective medium and the master plates do not, the positions of selected mutant colonies on the replica plates can be noted, and bacteria that were not exposed to the selective conditions can be isolated from the same positions on the master plate. Mutants of E coli resistant to bacteriophage T1 or to streptomycin have been isolated in this way, without exposing the wild-type bacteria to the bacteriophage or the antibiotic.

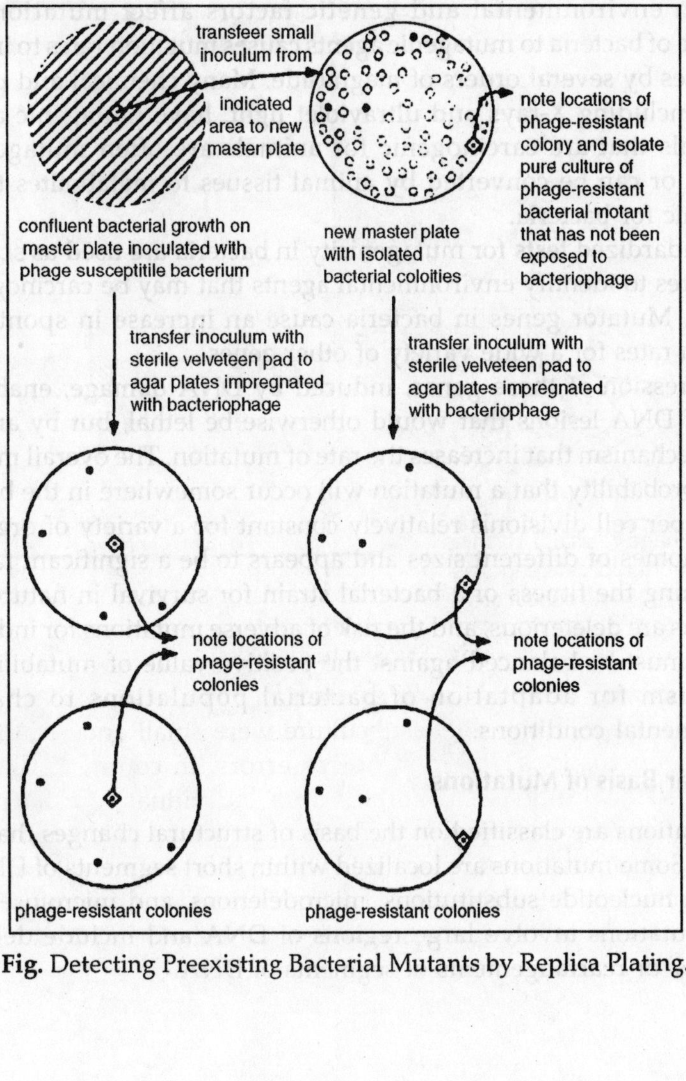


**Fig.** Detecting Preexisting Bacterial Mutants by Replica Plating.

Master plate was heavily inoculated with sample from pure cultures of phage-susceptible bacterium. After incubation, bacteria from master plate were transferred by replica plating to duplicate agar plates impregnated with bacteriophage. Phage-susceptible bacteria were killed by the bacteriophage. Colonies of phage-resistant bacteria appeared at identical positions on duplicate plates, indicating that phage-resistant bacteria had been transferred to each replica plate from the corresponding locations on master plate.

Bacterial inocula selected from appropriate locations on master plate contained a higher proportion of phage-resistant mutants than original bacterial culture. By repeating these procedures several times, it was possible to isolate pure cultures of phage-resistant bacterial mutants that had never been exposed to bacteriophage.

Both environmental and genetic factors affect mutation rates. Exposure of bacteria to mutagenic agents causes mutation rates to increase, sometimes by several orders of magnitude. Many chemical and physical agents, including X-rays and ultraviolet light, have mutagenic activity. Chemicals that are carcinogenic for animals are often mutagenic for bacteria, or can be converted by animal tissues to metabolites that are mutagenic for bacteria.

Standardized tests for mutagenicity in bacteria are used as screening procedures to identify environmental agents that may be carcinogenic in humans. Mutator genes in bacteria cause an increase in spontaneous mutation rates for a wide variety of other genes.

Expression of these genes, induced by DNA damage, enables the repair of DNA lesions that would otherwise be lethal, but by an error-prone mechanism that increases the rate of mutation. The overall mutation ratethe probability that a mutation will occur somewhere in the bacterial genome per cell divisionis relatively constant for a variety of organisms with genomes of different sizes and appears to be a significant factor in determining the fitness of a bacterial strain for survival in nature. Most mutations are deleterious, and the risk of adverse mutations for individual bacteria must be balanced against the positive value of mutability as a mechanism for adaptation of bacterial populations to changing environmental conditions.

### Molecular Basis of Mutations

Mutations are classified on the basis of structural changes that occur in DNA. Some mutations are localized within short segments of DNA (for example, nucleotide substitutions, microdeletions, and microinsertions). Other mutations involve large regions of DNA and include deletions, insertions, or rearrangements of segments of DNA.

**Table. Classification of Mutations**

| *Change in DNA* | *Effect on Polypeptide Structure* | *Effect on Polypeptide Function* | *Comments* |
|---|---|---|---|
| Nucleotide substitution | 1. None | 1. None | 1. Silent mulation (no phenotypic change) |
| | 2. Amino acid substitution | 2. Variable | 2. Missense mutation (usually CRM$^+$) |
| | 3. Premature termination | 3. Usually lost | 3. Nonsense mutations (CRM or CRM$^+$); extragenic suppression common |
| Microdeletion or microsertion | Frameshift mutation | Usually lost | Intragenic suppression common |
| Large insertions | Altered | Usually lost | See section on transposons |
| Large deletions | Altered | Usually | No reversion |

CRM, Cross-reacting material. Mutant polypeptides are CRM$^+$ if they share antigenic determinants with the corresponding wildtype polypeptides.

When a nucleotide substitution occurs in a region of DNA that codes for a polypeptide, one of the three nucleotides within a single codon of a corresponding mRNA molecule will be changed. Silent mutations cause no change in polypeptide structure or function, because one codon in mRNA is changed to another for the same amino acid. Other substitutions cause one amino acid to be replaced by another at the specific position within the polypeptide corresponding to the altered codon.

Mutations that result in replacement of one amino acid for another within a polypeptide chain are called missense mutations. The effects of amino acid replacements on the function of a polypeptide gene product vary and depend on the location and the identity of the amino acid replacement.

Mutant polypeptides containing amino acid replacements usually share antigenic determinants with the wild-type polypeptide and often have some residual biologic activity. Mutations that result in replacement of an amino acid codon with a termination codon are called nonsense mutations. This results in production of an amino-terminal fragment of the normal polypeptide when the mutant mRNA is translated. Nonsense mutations often result in complete loss of activity of the gene product.

Because of the triplet nature of the genetic code, the consequences of mutations caused by insertions or deletions of small numbers of nucleotides (microinsertions, microdeletions) depend on both the number and sequence of nucleotides involved. Deletion or addition of multiples of three nucleotide pairs does not affect the reading frame, but causes deletion or addition of appropriate numbers of amino acids at one site within the polypeptide. If a new chain-terminating codon is introduced, premature chain termination occurs within the polypeptide. In contrast, addition or deletion of other numbers of nucleotide pairs alters the reading frame for the entire segment of mRNA from the mutation to the distal end of the gene. Therefore, frameshift mutations are likely to cause drastic changes in the structure and activity of polypeptide gene products, and they are often classified as nonsense mutations.

**Complementation Tests**

To determine if mutations are located in the same gene or different genes, complementation tests are performed with partially diploid bacterial strains. Two copies of the region of the bacterial chromosome harboring a mutation are present in the same bacterium, with each copy containing a different mutation (mutations are in the trans arrangement). A wild-type phenotype indicates that the mutations are in different genes. This phenomenon is called complementation. If a mutant phenotype is observed, a control experiment should be performed with the mutations in the cis arrangement to exclude the possibility that the wild-type alleles cannot be expressed normally in a partially diploid bacterial strain. Complementation tests were originally called "cis-trans" tests, and the term cistron is sometimes used as a synonym for gene. Complementation tests can be performed and interpreted even if the specific biochemical functions of the gene products are unknown.

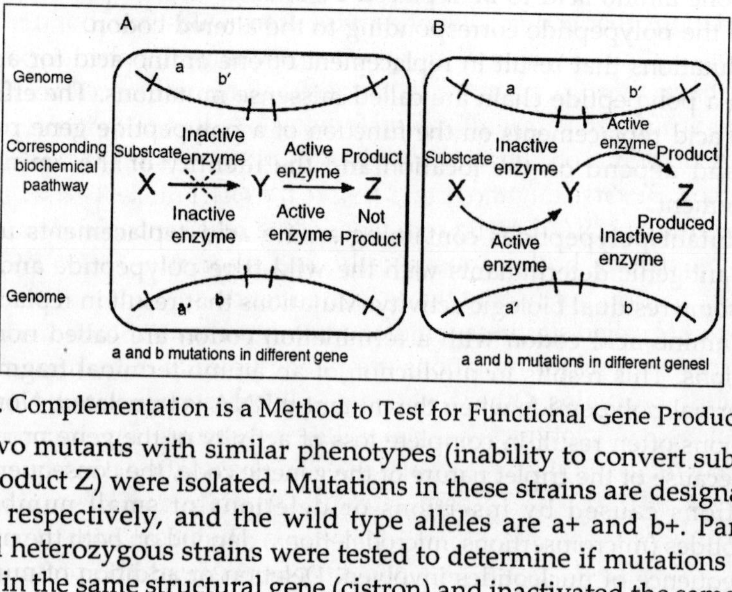


**Fig.** Complementation is a Method to Test for Functional Gene Products.

Two mutants with similar phenotypes (inability to convert substrate X to product Z) were isolated. Mutations in these strains are designated a and b, respectively, and the wild type alleles are a+ and b+. Partially diploid heterozygous strains were tested to determine if mutations a and b were in the same structural gene (cistron) and inactivated the same gene products. A), If a and b are in the same structural gene (e.g., encoding the enzyme that converts X to Y), neither the a+b nor the ab+ allele codes for an active enzyme, substrate X cannot be utilized, the mutant phenotype is expressed, and no complementation occurs. B), If a and b are in different cistrons (e.g., encoding the enzymes that convert X to Y and Y to Z), the a+ and b+ alleles encode active enzymes, substrate X is converted to product Z, the wild type phenotype is expressed, and complementation occurs.

As an example, consider using a complementation test to characterize two independently derived Lac-mutants of E coli. The biochemical pathway for utilization of lactose requires β-galactoside permease (genotypic symbol lacY) to transport lactose into the bacterial cell and β-galactosidase (genotypic symbol lacZ) to convert lactose into D-glucose and D-galactose. Mutants that lack b-galactoside permease or β-galactosidase cannot utilize lactose for growth. If the mutations in both Lac- mutants inactivated the same protein (e.g., b-galactoside) then a partial diploid strain containing the lacZ genes from both mutants in the trans arrangement would be unable to utilize lactose. In contrast, if the genotypes of the two mutants were lacZ+ lacY and lacZ lacY+, the partially diploid bacterium would produce active b-galactosidase from the lacZ+ determinant and active b-galactoside permease from the lacY+ determinant. Complementation would occur, and the partially diploid strain would utilize lactose.

## Reversion and Suppression

Mutations that convert the phenotype from wild-type to mutant are called forward mutations, and mutations that change the phenotype from mutant back to wild-type are called reverse mutations (reversions). Bacterial strains that contain reverse mutations are called revertants. Analysis of mutations that cause phenotypic reversion yields useful information. Reverse mutations that restore the exact nucleotide sequence of the wild-type DNA are true reversions. True revertants are identical to wild-type strains both genotypically and phenotypically. Reverse mutations that do not restore the exact nucleotide sequence of the wild-type DNA are called suppressor mutations (suppressors). Some revertants that harbor suppressor mutations are phenotypically indistinguishable from wild-type strains. Other revertants, called pseudorevertants, can be distinguished phenotypically from wild-type strains, for example, by subtle differences in the characteristics of an enzymatic activity that has been regained (such as specific activity, substrate specificity, kinetic constants, or susceptibility to thermal or chemical inactivation). Recognition of pseudorevertant phenotypes suggests the presence of suppressor mutations.

Suppressor mutations can be intragenic or extragenic. Intragenic suppressors are located in the same gene as the forward mutations that they suppress. The possible locations and nature of intragenic suppressors are determined by the original forward mutation and by the relationships between the primary structure of the gene product and its biologic activity. Extragenic suppressors are located in different genes from mutations whose effects they suppress. The ability of extragenic suppressors to suppress a variety of independent mutations can be tested.

Some extragenic suppressors are specific for particular genes, some are specific for particular codons, and some have other specificity patterns. Extragenic suppressors that reverse the phenotypic effects of chain-terminating codons have been well characterized and found to alter the structure of specific tRNAs.

A particular suppressor tRNA can permit a specific chain-terminating codon to be translated, resulting in incorporation of a specific amino acid into the nascent polypeptide at the position corresponding to the chain-terminating codon. In a bacterium that has a chain-terminating mutation and an appropriate extragenic suppressor, translation of the mRNA containing the mutant codon can therefore result in formation of a full-length polypeptide.

The biologic activity of the full-length polypeptide formed as a consequence of suppression depends both on the amount of protein made and on the functional consequences of the specific amino acid replacement determined by the suppressor tRNA.

## Exchange of Genetic Information

The biologic significance of sexuality in microorganisms is to increase the probability that rare, independent mutations will occur together in a single microbe and be subjected to natural selection. Genetic interactions between microbes enable their genomes to evolve much more rapidly than by mutation alone. Representative phenomena of medical importance that involve exchanges of genetic information or genomic rearrangements include the rapid emergence and dissemination of antibiotic resistance plasmids, flagellar phase variation in Salmonella, and antigenic variation of surface antigens in Neisseria and Borrelia.

Sexual processes in bacteria involve transfer of genetic information from a donor to a recipient and result either in substitution of donor alleles for recipient alleles or addition of donor genetic elements to the recipient genome. Transformation, transduction, and conjugation are sexual processes that use different mechanisms to introduce donor DNA into recipient bacteria. Because donor DNA cannot persist in the recipient bacterium unless it is part of a replicon, recombination between donor and recipient genomes is often required to produce stable, hybrid progeny. Recombination is most likely to occur when the donor and recipient bacteria are from the same or closely related species.

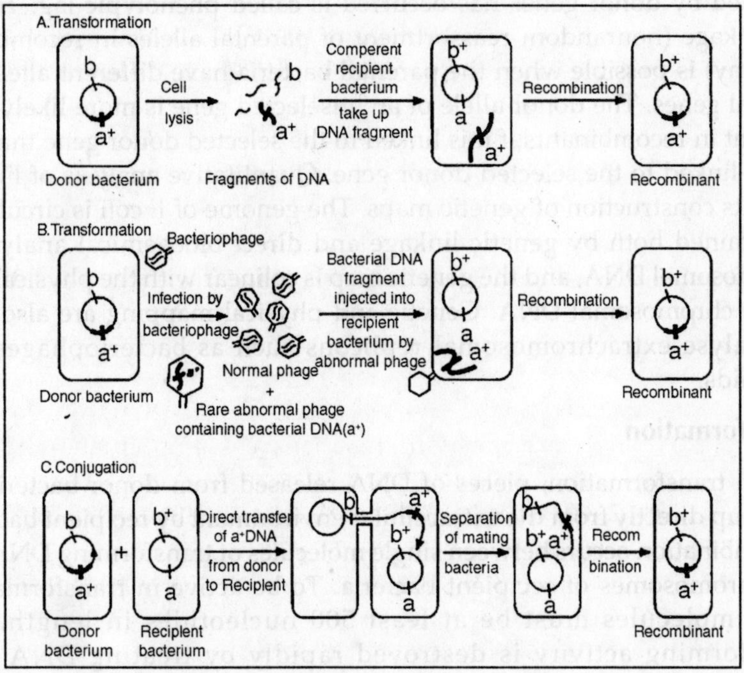


**Fig.** Exchange of Genetic Information in Bacteria.

Transformation, transduction, and conjugation differ in means for introducing DNA from donor cell into recipient cell. A) In transformation, fragments of DNA released from donor bacteria are taken up by competent recipient bacteria. B) In transduction, abnormal bacteriophage particles containing DNA from donor bacteria inject their DNA into recipient bacteria. C) Conjugation occurs by formation of cytoplasmic connections between donor and recipient bacteria, with direct transfer of newly synthesized donor DNA into the recipient cells. In all three cases, recombination between donor and recipient DNA molecules is required for formation of stable recombinant genomes. Bacterial genome is represented diagrammatically as a circular element in bacterial cells. Donor and recipient DNA are indicated by fine lines and heavy lines, respectively. In each recombinant genome, the a+ allele from donor strain has replaced the a allele from recipient strain, and the b+ allele is derived from recipient strain.

For a recombinant to be detected, its phenotype must be different from both parental phenotypes. Growth or cell division may be required before the recombinant phenotype is expressed. Delay in expression of a recombinant phenotype until a haploid recombinant genome has segregated is called segregation lag, and delay until synthesis of products encoded by donor genes has occurred is called phenotypic lag. Testing for linkage (nonrandom reassortment of parental alleles in recombinant progeny) is possible when the parental bacteria have different alleles for several genes. The donor allele of an unselected gene is more likely to be present in recombinants if it is linked to the selected donor gene than if it is not linked to the selected donor gene. Quantitative analysis of linkage permits construction of genetic maps. The genome of E coli is circular, as determined both by genetic linkage and direct biochemical analysis of chromosomal DNA, and the genetic map is colinear with the physical map of the chromosomal DNA. Genetic and physical mapping are also used to analyse extrachromosomal replicons such as bacteriophages and plasmids.

## Transformation

In transformation, pieces of DNA released from donor bacteria are taken up directly from the extracellular environment by recipient bacteria. Recombination occurs between single molecules of transforming DNA and the chromosomes of recipient bacteria. To be active in transformation, DNA molecules must be at least 500 nucleotides in length, and transforming activity is destroyed rapidly by treating DNA with deoxyribonuclease. Molecules of transforming DNA correspond to very small fragments of the bacterial chromosome. Cotransformation of genes is unlikely, therefore, unless they are so closely linked that they can be encoded on a single DNA fragment. Transformation was discovered in Streptococcus pneumoniae and occurs in other bacterial genera including Haemophilus, Neisseria, Bacillus, and Staphylococcus.

The ability of bacteria to take up extracellular DNA and to become transformed, called competence, varies with the physiologic state of the bacteria. Many bacteria that are not usually competent can be made to take up DNA by laboratory manipulations, such as calcium shock or exposure to a high-voltage electrical pulse (electroporation).

In some bacteria (including Haemophilus and Neisseria) DNA uptake depends on the presence of specific oligonucleotide sequences in the transforming DNA, but in others (including Streptococcus pneumoniae) DNA uptake is not sequence-specific. Competent bacteria may also take up intact bacteriophage DNA (transfection) or plasmid DNA, which can then replicate as extrachromosomal genetic elements in the recipient bacteria. In contrast, a piece of chromosomal DNA from a donor bacterium usually cannot replicate in the recipient bacterium unless it becomes part of a replicon by recombination. Historically, characterization of "transforming principle" from S pneumoniae provided the first direct evidence DNA is genetic material.

## Transduction

In transduction, bacteriophages function as vectors to introduce DNA from donor bacteria into recipient bacteria by infection. For some phages, called generalized transducing phages, a small fraction of the virions produced during lytic growth are aberrant and contain a random fragment of the bacterial genome instead of phage DNA. Each individual transducing phage carries a different set of closely linked genes, representing a small segment of the bacterial genome. Transduction mediated by populations of such phages is called generalized transduction, because each part of the bacterial genome has approximately the same probability of being transferred from donor to recipient bacteria.

When a generalized transducing phage infects a recipient cell, expression of the transferred donor genes occurs. Abortive transduction refers to the transient expression of one or more donor genes without formation of recombinant progeny, whereas complete transduction is characterized by production of stable recombinants that inherit donor genes and retain the ability to express them.

In abortive transduction the donor DNA fragment does not replicate, and among the progeny of the original transductant only one bacterium contains the donor DNA fragment. In all other progeny the donor gene products become progressively diluted after each generation of bacterial growth until the donor phenotype can no longer be expressed.

On selective medium upon which only bacteria with the donor phenotype can grow, abortive transductants produce minute colonies that can be distinguished easily from colonies of stable transductants. The frequency of abortive transduction is typically one to two orders of magnitude greater than the frequency of generalized transduction, indicating that most cells infected by generalized transducing phages do not produce recombinant progeny.

Specialized transduction differs from generalized transduction in several ways. It is mediated only by specific temperate phages, and only a few specific donor genes can be transferred to recipient bacteria. Specialized transducing phages are formed only when lysogenic donor bacteria enter the lytic cycle and release phage progeny. The specialized transducing phages are rare recombinants which lack part of the normal phage genome and contain part of the bacterial chromosome located adjacent to the prophage attachment site.

Many specialized transducing phages are defective and cannot complete the lytic cycle of phage growth in infected cells unless helper phages are present to provide missing phage functions. Specialized transduction results from lysogenization of the recipient bacterium by the specialized transducing phage and expression of the donor genes. Phage conversion and specialized transduction have many similarities, but the origin of the converting genes in temperate converting phages is unknown.

## Conjugation

In conjugation, direct contact between the donor and recipient bacteria leads to establishment of a cytoplasmic bridge between them and transfer of part or all of the donor genome to the recipient. Donor ability is determined by specific conjugative plasmids called fertility plasmids or sex plasmids.

The F plasmid (also called F factor) of E coli is the prototype for fertility plasmids in Gram-negative bacteria. Strains of E coli with an extrachromosomal F plasmid are called F+ and function as donors, whereas strains that lack the F plasmid are F- and behave as recipients. The conjugative functions of the F plasmid are specified by a cluster of at least 25 transfer (tra) genes which determine expression of F pili, synthesis and transfer of DNA during mating, interference with the ability of F+ bacteria to serve as recipients, and other functions.

Each F+ bacterium has 1 to 3 F pili that bind to a specific outer membrane protein (the ompA gene product) on recipient bacteria to initiate mating. An intercellular cytoplasmic bridge is formed, and one strand of the F plasmid DNA is transferred from donor to recipient, beginning at a unique origin and progressing in the 5' to 3' direction. The transferred strand is converted to circular double-stranded F plasmid DNA in the recipient bacterium, and a new strand is synthesized in the donor to replace the transferred strand. Both of the exconjugant bacteria are F+, and the F plasmid can therefore spread by infection among genetically compatible populations of bacteria. In addition to the role of the F pili in conjugation, they also function as receptors for donor-specific (male-specific) phages.

The F plasmid in E coli can exist as an extrachromosomal genetic element or be integrated into the bacterial chromosome. Because the F plasmid and the bacterial chromosome are both circular DNA molecules, reciprocal recombination between them produces a larger DNA circle consisting of F plasmid DNA inserted linearly into the chromosome. E coli contains multiple copies of several different genetic elements called insertion sequences on transposons for more detail), at various locations in its chromosome and in the F plasmid. Homologous recombination between insertion sequences in the chromosome and the F plasmid leads to preferential integration of the F plasmid at chromosomal sites where insertion sequences are located. The chromosomal sites where insertion sequences are found vary, however, among strains of E coli.

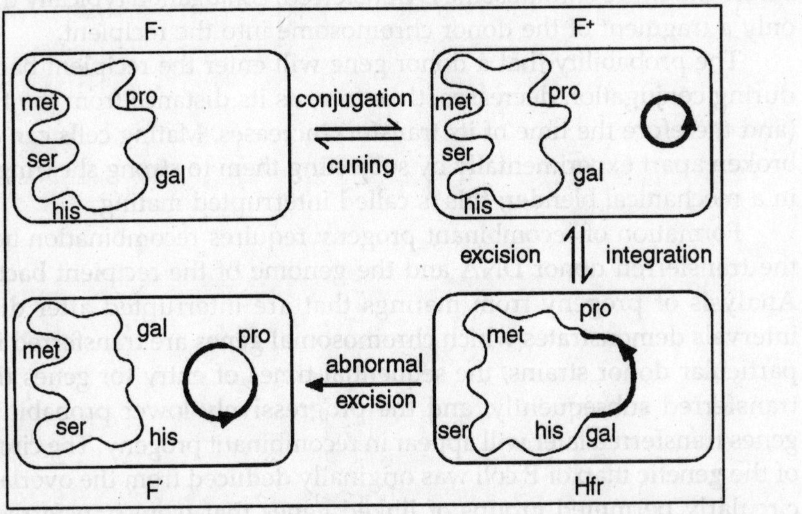


**Fig.** Role of F Plasmid in Determining Donor and Recipient States of E Coli.

The F plasmid is representative of specific conjugative plasmids that control donor ability in E coli. F-strains lack the F plasmid and are genetic recipients. F+ strains harbor the F plasmid as a cytoplasmic element, express F pili, and are genetic donors. The F plasmid can become integrated into bacterial chromosome at various locations to produce Hfr (high-frequency recombination) donor strains. Abnormal excision of F plasmid can result in formation of F′ plasmids that contain segments of bacterial chromosome and the corresponding bacterial genes. The arrowhead in F plasmid defines origin for transfer of DNA during conjugation. F plasmid and chromosomal DNA are indicated by heavy and fine lines, respectively. For additional data concerning the genomes of Hfr and F′ strains.

An E coli strain with an integrated F plasmid retains its ability to function as a donor in conjugal matings. Because donor strains with integrated F factors can transfer chromosomal genes to recipients with high efficiency, they are called Hfr (High frequency recombination) strains. Transfer of single-stranded DNA from an Hfr donor to a recipient begins from the origin within the F plasmid and proceeds as described above, except that the transferred DNA is the hybrid replicon consisting of F plasmid integrated into the bacterial chromosome. Transfer of this entire replicon, including the bacterial chromosome, requires approximately 100 minutes. The identity of the first chromosomal gene to be transferred and the polarity of chromosomal transfer are determined by the site of integration of the F plasmid and its orientation with respect to the bacterial chromosome. Because the mating bacteria usually separate spontaneously before the entire chromosome is transferred, conjugation typically transfers only a fragment of the donor chromosome into the recipient.

The probability that a donor gene will enter the recipient bacterium during conjugation decreases, therefore, as its distance from the F origin (and therefore the time of its transfer) increases. Mating cells can also be broken apart experimentally by subjecting them to strong shearing forces in a mechanical blender; this is called interrupted mating.

Formation of recombinant progeny requires recombination between the transferred donor DNA and the genome of the recipient bacterium. Analysis of progeny from matings that are interrupted after different intervals demonstrates which chromosomal genes are transferred first by particular donor strains, the sequential times of entry for genes that are transferred subsequently, and the progressively lower probability that genes transferred later will appear in recombinant progeny. The circularity of the genetic map of E coli was originally deduced from the overlapping, circularly permuted groups of linked genes that were transferred early by individual donor strains in which the F factor was integrated at different chromosomal locations.

In matings between F+ and F- bacteria, only the F plasmid is transferred with high efficiency to recipients. Chromosomal genes are transferred with very low efficiency, and it is the spontaneous Hfr mutants in F+ populations that mediate transfer of donor chromosomal genes. In matings between Hfr and F- strains, the segment of the F plasmid containing the tra region is transferred last, after the entire bacterial chromosome has been transferred. Most recombinants from matings between Hfr and F- cells fail to inherit the entire set of F plasmid genes and are phenotypically F-. In matings between F+ and F- strains, the F plasmid spreads rapidly throughout the bacterial population, and most recombinants are F+.

Integrated F plasmids in Hfr strains can sometimes be excised from the bacterial chromosome. If excision precisely reverses the integration process, F+ cells are produced. On rare occasions, however, excision occurs by recombinations involving insertion sequences or other genes on the bacterial chromosome that are located at some distance from the original integration site. In such cases segments of the bacterial chromosome can become incorporated into hybrid F plasmids that are called F′ plasmids. By similar processes, segments of the bacterial chromosome can sometimes become incorporated into R plasmids to produce hybrid R′ plasmids. Conjugative R′ plasmids can function as fertility plasmids because they can integrate into the bacterial chromosome by homologous recombination and mediate transfer of chromosomal genes during matings with recipient bacteria. F′ plasmids, R′ plasmids, specialized transducing phages, and recombinant plasmids or phages constructed by gene cloning are hybrid replicons that can include segments of the bacterial chromosome. Therefore, any of these genetic elements can be used to construct the partially diploid bacterial strains that are required for complementation tests and other purposes.

Conjugation also occurs in Gram-positive bacteria. Gram-positive donor bacteria produce adhesins that cause them to aggregate with recipient cells, but sex pili are not involved. In some Streptococcus species, recipient bacteria produce extracellular sex pheromones that cause the donor phenotype to be expressed by bacteria that harbor an appropriate conjugative plasmid, and the conjugative plasmid prevents the donor cells from producing the corresponding pheromone.

## Recombination

Recombination involves breakage and joining of parental DNA molecules to form hybrid, recombinant molecules. Several distinct kinds of recombination have been identified that depend on different features of the participating genomes and require the activities of different gene products. Specific enzymes that act on DNA (for example, exonucleases, endonucleases, polymerases, ligases) participate in recombination. Detailed discussion of the biochemical events in recombination is beyond the scope of this chapter.

Generalized recombination involves donor and recipient DNA molecules that have homologous nucleotide sequences. Reciprocal exchanges can occur between any homologous donor and recipient sites. In E coli, the product of the recA gene is essential for generalized recombination, but other gene products also participate.

Site-specific recombination involves reciprocal exchanges only between specific sites in donor and recipient DNA molecules. The recA gene product is not required for site-specific recombination. Integration of the temperate bacteriophage l into the chromosome of E coli is a well-studied example of site-specific recombination. The specific attachment (att) sites on the E coli chromosome and l phage DNA have a common core sequence of 15 nucleotides, within which reciprocal recombination occurs, flanked by adjacent sequences that are not homologous in the phage and bacterial genomes. In phage l the product of the int gene (integrase) is required for the site-specific integration event in lysogenization; the products of the int and xis (excisionase) genes are both needed for the complementary site-specific excision event that occurs during induction of lytic phage development in lysogenic cells. Illegitimate recombination is the term used to describe nonhomologous, aberrant recombination events such as those involved in formation of specialized transducing phages. The mechanisms of illegitimate recombination are unknown.

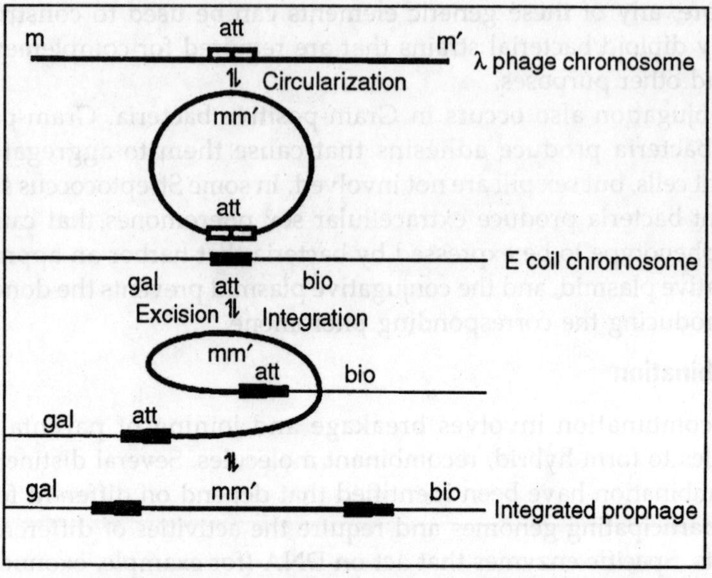


**Fig.** Integration and Excision of Bacteriophage l are Examples of Site-specific Recombination.

l DNA is shown by thin lines and chromosomal DNA by thick lines. Attachment (att) sites are closed boxes for the bacterial chromosome and open boxes for the l chromosome. The gal and bio operons, which determine utilization of galactose and biosynthesis of biotin, are located adjacent to the bacterial attachment site. In an infected E coli the l DNA becomes circular by joining ends m and m′, and site-specific recombination between phage and bacterial att sites results in insertion of the l genome into the bacterial chromosome. The arrangement of the prophage DNA (m and m′ located internally) is, therefore, a circular permutation of l virion DNA (m and m′ located terminally).

**Transposons**

Transposons are segments of DNA that can move from one site in a DNA molecule to other target sites in the same or a different DNA molecule. The process is called transposition and occurs by a mechanism that is independent of generalized recombination. Transposons are important genetic elements because they cause mutations, mediate genomic rearrangements, function as portable regions of genetic homology, and acquire new genes and contribute to their dissemination within bacterial populations. Insertion of a transposon often interrupts the linear sequence of a gene and inactivates it. Transposons have a major role in causing deletions, duplications, and inversions of DNA segments as well as fusions between replicons. Transposons are not self-replicating genetic elements, however, and they must integrate into other replicons to be maintained stably in bacterial genomes.

Most transposons share a number of common features. Each transposon encodes the functions necessary for its transposition, including a transposase enzyme that interacts with specific sequences at the ends of the transposon. During transposition a short sequence of target DNA is duplicated, and the transposon is inserted between the directly repeated target sequences. The length of this short duplication varies, but is characteristic for each transposon. The duplication is presumed to involve asymmetric cleavage of DNA at the target site, followed by synthesis of new complementary strands corresponding to the region between the cleavage sites. Some transposons insert into almost any target sequence, whereas others have relatively stringent target specificity. Two types of transposition are recognized. Excision of the transposon from a donor site followed by its insertion into a target site is called nonreplicative transposition. If the transposon at a donor site is replicated and a copy is inserted into the target site, however, the process is called replicative transposition. The process of replicative transposition can involve formation of a cointegrate, a single circular DNA molecule consisting of two replicons joined with copies of the transposon in an alternating sequence. Resolution of the cointegrate into its component replicons is often accomplished by a transposon-encoded resolvase that catalyzes site-specific recombination between the transposons. Generalized recombination between homologous transposons can also lead to the formation or resolution of cointegrates. Transposition differs from site-specific recombination by duplicating a segment of the target sequence and by using a variety of different target sequences for a single donor sequence.

Most transposons in bacteria can be separated into three major classes. Insertion sequences and related composite transposons comprise the first class. Insertion sequences are simplest in structure and encode only the functions needed for transposition. The known insertion sequences vary in length from approximately 780 to 1500 nucleotide pairs, have short (15-25 base pair) inverted repeats at their ends, and are not closely related to each other. The DNA between the inverted terminal repeats contains one (or rarely two) transposase genes and does not encode a resolvase.

Complex transposons vary in length from about 2,000 to more than 40,000 nucleotide pairs and contain insertion sequences (or closely related sequences) at each end, usually as inverted repeats. The entire complex element can transpose as a unit. The DNA between the terminal insertion sequences of complex transposons encodes multiple functions that are not essential for transposition. In medically important bacteria, genes that determine production of adherence antigens, toxins, or other virulence factors, or specify resistance to one or more antibiotics, are often located in complex transposons. Well-known examples of complex transposons are Tn5 and Tn10, which determine resistance to kanamycin and tetracycline, respectively. The complex transposons probably evolve by transposition of homologous insertion sequences to nearby sites within a DNA molecule.

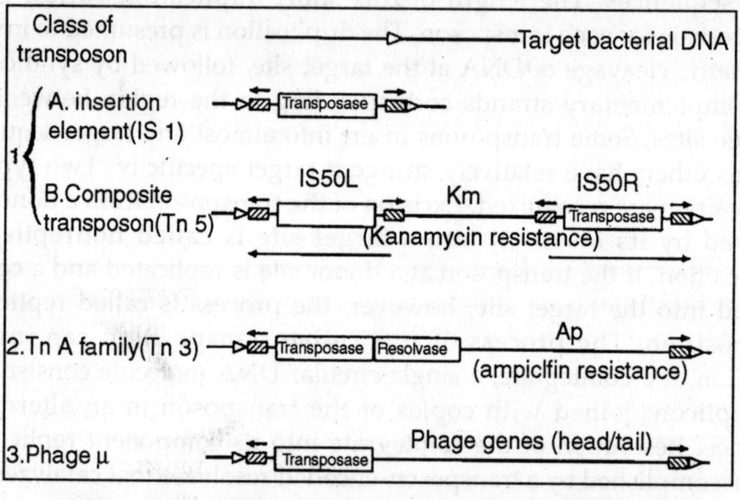


**Fig.** Features of Representative Transposons (heavy lines) Integrated into the Bacterial Chromosome (fine lines).

Transposons are important genetic elements because they cause mutations, mediate genomic rearrangements, function as portable regions of genetic homology, and acquire new genes and contribute to their dissemination within bacterial populations. 1A) IS1 insertion sequence (786 base pairs) has transposase gene flanked by inverted terminal repeats. The IS1 element is flanked by copies of target site (open arrows) with same orientation. 1B) Composite transposon Tn5 (5816 base pairs) consists of kanamycin resistance determinant flanked by inverted copies of IS50 insertion element. 2), Transposon TnA (4957 base pairs) contains ampicillin resistance determinant, transposase and resolvase genes between terminal inverted repeat sequences, flanked by direct repeats of target site (open arrows). 3) Phage Mu (37 kilobase pairs) encodes transposase that catalyzes recombination between the ends of Mu DNA and target DNA. Direct repeats of the target site (open arrows) flank the integrated Mu genome. Mu virion DNA is longer than Mu prophage and contains chromosomal sequences at both ends, reflecting the process by which prophage Mu is excised and packaged.

The second class of transposons consists of the highly homologous TnA family. These transposons have longer (35 to 40 base pair) terminal inverted repeats than the complex transposons described above, but they lack terminal insertion sequences. All members of the family encode both transposase and resolvase functions. Well known examples from the TnA transposon family include the ampicillin resistance transposon Tn3 and Tn1000 (the gamma-delta transposon) found in the F plasmid. The TnA family has an important place in the history of medical microbiology. The development of high-level resistance to ampicillin in Haemophilus influenzae and Neisseria gonorrhoeae during the 1970s, which severely limited the usefulness of ampicillin for treatment of gonorrhea and Haemophilus infections in areas where such strains became prevalent, was caused by dissemination of ampicillin resistance determinants from TnA transposons in plasmids of the Enterobacteriaceae to plasmids in Haemophilus and Neisseria.

The third class of transposons consists of bacteriophage Mu and related temperate phages. The entire phage genome functions as a transposon, and replication of the phage DNA during vegetative growth occurs by replicative transposition. Prophage integration can occur at many different sites in the bacterial chromosome and often causes mutations. For that reason Mu and related phages are sometimes called mutator phages.

A fourth class of transposons, discovered in Gram-positive bacteria and represented by Tn917, consists of conjugative transposons that are completely different from the transposons described above. The conjugative transposon does not generate a duplication of the target sequence into which it inserts, and in Gram-positive bacteria the host strain carrying the transposon can act as a conjugal donor. Recipient bacteria need not be closely related to the donor bacterium. The transposon is excised from the chromosome of the donor and transmitted by conjugation to the recipient, where it integrates randomly into the chromosome. Tn917 encodes tetracycline resistance, but other larger conjugative transposons may encode additional antibiotic resistances. Conjugative transposons appear to be a major cause of the spread of antibiotic resistance in Gram-positive bacteria.

Some roles of transposons in bacterial evolution are illustrated by considering enteric Gram-negative bacteria and the structure of their plasmids. Bacteria collected during the pre-antibiotic era contained many plasmids, but they usually lacked resistance determinants. Many of the R plasmids from current clinical isolates belong to the same incompatibility groups as plasmids found previously, but they also determine resistance to multiple antibiotics.

The close relationships between their replicons provide strong evidence that many current R plasmids evolved from the older plasmids by acquisition of resistance determinants. Some of the multiple antibiotic resistant plasmids have individual transposons with several resistance determinants, others have multiple resistance transposons located at separate sites, and still others contain complex hybrid resistance transposons formed by integration of one transposon into another. The stepwise acquisition of resistance determinants can lead, in some cases, to the formation of composite transposons that encode multiple resistance determinants.

Therapeutic use of antibiotics and their incorporation into animal feeds provide selective advantages for bacteria with R plasmids, whereas conjugation, transformation and transfection provide means for dissemination of R plasmids within and between bacterial species. After a plasmid carrying a transposon is introduced into a new bacterial host, the transposon and its determinants can jump into the chromosome or indigenous plasmids of the new host. Therefore, stability of the mobilizing plasmid in a new bacterial host is not essential for persistence of genetic determinants located on a transposon.

## RECOMBINATION DNA AND GENE CLONING

Many methods are available to make hybrid DNA molecules in vitro (recombinant DNA) and to characterize them. Such methods include isolating specific genes in hybrid replicons, determining their nucleotide sequences, and creating mutations at designated locations (site-directed mutagenesis). A clone is a population of organisms or molecules derived by asexual reproduction from a single ancestor. Gene cloning is the process of incorporating foreign genes into hybrid DNA replicons. Cloned genes can be expressed in appropriate host cells, and the phenotypes that they determine can be analyzed. Some key concepts underlying representative methods are summarized here.

The first step in gene cloning is to make fragments of the donor DNA by mechanical or enzymatic methods. Certain restriction endonucleases, designated as class II, are particularly useful for preparing defined fragments of DNA molecules. They cleave both strands of double-stranded DNA molecules at specific, palindromic sequences (restriction sites) that usually vary from four to eight nucleotides in length, and the resulting

DNA fragments are called restriction fragments. Some restriction endonucleases cleave at coincident sites to create blunt-ended DNA fragments, and others cut at staggered positions to create DNA fragments with short, self-complementary, single-stranded 5' or 3' ends.

The random probability that n adjacent nucleotides in a DNA strand will correspond to a specific restriction site is approximately 1/4n. Sites for enzymes that recognize unique 4, 6, or 8 nucleotide targets are likely to occur about once in every 256, 4096, or 65,536 nucleotides, respectively. By choosing appropriate restriction enzymes, specific DNA molecules, including bacterial chromosomes, plasmids, and phage genomes, can be digested into sets of restriction fragments that have appropriate sizes for specific applications.

**Table. Specificities of Representative Class II Restriction Endonucleases**

| *Enzyme* | *Isolated from* | *Recognition site Length (bp)* | *Sequence* (5' – 3') | *End structure of Restriction Fragment* |
|---|---|---|---|---|
| Sau3A | Staphylococcus aureus | 4 | ↓GATC | 4-base 5' extension |
| Nlalll | Neisseria lactamica | 4 | CATG↓ | 4-base 3' extension |
| Dpnl | Diplococcus pneumoniae | 4 | GA↓TC | Blunt |
| Sspl | Sphaerotilus Natans | 6 | AAT↓ATT | Blunt |
| Pstl | Providencla stuartii | 6 | CTGCA↓G | 4-base 3' extension |
| EcoRI | Escherichia coli | 6 | G↓AATTC | 4-base 5' extension |
| ClaI | Caryophanon jatum | 6 | AT↓CGAT | 2-base 5' extension |
| NorI | Nocardia otitidis -caviarum | 8 | GC↓GGCCGC | 4-base 5' extension |

Each recognition site in double-stranded DNA is a palindrome, and the 5' to 3' sequence is identical for each of the antiparallel strands. Both strands are cleaved at the site indicated by the vertical arrow.

A restriction map identifies the positions of target sites for specific restriction endonucleases in a DNA molecule. Restriction maps are available for many cloned DNA fragments, plasmids and phage genomes, as well as for the entire chromosome of E coli and several other bacteria.

The second step in gene cloning is to create hybrid replicons consisting of donor DNA fragments and a cloning vector. Cloning vectors are small plasmid or phage replicons that have one or more restriction sites into which foreign DNA can be inserted. Hybrid replicons are produced by using DNA ligase to join the restricted vector DNA with donor DNA fragments that have compatible ends, or, alternatively, synthetic oligonucleotides are used as linkers to create compatibility between donor and vector DNA molecules with different ends.

Ligating a vector to a heterogeneous set of DNA fragments from a donor genome is called shotgun cloning, and the collection of recombinant DNA molecules that contains the various fragments is called a genomic library. If a specific DNA fragment is available, it can be incorporated into a recombinant replicon by direct cloning into an appropriate vector chosen from the wide variety of vectors available.

Plasmid and phage vectors are used mainly to clone small inserts usually less than 10 kbp. Examples of more special purpose vectors include cosmids, which are plasmid vectors that can be packaged into phage capsids (lambda cosmids accept inserts up to 30-40 kbp), and phagemids, which are plasmid-phage hybrid replicons that can exist either as plasmids or as single-stranded DNA phages under different experimental conditions.

Phage P1 cosmids can accept inserts up to 100 kbp, and still larger DNA molecules can be cloned in yeast artificial chromosomes (YACs) which can stably maintain inserts up to and exceeding 1 Mbp in size. Other specialized vectors detect promoters, transcription termination signals, or other regulatory elements within foreign DNA inserts or, conversely, provide promoters from which transcription of cloned genes can be initiated.

Plasmid cloning vector pBR322 is 4.36 kilobase pairs in size, has genes for resistance to ampicillin (ampr) and to tetracycline (tetr), and has only one HindIII restriction site that is located within the tetr locus. HindIII is used to treat samples of DNA from plasmid pBR322 and from a donor organism with a gene, designated a+, to be cloned.

The donor can be a prokaryotic or a eukaryotic organism. If HindIII restriction sites are located adjacent to a+ in donor DNA, but do not occur within a+, a restriction fragment carrying intact a+ marker can be generated from donor DNA. Hybrid plasmids can be formed by random association and ligation of the HindIII-treated donor and vector DNA fragments. Although pBR322 is tetr, hybrid plasmids will be tets because the donor DNA fragments are inserted at the HindIII restriction site within the tetr locus. After transformation of amps-recipient bacteria that also lack a+, transconjugants with hybrid plasmids can be selected by their ampr tets phenotypes. Strains in which the a+ gene is present can then be identified by expresion of a+ or by testing for the polynucleotide sequence corresponding to a+. The pBR322 plasmid contains other uniuqe restriction sites that can also be used for cloning (e.g., PstI in ampr and BamHI in tetr). Many other cloning vectors and restriction endonucleases have also been used for gene cloning experiments.

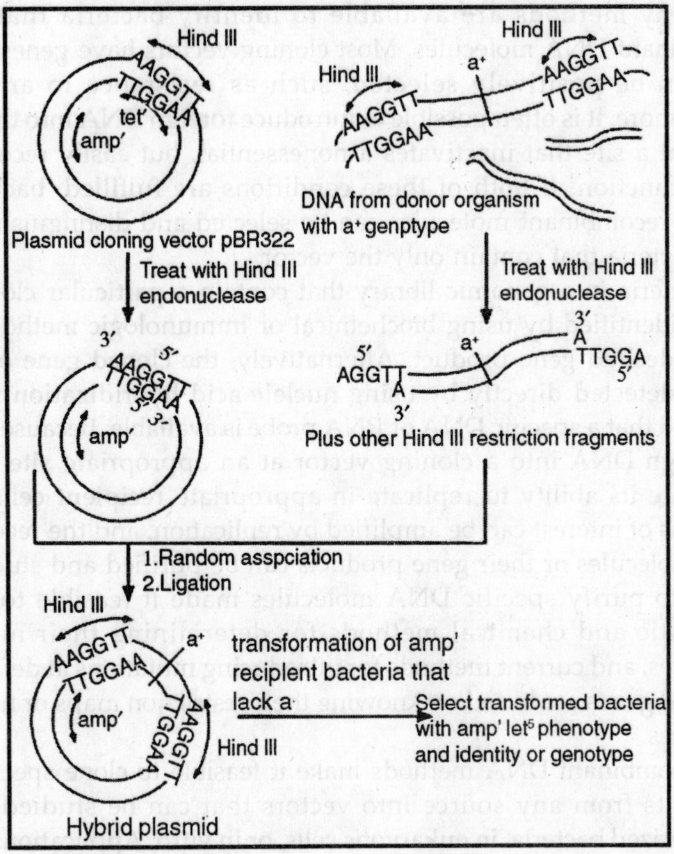


**Fig.** Diagrammatic Representation of Gene Cloning Experiment.

The final steps in gene cloning are to introduce hybrid replicons into appropriate recipient cells and test them for expression of donor genes of interest. Prokaryotic cells (including bacteria) or eukaryotic cells (including yeast, animal or plant cells) can be used as recipients, but they differ with respect to their permissiveness for specific replicons, the transcriptional signals that they recognize, and the post-translational modifications of protein structure that they can accomplish. Recombinant DNA molecules produced in vitro can be introduced directly into recipient cells by transformation or transfection. In addition, clones in cosmid or phage vectors can be packaged into phage coats and introduced into susceptible recipient cells by transduction. By using specialized vectors (shuttle vectors) that can replicate in multiple cell types, genes from any organism can be cloned and manipulated in a convenient bacterial system and subsequently reintroduced into cells of the original organism for analysis in their natural environment.

Many methods are available to identify bacteria that contain recombinant DNA molecules. Most cloning vectors have genes for traits that can be positively selected, such as resistance to antibiotics. Furthermore, it is often possible to introduce foreign DNA into the cloning vector at a site that inactivates a nonessential, but easily recognizable, vector function. If both of these conditions are fulfilled, bacteria that contain recombinant molecules can be selected and distinguished easily from bacteria that contain only the vector.

Bacteria in a genomic library that contain a particular cloned gene can be identified by using biochemical or immunologic methods to test for the desired gene product. Alternatively, the cloned gene of interest can be detected directly by using nucleic acid hybridization methods, provided that a specific DNA or RNA probe is available. Because insertion of foreign DNA into a cloning vector at an appropriate site does not inactivate its ability to replicate in appropriate recipient cells, hybrid replicons of interest can be amplified by replication, and the recombinant DNA molecules or their gene products can be purified and studied. The ability to purify specific DNA molecules made it feasible to develop enzymatic and chemical methods for determining their nucleotide sequences, and current methods for introducing mutations at defined sites in cloned genes are based on knowing their restriction maps or nucleotide sequences.

Recombinant DNA methods make it feasible to clone specific DNA fragments from any source into vectors that can be studied in well-characterized bacteria, in eukaryotic cells, or in vitro. Applications of DNA cloning are expanding rapidly in all fields of biology and medicine. In medical genetics such applications range from the prenatal diagnosis of inherited human diseases to the characterization of oncogenes and their roles in carcinogenesis. Pharmaceutical applications include large-scale production from cloned human genes of biologic products with therapeutic value, such as polypeptide hormones, interleukins, and enzymes. Applications in public health and laboratory medicine include development of vaccines to prevent specific infections and probes to diagnose specific infections by nucleic acid hybridization or polymerase chain reaction (PCR). The latter process uses oligonucleotide primers and DNA polymerase to amplify specific target DNA sequences during multiple cycles of synthesis in vitro, making it possible to detect rare target DNA sequences in clinical specimens with great sensitivity.

**Regulation of Gene Expression**

The phenotypic properties of bacteria are determined by their genotypes and growth conditions. For bacteria in pure culture, changes in growth conditions often result in predictable physiological adaptations in all members of the population. Typically, essential gene products are made in amounts that permit fastest growth in the given environment, and products required under special circumstances are made only when they are needed.

Physiological adaptations are often associated with changes in metabolic activities. The flow of metabolites through particular biochemical pathways can be controlled both by regulating the synthesis of specific enzymes and by altering the activities of existing enzymes. Mechanisms that regulate expression of genes by affecting synthesis of specific gene products are discussed here.

Specific regulation involves a gene or group of genes involved in a particular metabolic process. Induction and repression enable bacteria to regulate production of specific gene products in response to appropriate signals. Generally catabolic enzymes are induced when the substrate for the pathway is present in the growth medium, and biosynthetic enzymes are repressed by the product of the pathway. Enzymes that participate in a single biochemical pathway often occupy adjacent positions on the bacterial chromosome and are coordinately induced or repressed. They form an operon, a group of contiguous genes that is transcribed as a single unit and translated to produce the corresponding gene products. Organization into an operon is an important strategy for coordinately regulating the expression of genes in bacteria. Operons that can be induced or repressed are controlled by binding of specific regulatory proteins to particular nucleotide sequences that function as regulatory sites within the operon. Comparison of the amino acid sequences of many of these different regulatory proteins showed that they could be grouped together into families of regulators (e.g. the lysR family of proteins) that may have evolved from common ancestoral genes. Members of the lysR family include regulators of such diverse phenomena as lysine, cysteine and methionine metabolism in E coli and iron repression in V cholerae.

Global regulation simultaneously alters expression of a group of genes and operons, collectively called a regulon, that are controlled by the same regulatory signal. Global regulation determines responses of bacteria to basic nutrients such as carbon, nitrogen or phosphate, reactions to stresses such as DNA damage or heat shock, and synthesis by pathogens of specific virulence factors during growth in their host animals.

The amount of a specific protein in a bacterial cell can vary from none to many thousands of molecules. This wide range is often determined by the combined action of several regulatory mechanisms that affect expression of the corresponding structural gene. Regulation is achieved by determining how often a gene is transcribed into functional mRNA, how efficiently the mRNA is translated into protein, how rapidly the mRNA is degraded, how rapidly the protein product turns over, and whether the activity of the protein product can be altered by allosteric effects or covalent modifications.

### mRNAs as Transcriptional Units

Gene expression begins with DNA-dependent RNA polymerase (RNA polymerase) catalyzing the transcription of specific mRNA from one strand of a DNA template. Binding of RNA polymerase to DNA occurs at specific sites called promoters, and transcription begins adjacent to the promoter. Strong promoters can interact efficiently with RNA polymerase and initiate transcription at a high rate; weak promoters initiate transcription at slow rates. In either case, mRNA is synthesized from its 5' end toward its 3' end at an approximately constant rate until the RNA polymerase recognizes another specific site called a terminator. RNA polymerase then dissociates from the template, and transcription of the mRNA is completed.

Individual mRNA molecules may code for one or more polypeptides. Transcription of an operon produces a polycistronic mRNA that codes for several polypeptides. Translation of polycistronic mRNAs leads to coordinate synthesis of the encoded polypeptides, but each polypeptide is synthesized as a separate molecule. A specific ribosome binding site is located just upstream from the start of each coding sequence on the mRNA molecule.

Messenger RNAs in bacteria are degraded rapidly with an average half life of several minutes, in contrast to tRNAs and rRNAs which are much more stable. Although mRNAs represent about half of the newly synthesized RNA, they represent only a small fraction of the total RNA. The short half-life of mRNAs has important consequences for gene expression. If the synthesis of a specific mRNA is prevented, production of the corresponding polypeptides declines rapidly.

Control of gene expression occurs by regulating one or more of the steps in the pathway from the DNA template to the active gene product. Simultaneous regulation at several levels permits greater control over gene expression than would be possible with a single regulatory mechanism. The most common way to regulate gene expression in bacteria is to control the production of specific mRNAs. Since the rate of elongation of an RNA molecule is approximately constant, the major factors that control mRNA synthesis are the rate of initiation and the probability that a full length transcript will be produced.

## Regulation of Transcription Initiation

Some mRNAs in bacteria are synthesized at constant rates, resulting in constitutive production of the encoded polypeptides. The amounts of specific mRNAs and polypeptides produced from different constitutive genes vary greatly, however, and often reflect differences in strength of the promoters for those genes.

Transcription of many operons is regulated in response to changing environmental conditions. The promoters determine the maximum rate of transcription initiation for such operons, but regulatory proteins participate in controlling transcription. Nucleotide sequences in operons to which specific regulatory proteins bind are called regulatory sites or operators. Operators and promoters are located close together within operons and may have overlapping DNA sequences. The binding of regulatory proteins to operators can either increase (positive regulation) or decrease (negative regulation) the frequency of transcription initiation. Proteins that function as negative regulators are usually called repressors. Because regulatory proteins can diffuse through the cytoplasm, the structural genes for regulatory proteins do not have to be linked to the target operons.

The ability to sense the presence or absence of specific compounds and change the rates of synthesis of appropriate gene products are central to the control of gene expression. Regulatory proteins offer one solution to this problem of stimulus-response coupling. Many regulatory proteins are bifunctional and bind not only to appropriate operators but also to specific effectors, which are small molecules such as particular sugars, amino acids, and other metabolites. Furthermore, regulatory proteins are allosteric, meaning that they can exist in different conformations which exhibit different binding affinities for their cognate operators and effectors. A sufficient concentration of effector favors formation of the regulatory protein-effector complex, which has either high or low affinity for the operator in any specific case. In negatively regulated systems the effector functions as a corepressor if the regulatory protein-effector complex is the active repressor, and the effector functions as an inducer and causes derepression if the free regulatory protein is the active repressor. Conversely, in positively regulated systems, the effector stimulates expression of the operon if the regulatory protein-effector complex is the positive regulator, and the effector inhibits expression of the operon if the free regulatory protein is the positive regulator.

The lactose (lac) operon of E coli is an example of an inducible, negatively regulated operon. The lacI gene codes for a repressor that binds to the lac operator and prevents transcription from the lac promoter. The structural gene for this repressor is separate from the lac operon, and the repressor is synthesized constitutively at a low rate. When inducer binds to the lac repressor, the complex cannot bind to the operator and cannot prevent binding of the RNA polymerase to the promoter. If other conditions are favorable, the lac operon is expressed, resulting in synthesis of β-galactosidase, β-galactoside permease and β-galactoside transacetylase. The lac operon can be induced by lactose or by structurally related compounds such as isopropyl-b-D-thiogalactoside (IPTG). IPTG is called a gratuitous inducer because it induces the lac operon, but is not a substrate for b-galactosidase. Negative regulation also occurs in many biosynthetic operons in E coli. In such operons a product of the biosynthetic pathway functions as the effector for the negative regulatory system.

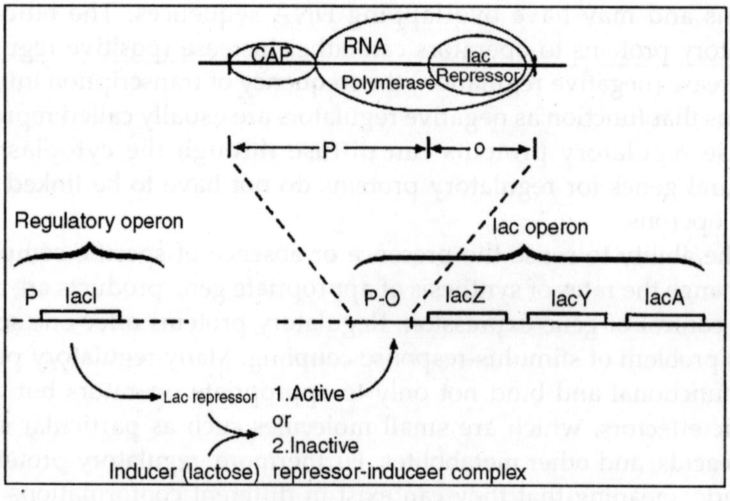


**Fig.** Regulation of Lac Operon in E Coli.

Structural genes lacZ, lacY, and lacA code for b-galactosidase, b-galactoside permease, and b-galactoside transacetylase, respectively. The physiologic role of lacA is unknown. The lac repressor is product of lacI gene in separate regulatory operon. Transcription of mRNA encoding lacZ, lacY, and lacA is negatively regulated. and binding of lac repressor to operator lacO prevents initiation of transcription at promoter lacP. Inducer binds to lac repressor and inactivates it. Catabolite activator protein (CAP) forms a complex with cyclic AMP, and binding of the complex to a site immediately adjacent to the lac promoter stimulates transcription of the lac operon by RNA polymerase. An expanded diagram of the lac operator-promoter region shows the binding sites for CAP, RNA polymerase, and lac repressor.

The arabinose (ara) operon in E coli is both positively and negatively regulated. In the presence of arabinose the regulatory protein stimulates transcription of the ara operon. In the absence of arabinose, however, the regulatory protein represses the ara operon.

Operons are often controlled by more than one mechanism. When E coli is grown in a medium containing glucose and an alternative carbon source such as lactose or arabinose, induction of the lac or ara operon and utilization of the lactose or arabinose are delayed until the glucose has been consumed. This phenomenon is called diauxic growth. The failure to induce the lac or ara operon in the presence of glucose is an example of catabolite repression. The lac and ara operons are positively regulated by cyclic-3',5'-adenosine monophosphate (cAMP) and the catabolite gene activator (CAP) protein (the product of the crp gene). The cAMP-CAP complex interacts with CAP binding sites in the regulatory regions of some operons, including the lac and ara operons, and stimulates transcription from the corresponding promoters. The level of intracellular cAMP in E coli is high during growth in the absence of glucose, and low during growth in the presence of glucose. Catabolite repression is due, therefore, to lack of activation of cAMP-dependent operons when the bacteria are grown in the presence of glucose or certain other rapidly metabolizable carbon sources.

## Regulation of Transcription Termination

Attenuation is a mechanism for regulating operons by terminating transcription of mRNA prematurely. Attenuation is common in biosynthetic operons, including the trp, histidine (his), threonine (thr), isoleucine-valine (ilv), and phenylalanine (phe) operons. The trp operon in E coli is controlled both by repression and attenuation. In the presence of excess tryptophan, initiation of transcription from the trp promoter is repressed. In addition, however, those transcripts that are initiated from the trp promoter are usually terminated before any of the structural genes of the trp operon are transcribed. The concentration of intracellular tryptophan required to maintain repression exceeds that needed for attenuation. Such dual control enables the cell to fine tune the expression of the trp operon in response to decreasing concentrations of tryptophan.

The secondary structure of mRNA has an important role in the mechanism of attenuation. All mRNAs have a leader sequence between the transcriptional start site and the beginning of the coding sequence for the first structural gene. For amino acid biosynthetic operons that are subject to attenuation, the mRNA leader sequence has two distinctive features. It encodes a short peptide containing the amino-acid produced by the regulated pathway, and it can form alternative, mutually incompatible, double-stranded RNA structures that participate in regulatory events.

For example, the peptide encoded by the trp mRNA leader sequence contains two adjacent tryptophan residues, and the peptide encoded by the his mRNA leader sequence has a series of seven consecutive histidine residues. Fig.shows the trp operon and illustrates alternative secondary structures in the leader sequence of trp mRNA.

There are three possible secondary structures for this region, called the pause site (segments 1+2), the anti-terminator (segments 2+3), and the attenuator (segments 3+4). Segment 1 of the pause site overlaps with the coding region for the trpL peptide. Which secondary structures are formed depends on efficiency of translation of the trpL peptide.

When segments 1 and 2 are transcribed, they immediately anneal and cause the RNA polymerase to pause temporarily. Subsequent initiation of translation of the trpL peptide disrupts the pause site and allows RNA polymerase to continue transcription. If tryptophan is present, transcription of segments 3 and 4 and formation of the attenuator structure occurs while the ribosome is blocking segment 2, causing the RNA polymerase to terminate transcription.

If tryptophan is deficient, however, tryptophanyl-tRNA is also deficient, and the ribosome stalls at the tryptophan codons in segment 1. This allows segment 2 to anneal with newly synthesized segment 3 to form the antiterminator, thereby making segment 3 unavailable to anneal with segment 4. Formation of the attenuator is therefore prevented, and the RNA polymerase transcribes the entire trp operon. In this manner depletion of tryptophan (actually the supply of tryptophanyl-tRNA) is coupled to regulation of transcription of the biosynthetic operon for tryptophan.

The organization of the trp operon is shown at the top of the figure. The five structural genes trpE, trpD, trpC, trpB, and trpA encode enzymes that catalyze terminal sequence of reactions in tryptophan formation. Transcription initiation is controlled at the promoter-operator (p-o) locus, and signals within the 162 nucleotide trp mRNA leader sequence control termination of transcription by attenuation. The leader sequence of trp mRNA is expanded to show locations of the trpL coding sequence, the complementary segments 1, 2, 3, and 4, and their possible alternative secondary structures which function as pause site, anti-terminator, or attenuator. In E coli transcription and translation are functionally coupled. Nonsense mutations that cause premature termination of translation often cause decreased transcription of more distal genes in the same operon. This phenomenon is called polarity. Ribosomes usually initiate translation of a growing mRNA molecule prior to completion of transcription, and such translation masks sites that would otherwise cause the RNA polymerase to terminate transcription. Premature termination of translation by a nonsense codon dissociates the ribosomes from the mRNA and enables RNA polymerase to interact with the unmasked transcription termination sites.

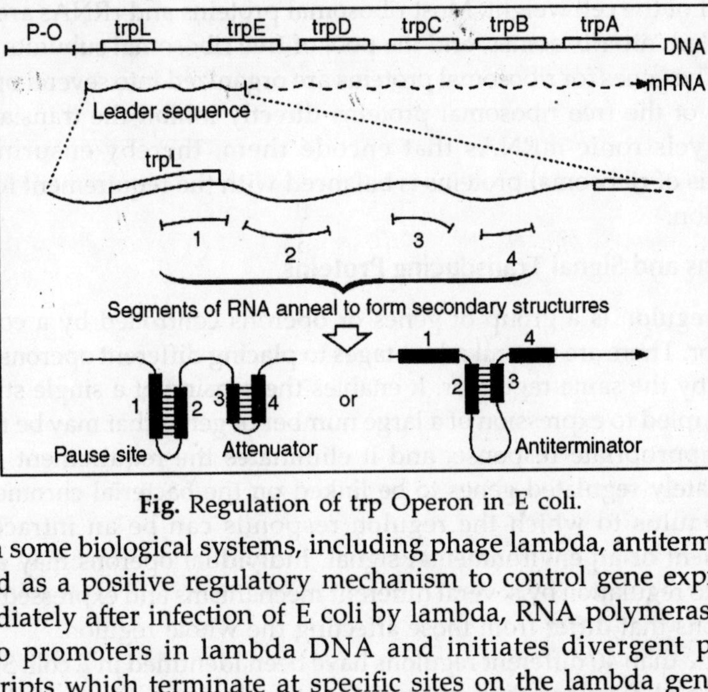


**Fig.** Regulation of trp Operon in E coli.

In some biological systems, including phage lambda, antitermination is used as a positive regulatory mechanism to control gene expression. Immediately after infection of E coli by lambda, RNA polymerase binds to two promoters in lambda DNA and initiates divergent primary transcripts which terminate at specific sites on the lambda genome. A protein encoded by one of the primary transcripts interacts with RNA polymerase and enables it to continue transcription through the primary termination sites, thereby expressing a second set of lambda genes. One of the products encoded by a secondary transcript blocks termination of another mRNA and activates expression of a third set of genes. Antitermination has a key role, therefore, in controlling the cascade of gene expression during lytic growth of phage lambda. Antitermination is also involved in the regulation of E coli rRNA operons.

## Regulation of Translation

The ribosome binding site on mRNA is complementary to a sequence at the 3' end of 16S rRNA. Interaction between these sequences facilitates formation of the initiation complex for protein synthesis. Both the extent of homology with 16S rRNA and the spacing of the ribosome binding site from the initiation codon affect the efficiency of translation initiation. Codon usage in mRNA also influences translation efficiency. Messenger RNAs for proteins that are required in large amounts tend to use codons that are translated by the most abundant species of tRNA, and the converse is also true.

Translational control is important for regulation of synthesis of ribosomal proteins. Production of ribosomes involves a high metabolic cost for bacteria, and at high growth rates ribosomes can constitute nearly one-half of the cell weight. Most ribosomal proteins and rRNAs are found assembled into ribosomes, and the pool of free ribosomal subunits is very small. The genes for ribosomal proteins are organized into several operons. Certain of the free ribosomal proteins directly inhibit the translation of the polycistronic mRNAs that encode them, thereby ensuring that synthesis of ribosomal proteins is balanced with the requirement for their utilization.

## Regulons and Signal Transducing Proteins

A regulon is a group of genes or operons controlled by a common regulator. There are several advantages to placing different operons under control by the same regulator. It enables the sensing of a single stimulus to be coupled to expression of a large number of genes that may be needed for an appropriate response, and it eliminates the requirement for the coordinately regulated genes to be linked on the bacterial chromosome. The stimulus to which the regulon responds can be an intracellular component or an environmental signal. Individual operons may also be subject to regulation by several different mechanisms and expressed under conditions that differ from those affecting the whole regulon.

More than 40 different regulons have been identified in E coli. Specific examples of regulons that respond to intracellular components include the cAMP-CAP regulon described previously and the regulons controlled by the stringent response and the SOS response. When ribosomes encounter uncharged tRNA molecules during protein synthesis, the stringent response is activated and results in prompt cessation of rRNA synthesis.

A novel nucleotide called guanosine-3'-diphosphate-5'-diphosphate (ppGpp) accumulates during amino-acid starvation. The ppGpp produced by idling ribosomes appears to be a mediator of the stringent response, but the precise mechanism causing inhibition of rRNA synthesis is unknown. The SOS response is associated with damage to DNA and involves induction of more than 20 genes involved in several DNA repair pathways. The product of the recA gene detects inhibition of DNA synthesis and initiates events leading to proteolytic cleavage and inactivation of the repressor for the SOS pathway, encoded by the lexA gene.

Some regulons are induced by specific environmental stimuli, such as nutrient limitation or osmotic stress. Often operons from more than one regulon may be induced, and the term stimulon has been used to describe the set of genes so induced. Typically, bacteria sense such environmental conditions by two component systems.

The first component is a membrane-spanning protein with extracellular and intracellular domains. Its extracellular domain detects the environmental stimulus, and its cytoplasmic domain transmits the signal.

The second component is a bifunctional cytoplasmic protein. It has a receiver domain that interacts with the transmitter module of the first component, as well as an effector domain that controls expression of the corresponding regulon. The transmitter and receiver modules of the two component regulatory systems from a wide variety of regulons are genetically related and share amino-acid homology. The signal-detecting and effector domains of the proteins from different regulons vary, however, and determine the signal that is detected and the operons that are activated or repressed in response to that signal.

Global regulation has an important role in the physiology of pathogenic bacteria. For example, Vibrio cholerae and Bordetella pertussis express many of their virulence determinants under the control of signal transducing systems that are related to the two component systems described above.

The expression of proteins needed for the invasive phenotype is controlled by temperature in Shigella. Yersinia enterocolitica senses both the environmental temperature and the concentration of calcium ions and couples these signals to the expression of genes and cellular location of the gene products that are appropriate for an intracellular or extracellular environment.

In host tissues the concentration of free iron is extremely low, and most pathogenic bacteria have high affinity iron transport systems that are induced under low-iron conditions.

The synthesis of diphtheria toxin by C diphtheriae, Shiga toxin by Shigella dysenteriae, exotoxin A by Pseudomonas aeruginosa, and other specific proteins in many pathogenic bacteria is induced under conditions of iron-limited growth. These examples illustrate how environmental factors can regulate the expression of virulence genes in pathogenic bacteria.

# Chapter 6

# Normal Flora

## INTRODUCTION

A diverse microbial flora is associated with the skin and mucous membranes of every human being from shortly after birth until death. The human body, which contains about $10^{13}$ cells, routinely harbors about $10^{14}$ bacteria. This bacterial population constitutes the normal microbial flora. The normal microbial flora is relatively stable, with specific genera populating various body regions during particular periods in an individual's life.

Microorganisms of the normal flora may aid the host (by competing for microenvironments more effectively than such pathogens as Salmonella spp or by producing nutrients the host can use), may harm the host (by causing dental caries, abscesses, or other infectious diseases), or may exist as commensals (inhabiting the host for long periods without causing detectable harm or benefit).

Even though most elements of the normal microbial flora inhabiting the human skin, nails, eyes, oropharynx, genitalia, and gastrointestinal tract are harmless in healthy individuals, these organisms frequently cause disease in compromised hosts. Viruses and parasites are not considered members of the normal microbial flora by most investigators because they are not commensals and do not aid the host.

Numbers represent the number of organisms per gram of homogenized tissue or fluid or per square centimeter of skin surface.

### Significance of the Normal Flora

The fact that the normal flora substantially influences the well-being of the host was not well understood until germ-free animals became available. Germ-free animals were obtained by cesarean section and maintained in special isolators; this allowed the investigator to raise them in an environment free from detectable viruses, bacteria, and other organisms. Two interesting observations were made about animals raised under germ free conditions. First, the germ-free animals lived almost twice as long as their conventionally maintained counterparts, and second, the major causes of death were different in the two groups. Infection often caused death in conventional animals, but intestinal atonia frequently killed germ-free animals.

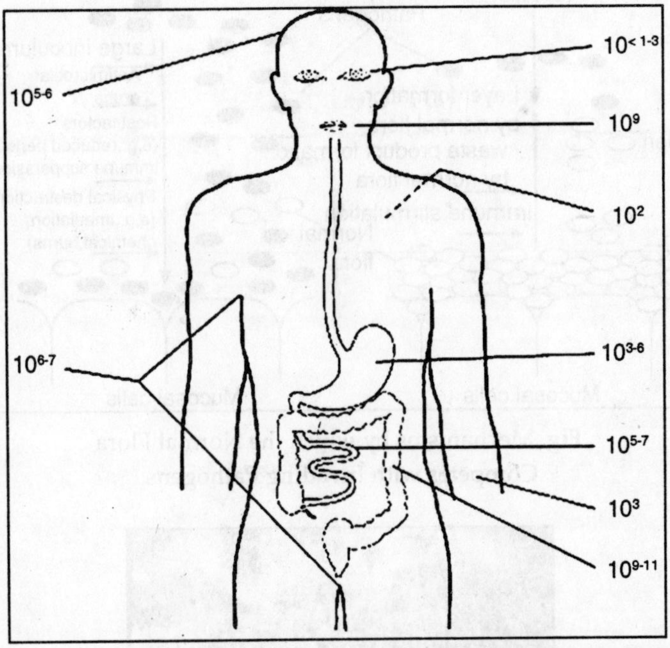


**Fig.** Numbers of Bacteria that Colonize Different Parts of the Body.

Other investigations showed that germ-free animals have anatomic, physiologic, and immunologic features not shared with conventional animals.

For example, in germ-free animals, the alimentary lamina propria is underdeveloped, little or no immunoglobulin is present in sera or secretions, intestinal motility is reduced, and the intestinal epithelial cell renewal rate is approximately one-half that of normal animals (4 rather than 2 days).

Although the foregoing indicates that bacterial flora may be undesirable, studies with antibiotic treated animals suggest that the flora protects individuals from pathogens. Investigators have used streptomycin to reduce the normal flora and have then infected animals with streptomycin-resistant Salmonella. Normally, about $10^6$ organisms are needed to establish a gastrointestinal infection, but in streptomycin-treated animals whose flora is altered, fewer than 10 organisms were needed to cause infectious disease. Further studies suggested that fermentation products (acetic and butyric acids) produced by the normal flora inhibited

Salmonella growth in the gastrointestinal tract. Figure shows some of the factors that are important in the competition between the normal flora and bacterial pathogens.

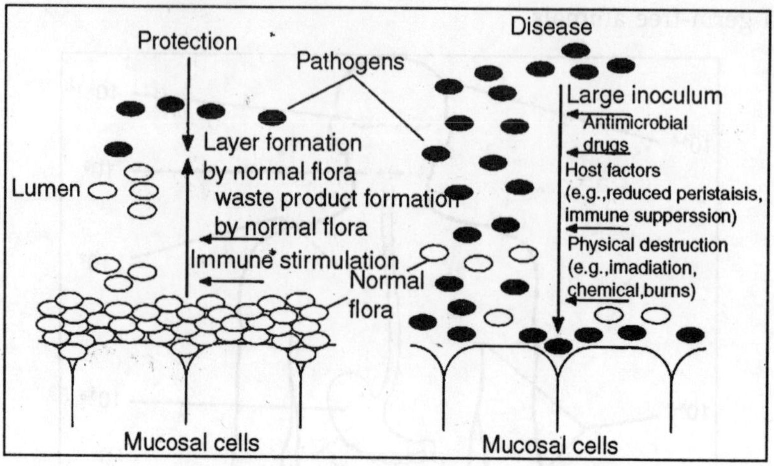


**Fig.** Mechanisms by which the Normal Flora Competes with Invading Pathogens.

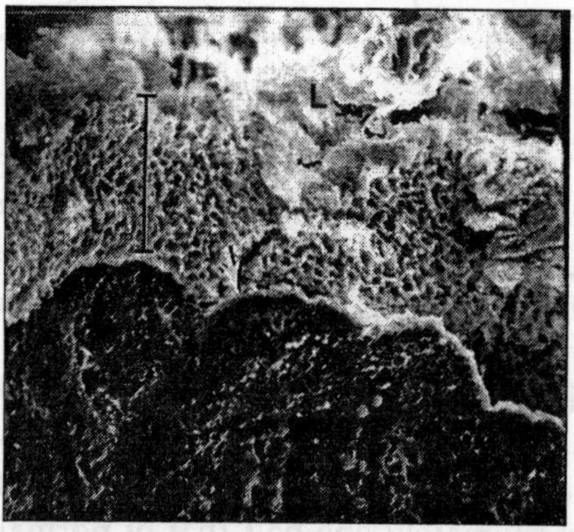


**Fig.** B Scanning Electron Micrograph of a Cross-section of Rat Colonic Mucosa.

The bar indicates the thick layer of bacteria between the mucosal surface and the lumen A, showing a mass of bacteria (B) immediately adjacent to colonized intestinal tissue. The normal flora in humans usually develops in an orderly sequence, or succession, after birth, leading to the stable populations of bacteria that make up the normal adult flora. The main factor determining the composition of the normal flora in a body region is the nature of the local environment, which is determined by pH, temperature, redox potential, and oxygen, water, and nutrient levels. Other factors such as peristalsis, saliva, lysozyme secretion, and secretion of immunoglobulins also play roles in flora control.

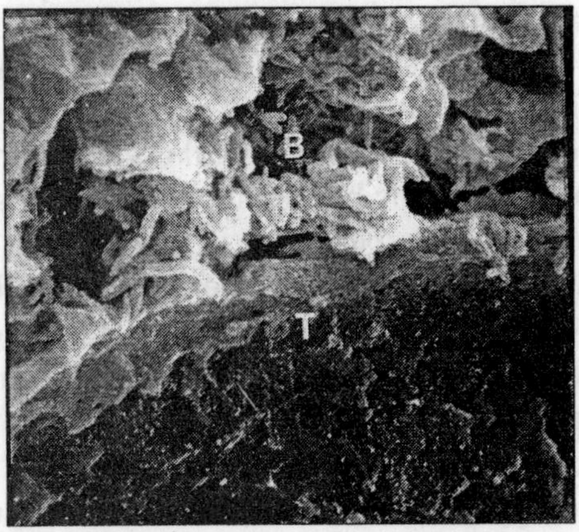


**Fig.** (B) Higher magnification of the Area Indicated by the Arrow in Fig.

The local environment is like a concerto in which one principal instrument usually dominates. For example, an infant begins to contact organisms as it moves through the birth canal. A Gram-positive population (bifidobacteria arid lactobacilli) predominates in the gastrointestinal tract early in life if the infant is breast-fed. This bacterial population is reduced and displaced somewhat by a Gram-negative flora (Enterobacteriaceae) when the baby is bottle-fed. The type of liquid diet provided to the infant is the principal instrument of this flora control; immunoglobulins and, perhaps, other elements in breast milk may also be important.

What, then, is the significance of the normal flora? Animal and some human studies suggest that the flora influences human anatomy, physiology, lifespan, and, ultimately, cause of death. Although the causal relationship of flora to death and disease in humans is accepted, of her roles of the human microflora need further study.

## NORMAL FLORA OF SKIN

Skin provides good examples of various microenvironments. Skin regions have been compared to geographic regions of Earth: the desert of the forearm, the cool woods of the scalp, and the tropical forest of the armpit. The composition of the dermal microflora varies from site to site according to the character of the microenvironment. A different bacterial flora characterizes each of three regions of skin: (1) axilla, perineum, and toe webs; (2) hand, face and trunk; and (3) upper arms and legs. Skin sites with partial occlusion (axilla, perineum, and toe webs) harbor more microorganisms than do less occluded areas (legs, arms, and trunk). These quantitative differences may relate to increased amount of moisture, higher body temperature, and greater concentrations of skin surface lipids. The axilla, perineum, and toe webs are more frequently colonized by Gram-negative bacilli than are drier areas of the skin.

The number of bacteria on an individual's skin remains relatively constant; bacterial survival and the extent of colonization probably depend partly on the exposure of skin to a particular environment and partly on the innate and species-specific bactericidal activity in skin.

Also, a high degree of specificity is involved in the adherence of bacteria to epithelial surfaces. Not all bacteria attach to skin; staphylococci, which are the major element of the nasal flora, possess a distinct advantage over viridans streptococci in colonizing the nasal mucosa. Conversely, viridans streptococci are not seen in large numbers on the skin or in the nose but dominate the oral flora.

The microbiology literature is inconsistent about the density of bacteria on the skin; one reason for this is the variety of methods used to collect skin bacteria.

The scrub method yields the highest and most accurate counts for a given skin area. Most microorganisms live in the superficial layers of the stratum corneum and in the upper parts of the hair follicles. Some bacteria, however, reside in the deeper areas of the hair follicles and are beyond the reach of ordinary disinfection procedures. These bacteria are a reservoir for recolonization after the surface bacteria are removed.

## Staphylococcus Epidermidis

S epidermidis is a major inhabitant of the skin, and in some areas it makes up more than 90 per cent of the resident aerobic flora.

## Staphylococcus Aureus

The nose and perineum are the most common sites for S aureus colonization, which is present in 10 per cent to more than 40 per cent of normal adults. S aureus is prevalent (67 per cent) on vulvar skin. Its occurrence in the nasal passages varies with age, being greater in the newborn, less in adults. S aureus is extremely common (80 to 100 per cent) on the skin of patients with certain dermatologic diseases such as atopic dermatitis, but the reason for this finding is unclear.

## Micrococci

Micrococci are not as common as staphylococci and diphtheroids; however, they are frequently present on normal skin. Micrococcus luteus, the predominant species, usually accounts for 20 to 80 per cent of the micrococci isolated from the skin.

## Diphtheroids (Coryneforms)

The term diphtheroid denotes a wide range of bacteria belonging to the genus Corynebacterium. Classification of diphtheroids remains unsatisfactory; for convenience, cutaneous diphtheroids have been categorized into the following four groups: lipophilic or nonlipophilic diphtheroids; anaerobic diphtheroids; diphtheroids producing porphyrins (coral red fluorescence when viewed under ultraviolet light); and those that possess some keratinolytic enzymes and are associated with trichomycosis axillaris (infection of axillary hair). Lipophilic diphtheroids are extremely common in the axilla, whereas nonlipophilic strains are found more commonly on glabrous skin. Anaerobic diphtheroids are most common in areas rich in sebaceous glands. Although the name Corynebacterium acnes was originally used to describe skin anaerobic diphtheroids, these are now classified as Propionibacterium acnes and as P granulosum. P acnes is seen eight times more frequently than P granulosum in acne lesions and is probably involved in acne pathogenesis. Children younger than 10 years are rarely colonized with P acnes. The appearance of this organism on the skin is probably related to the onset of secretion of sebum (a semi-fluid substance composed of fatty acids and epithelial debris secreted from sebaceous glands) at puberty. P avidum, the third species of cutaneous anaerobic diphtheroids, is rare in acne lesions and is more often isolated from the axilla.

## Streptococci

Streptococci, especially β-hemolytic streptococci, are rarely seen on normal skin. The paucity of β-hemolytic streptococci on the skin is attributed at least in part to the presence of lipids on the skin, as these lipids are lethal to streptococci. Other groups of streptococci, such as α-hemolytic streptococci, exist primarily in the mouth, from where they may, in rare instances, spread to the skin.

## Gram-Negative Bacilli

Gram-negative bacteria make up a small proportion of the skin flora. In view of their extraordinary numbers in the gut and in the natural environment, their scarcity on skin is striking. They are seen in moist intertriginous areas, such as the toe webs and axilla, and not on dry skin. Desiccation is the major factor preventing the multiplication of Gram-negative bacteria on intact skin. Enterobacter, Klebsiella, Escherichia coli, and Proteus spp are the predominant Gram-negative organisms found on the skin. Acinetobacter spp also occurs on the skin of normal individuals and, like other Gram-negative bacteria, is more common in the moist intertriginous areas.

## Nail Flora

The microbiology of a normal nail is generally similar to that of the skin. Dust particles and other extraneous materials may get trapped under the nail, depending on what the nail contacts. In addition to resident skin flora, these dust particles may carry fungi and bacilli. Aspergillus, Penicillium, Cladosporium, and Mucor are the major types of fungi found under the nails.

## Oral and Upper Respiratory Tract Flora

The oral flora is involved in dental caries and periodontal disease, which affect about 80 per cent. of the population in the Western world. The oral flora, its interactions with the host, and its response to environmental factors are thoroughly discussed in another Chapter. Anaerobes in the oral flora are responsible for many of the brain, face, and lung infections that are frequently manifested by abscess formation.

The pharynx and trachea contain primarily those bacterial genera found in the normal oral cavity (for example, alpha-and β-hemolytic streptococci); however, anaerobes, staphylococci, neisseriae, diphtheroids, and others are also present. Potentially pathogenic organisms such as Haemophilus, mycoplasmas, and pneumococci may also be found in the pharynx.

Anaerobic organisms also are reported frequently. The upper respiratory tract is so often the site of initial colonization by pathogens (Neisseria meningitides, C diphtheriae, Bordetella pertussis, and many others) and could be considered the first region of attack for such organisms. In contrast, the lower respiratory tract (small bronchi and alveoli) is usually sterile, because particles the size of bacteria do not readily reach it. If bacteria do reach these regions, they encounter host defense mechanisms, such as alveolar macrophages, that are not present in the pharynx.

## Gastrointestinal Tract Flora

The stomach is a relatively hostile environment for bacteria. It contains bacteria swallowed with the food and those dislodged from the mouth. Acidity lowers the bacterial count, which is highest (approximately $10^3$ to $10^6$ organisms/g of contents) after meals and lowest (frequently undetectable) after digestion. Some Helicobacter species can colonize the stomach and are associated with type B gastritis and peptic ulcer disease. Aspirates of duodenal or jejunal fluid contain approximately $10^3$ organisms/ml in most individuals. Most of the bacteria cultured (streptococci, lactobacilli, Bacteroides) are thought to be transients. Levels of $10^5$ to about $10^7$ bacteria/ml in such aspirates usually indicate an abnormality in the digestive system (for example, achlorhydria or malabsorption syndrome). Rapid peristalsis and the presence of bile may explain in part the paucity of organisms in the upper gastrointestinal tract. Further along the jejunum and into the ileum, bacterial populations begin to increase, and at the ileocecal junction they reach levels of $10^6$ to $10^8$ organisms/ml, with streptococci, lactobacilli, Bacteroides, and bifidobacteria predominating.

Concentrations of $10^9$ to $10^{11}$ bacteria/g of contents are frequently found in human colon and feces. This flora includes a bewildering array of bacteria (more than 400 species have been identified); nonetheless, 95 to 99 per cent belong to anaerobic genera such as Bacteroides, Bifidobacterium, Eubacterium, Peptostreptococcus, and Clostridium. In this highly anaerobic region of the intestine, these genera proliferate, occupy most available niches, and produce metabolic waste products such as acetic, butyric, and lactic acids. The strict anaerobic conditions, physical exclusion (as is shown in many animal studies), and bacterial waste products are factors that inhibit the growth of other bacteria in the large bowel.

Although the normal flora can inhibit pathogens, many of its members can produce disease in humans. Anaerobes in the intestinal tract are the primary agents of intra-abdominal abscesses and peritonitis. Bowel perforations produced by appendicitis, cancer, infarction, surgery, or gunshot wounds almost always seed the peritoneal cavity and adjacent organs with the normal flora. Anaerobes can also cause problems within the gastrointestinal lumen. Treatment with antibiotics may allow certain anaerobic species to become predominant and cause disease. For example, *Clostridium difficile,* which can remain viable in a patient undergoing antimicrobial therapy, may produce pseudomembranous colitis. Other intestinal pathologic conditions or surgery can cause bacterial overgrowth in the upper small intestine. Anaerobic bacteria can then deconjugate bile acids in this region and bind available vitamin B12 so that the vitamin and fats are malabsorbed. In these situations, the patient usually has been compromised in some way; therefore, the infection caused by the normal intestinal flora is secondary to another problem.

More information is available on the animal than the human microflora. Research on animals has revealed that unusual filamentous microorganisms attach to ileal epithelial cells and modify host membranes with few or no harmful effects. Microorganisms have been observed in thick layers on gastrointestinal surfaces and in the crypts of Lieberkuhn. Other studies indicate that the immune response can be modulated by the intestinal flora. Studies of the role of the intestinal flora in biosynthesis of vitamin K and other host-utilizable products, conversion of bile acids (perhaps to cocarcinogens), and ammonia production (which can play a role in hepatic coma) show the dual role of the microbial flora in influencing the health of the host. More basic studies of the human bowel flora are necessary to define their effect on humans.

### Urogenital Flora

The type of bacterial flora found in the vagina depends on the age, pH, and hormonal levels of the host. Lactobacillus spp predominate in female infants (vaginal pH, approximately 5) during the first month of life. Glycogen secretion seems to cease from about I month of age to puberty. During this time, diphtheroids, *S epidermidis*, streptococci, and *E coli* predominate at a higher pH (approximately pH 7). At puberty, glycogen secretion resumes, the pH drops, and women acquire an adult flora in which L acidophilus, corynebacteria, peptostreptococci, staphylococci, streptococci, and Bacteroides predominate. After menopause, pH again rises, less glycogen is secreted, and the flora returns to that found in prepubescent females. Yeasts (Torulopsis and Candida) are occasionally found in the vagina (10 to 30 per cent of women); these sometimes increase and cause vaginitis.

In the anterior urethra of humans, *S epidermidis*, enterococci, and diphtheroids are found frequently; *E coli*, Proteus, and Neisseria (nonpathogenic species) are reported occasionally (10 to 30 per cent). Because of the normal flora residing in the urethra, care must be taken in clinically interpreting urine cultures; urine samples may contain these organisms at a level of $10^4$/ml if a midstream (clean-catch) specimen is not obtained.

### Conjunctival Flora

The conjunctival flora is sparse. Approximately 17 to 49 per cent of culture samples are negative. Lysozyme, secreted in tears, may play a role in controlling the bacteria by interfering with their cell wall formation. When positive samples show bacteria, corynebacteria, neisseriae, and moraxellae are cultured. Staphylococci and streptococci are also present, and recent reports indicate that Haemophilus parainfluenzae is present in 25 per cent of conjunctival samples.

## HOST INFECTION BY ELEMENTS OF THE NORMAL FLORA

This chapter has briefly described the normal human flora; however, the pathogenic mechanisms of various genera or the clinical syndromes in which they are involved was not discussed. Although such material is presented in other chapters, note that a breach in mucosal surfaces often results in infection of the host by members of the normal flora. Caries, periodontal disease, abscesses, foul-smelling discharges, and endocarditis are hallmarks of infections with members of the normal human flora.

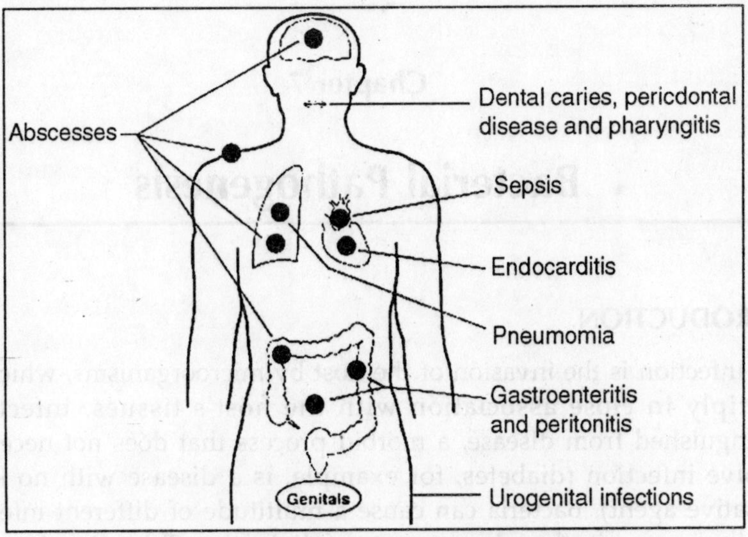


Fig. Clinical Conditions that may be Caused by Members of the Normal Flora.

In addition, impairment of the host (for example, those with heart failure or leukemia) or host defenses (due to immunosuppression, chemotherapy, or irradiation) may result in failure of the normal flora to suppress transient pathogens or may cause members of the normal flora to invade the host themselves. In either situation, the host may die.

# Chapter 7

# Bacterial Pathogenesis

## INTRODUCTION

Infection is the invasion of the host by microorganisms, which then multiply in close association with the host's tissues. Infection is distinguished from disease, a morbid process that does not necessarily involve infection (diabetes, for example, is a disease with no known causative agent). Bacteria can cause a multitude of different infections, ranging in severity from inapparent to fulminating. Table lists these types of infections.

**Table. Types of Bacterial Infections**

| *Type of Infection* | *Description* | *Examples* |
|---|---|---|
| Inapparent (subclinicat) | No defectable clinical symptoms of infection | Asymptomic gonorrhea in women and men |
| Dormant (latent) | Camer state | Typhoid carrier |
| Accidental | Zoorosis or environment or inadvertent exposures | Anthrax, cryptococcal infection, and laboratory, respectively |
| Opportunistic | Infection caused by normal flora or transient bacteria when normal host defenses are compromised | Serrati or Candida infection of the genitourinary tract |
| Primary | Clinically apparent (e.g., invasion and multiplication of microbes in body tissues, causing local tissue injury) | Shigella dysentery |
| Secondary | Microbial invasion subsequent to primary infection | Bacterial pneumonia following viral long infection |
| Mixed | Two or more microbes infecting the same tissue | Anaerobic abscess (E coli and Bacteroides tragifis) |
| Acute | Rapid onset (hours or days); one duration (days or weeks) | Diphtheria |

| | | |
|---|---|---|
| Chronic | Prolonged duration (months or years) | Mycobacterial diseases (tuberculosis and teprosy) |
| Localized | Confined to a small area or to an organ | Staphyiococcal boil |
| Generalized | Disseminated to many body regions (gonococcemia) | Gram-negative bacteremia |
| Pyogenic | Pus-forming | Staphyloccal and streptococcal infection |
| Retrograde | Microbes ascending in a duct or tube against the flow of secretions or excretions | E coli urinary tract infection |
| Fulminant | Infections that occur suddenly and intensely | Airbome Yersinia pestis (pneumonic plague) |

The capacity of a bacterium to cause disease reflects its relative pathogenicity. On this basis, bacteria can be organized into three major groups. When isolated from a patient, frank or primary pathogens are considered to be probable agents of disease (e.g., when the cause of diarrheal disease is identified by the laboratory isolation of Salmonella spp from feces).

Opportunistic pathogens are those isolated from patients whose host defense mechanisms have been compromised. They may be the agents of disease (e.g., in patients who have been predisposed to urinary tract infections with Escherichia coli by catheterization).

Finally, some bacteria, such as Lactobacillus acidophilus, are considered to be nonpathogens, because they rarely or never cause human disease. Their categorization as nonpathogens may change, however, because of the adaptability of bacteria and the detrimental effect of modern radiation therapy, chemotherapy, and immunotherapy on resistance mechanisms. In fact, some bacteria previously considered to be nonpathogens are now known to cause disease. Serratia marcescens, for example, is a common soil bacterium that causes pneumonia, urinary tract infections, and bacteremia in compromised hosts.

Virulence is the measure of the pathogenicity of an organism. The degree of virulence is related directly to the ability of the organism to cause disease despite host resistance mechanisms; it is affected by numerous variables such as the number of infecting bacteria, route of entry into the body, specific and nonspecific host defense mechanisms, and virulence factors of the bacterium. Virulence can be measured experimentally by determining the number of bacteria required to cause animal death, illness, or lesions in a defined period after the bacteria are administered by a designated route. Consequently, calculations of a lethal dose affecting 50 per cent of a population of animals ($LD_{50}$) or an effective dose causing a disease symptom in 50 per cent of a population of animals ($ED_{50}$) are useful in comparing the relative virulence of different bacteria.

Pathogenesis refers both to the mechanism of infection and to the mechanism by which disease develops. The purpose of this chapter is to provide an overview of the many bacterial virulence factors and, where possible, to indicate how they interact with host defense mechanisms and to describe their role in the pathogenesis of disease. It should be understood that the pathogenic mechanisms of many bacterial diseases are poorly understood, while those of others have been probed at the molecular level. The relative importance of an infectious disease to the health of humans and animals does not always coincide with the depth of our understanding of its pathogenesis. This information is best acquired by reading each of the ensuing chapters on specific bacterial diseases, infectious disease texts, and public health bulletins.

## Host Susceptibility

Susceptibility to bacterial infections depends on the physiologic and immunologic condition of the host and on the virulence of the bacteria. Before increased amounts of specific antibodies or T cells are formed in response to invading bacterial pathogens, the "nonspecific" mechanisms of host resistance (such as polymorphonuclear neutrophils and macrophage clearance) must defend the host against the microbes. Development of effective specific immunity (such as an antibody response to the bacterium) may require several weeks. The normal bacterial flora of the skin and mucosal surfaces also serves to protect the host against colonization by bacterial pathogens. In most healthy individuals, bacteria from the normal flora that occasionally penetrate the body (e.g., during tooth extraction or routine brushing of teeth) are cleared by the host's cellular and humoral mechanisms. In contrast, individuals with defective immune responses are prone to frequent, recurrent infections with even the least virulent bacteria. The best-known example of such susceptibility is acquired immune deficiency syndrome (AIDS), in which the $CD4^+$ helper lymphocytes are progressively decimated by human immunodeficiency virus (HIV). However, resistance mechanisms can be altered by many other processes. For example, aging often weakens both nonspecific and specific defense systems so that they can no longer effectively combat the challenge of bacteria from the environment. Infants are also especially susceptible to certain pathogens (such as group B streptococci because their immune systems are not yet fully developed and cannot mount a protective immune response to important bacterial antigens. In addition, some individuals have genetic defects of the complement system or cellular defenses (e.g., inability of polymorphonuclear neutrophils to kill bacteria). Finally, a patient may develop granulocytopenia as a result of a predisposing disease, such as cancer, or immunosuppressive chemotherapy for organ transplants or cancer.

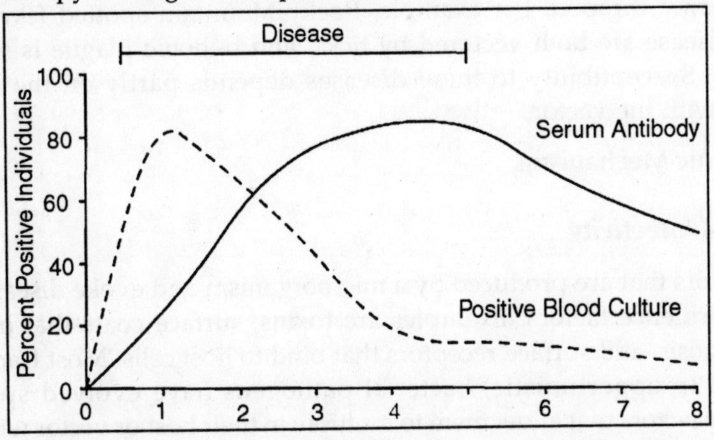


**Fig.** Serum Antibody Response to Salmonella Typhi During Typhoid Fever and Its Relationship to Septicemia.

Host resistance can be compromised by trauma and by some underlying diseases. An individual becomes susceptible to infection with a variety of bacteria if the skin or mucosa is breached, particularly in the case of severe wounds such as burns or contaminated surgical wounds. Cystic fibrosis patients, who have poor ciliary function and consequently cannot clear mucus efficiently from the respiratory tract, are abnormally susceptible to infection with mucoid strains of *Pseudomonas aeruginosa*, resulting in serious respiratory distress. Ascending urinary tract infections with *Escherichia coli* are common in women and are particularly troublesome in patients with urinary tract obstructions. A variety of routine medical procedures, such as tracheal intubation and catheterization of blood vessels and the urethra, increase the risk of bacterial infection. The plastic devices used in these procedures are readily colonized by bacteria from the skin, which migrate along the outside of the tube to infect deeper tissues or enter the bloodstream. Because of this problem, it is standard practice to change catheters frequently (e.g., every 72 hours for peripheral intravenous catheters).

Many drugs have been developed to treat bacterial infections. Antimicrobial agents are most effective, however, when the infection is also being fought by healthy phagocytic and immune defenses. Some reasons for this situation are the poor diffusion of antibiotics into certain sites (such as the prostate gland), the ability of many bacteria to multiply or survive inside cells (where many antimicrobial agents have little or no effect), the bacteriostatic rather than bactericidal action of some drugs, and the capacity of some organisms to develop resistance to multiple antibiotics.

Many bacterial pathogens are transmitted to the host by a vector, usually an arthropod. For example, Rocky Mountain spotted fever and Lyme disease are both vectored by ticks, and bubonic plague is spread by fleas. Susceptibility to these diseases depends partly on the host's contact with the vector.

## Pathogenic Mechanisms

### *Bacterial Infectivity*

Factors that are produced by a microorganism and evoke disease are called virulence factors. Examples are toxins, surface coats that inhibit phagocytosis, and surface receptors that bind to host cells. Most frank (as opposed to opportunistic) bacterial pathogens have evolved specific virulence factors that allow them to multiply in their host or vector without being killed or expelled by the host's defenses. Many virulence factors are produced only by specific virulent strains of a microorganism. For example, only certain strains of *E coli* secrete diarrhea-causing enterotoxins.

Virulence factors should never be considered independently of the host's defenses; the clinical course of a disease often depends on the interaction of virulence factors with the host's response. An infection begins when the balance between bacterial pathogenicity and host resistance is upset.

In essence, we live in an environment that favors the microbe, simply because the growth rate of bacteria far exceeds that of most eukaryotic cells. Furthermore, bacteria are much more versatile than eukaryotic cells in substrate utilization and biosynthesis. The high mutation rate of bacteria combined with their short generation time results in rapid selection of the best-adapted strains and species. In general, bacteria are much more resistant to toxic components in the environment than eukaryotes, particularly when the major barriers of eukaryotes (skin and mucous membranes) are breached.

From a practical standpoint, bacteria can be said to have a single objective: to multiply. Only a few of the vast number of bacterial species in the environment consistently cause disease in a given host. From a teleologic standpoint, it is not in the best interest of the pathogen to kill the host, because in most cases the death of the host means the death of the pathogen. The most highly evolved or adapted pathogens are the ones that acquire the necessary nutritional substances for growth and dissemination with the smallest expenditure of energy and least damage to the host. For example, *Rickettsia akari*, the etiologic agent of rickettsialpox, causes a mild, self-limited infection consisting of headache, fever, and a papulovesicular rash. Other members of the rickettsial group, such as *R rickettsii*, the agent of Rocky Mountain spotted fever, elicit more severe, life-threatening infections. Some bacteria that are poorly adapted to the host synthesize virulence factors (e.g., tetanus and diphtheria toxin) so potent that they threaten the life of the host.

### Host Resistance

Although easily damaged, the skin represents one of the most important barriers of the body to the microbial world, which contains a diverse array of bacteria in enormous numbers. Fortunately, most bacteria in the environment are relatively benign to individuals with normal immune systems. However, patients who are immunosuppressed, such as individuals receiving cancer chemotherapy or have AIDS, opportunistic microbial pathogens can establish life-threatening infections. Normally, microbes in the environment are prevented from entering the body by the skin and mucous membranes. The outermost surface of the skin consists of squamous cell epithelium, largely comprised of dead cells that are sloughed off as new cells are formed below them.

In addition to the skin barrier, mucous membranes of the respiratory, gastrointestinal, and urogenital systems represent other portals through which bacteria can gain access to the body. Like the squamous epithelial cells of the skin, the mucosal epithelial cells divide rapidly, and as the cells mature, they are pushed laterally toward the intestinal lumen and shed. The entire process is reported to require only 36–48 hours for complete replacement of the epithelium, which diminishes the number of bacteria associated with the epithelium. The skin surface is a dry, acidic environment, and the temperature is less than 37° C.

The pores and crevices of the skin also are colonized by the "normal bacterial flora", which ensure competition for pathogens to which the skin is exposed. Similarly, the mucous layer that covers the epithelia contains hostile substances to microbial colonization. Protective levels of lysozyme, lactoferrin, and lactoperoxidase in the mucus either kill bacteria or restrict their growth. In addition, the mucus contains secretory immunoglobulins (predominantly sIgA) synthesized by plasma cells resident in the submucosal tissue During the normal course of life, individuals develop local antibodies specific for a variety of intestinal bacteria that colonize mucosal surfaces.

Another mechanism of restricting growth of bacteria that penetrate the skin and mucous membranes is competition for iron. Typically, the amount of free iron in tissues and blood available to bacteria is very low, since plasma transferrin binds virtually all iron in the blood. Similarly, hemoglobin in the erythrocytes binds iron.

Without free iron, bacterial growth is restricted unless the bacteria synthesize siderophores or receptors for iron containing molecules that compete for transferrin-bound iron. Such siderophores strip iron from transferrin and present it to the bacteria, which enables them to grow. The phagocytic cells of the body patrol the blood and tissues for foreign substances, including bacteria.

This task is assumed predominantly by polymorphonuclear neutrophils; however, monocytes, macrophages, and eosinophils also participate. After phagocytosis, these bacterial cells usually are killed unless their numbers are excessive or they possess virulence factors, that enable them to survive the lysosomal enzymes and acidic pH.

In some instances, the bacteria kill the phagocyte or multiply within the macrophage, escaping the hostile extracellular environment. When inflammation occurs, phagocytic cells, along with lymphocytes, play an important role in innate immunity to bacterial infections. During the interaction of bacterial cells with macrophages, T cells, and B cells, specific antibody responses and/or cell-mediated immunity develop to protect against reinfection.

### *Genetic and Molecular Basis for Virulence*

Virulence factors in bacteria may be encoded on chromosomal DNA, bacteriophage DNA, plasmids, or transposons in either plasmids or the bacterial chromosome. For example, the capacity of the Shigella species to invade cells is a property encoded in part on a 140-mega-dalton plasmid.

Similarly, the heat-labile enterotoxin (LTI) of *E coli* is plasmid encoded, whereas the heat-labile toxin (LTII) is encoded on the chromosome. Other virulence factors are acquired by bacteria following infection by a particular bacteriophage, which integrates its genome into the bacterial chromosome by the process of lysogeny.

Temperate bacteriophages often serve as the basis of toxin production in pathogenic bacteria.

Examples include diphtheria toxin production by Corynebacterium diphtheriae, erythrogenic toxin formation by Streptococcus pyogenes, Shiga-like toxin synthesis by *E coli*, and production of botulinum toxin (types C and D) by *Clostridium botulinum*. Other virulence factors are encoded on the bacterial chromosome (e.g., cholera toxin, *Salmonella enterotoxin*, and *Yersinia* invasion factors).

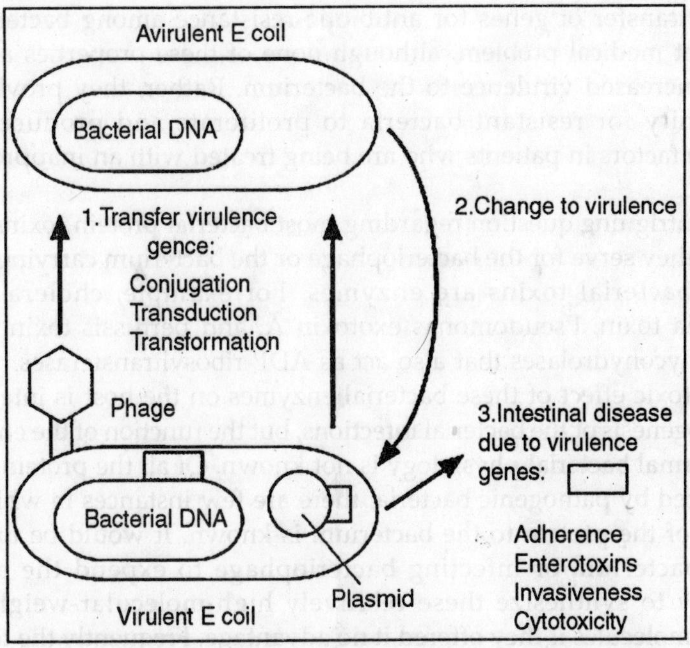


**Fig.** Mechanisms of Acquiring Bacterial Virulence Genes.

**Table. Genetic Basis for Virulence of Selected Bacterial Pathogens.**

| *Gene(s) Encoded on* | *Bacterial Pathogen* | *Virulence Factor* |
|---|---|---|
| Chromosome | Vibrio cholerae | Enterotoxin |
| | Salmonella typhimunium | Enterotoxin, invasion factors |
| | Shigella spp | Enterotoxin, invasion factors |
| | Aeromonas hydrophila | Enterotoxin, aerolysin |
| | Pseudomonas aeruginosa | Exotoxin A |
| | Staphylococcus aureus | Enterotoxin B |
| | Yersinia enterocolitica | Invasion factors |
| | Yersinia pseudofuberculosis | Invasion factors |
| | Escherichia coli | Enterotoxin (LTII) |
| Plasmid | Shigella spp | Invasion factors |
| | Escherichia coli | Invasion factors, colonization factor, and enterotoxin (LTI) |
| | Staphylococcus aureus | Exfoliative toxin |
| | Bacillus anthracis | Anthrax toxin |
| Bacteriophage | Corynebacterium diphtheriae | Diphtheria toxin |
| | Streptococcus pyogenes | Erythrogenic toxin |
| | Escherichia coli | Shiga-like enterotoxin |
| | Clostridium botulinum | Botulinum toxin (C.D) |
| Transposons | Eschenchia coli | Enterotoxins (STA and STB), iron acquistion, hemolysin |

The transfer of genes for antibiotic resistance among bacteria is a significant medical problem, although none of these properties actually confers increased virulence to the bacterium. Rather, they provide the opportunity for resistant bacteria to proliferate and produce other virulence factors in patients who are being treated with an inappropriate antibiotic.

An intriguing question regarding most bacterial protein toxins is the purpose they serve for the bacteriophage or the bacterium carrying them. Several bacterial toxins are enzymes. For example, cholera toxin, diphtheria toxin, Pseudomonas exotoxin A, and pertussis toxin all are NAD + glycohydrolases that also act as ADP-ribosyltransferases.

The toxic effect of these bacterial enzymes on the host is integral to the pathogenesis of the bacterial infections, but the function of the enzymes in the normal bacterial physiology is not known. Of all the protein toxins synthesized by pathogenic bacteria, there are few instances in which the function of the protein to the bacterium is known. It would be unlikely for the bacterium or infecting bacteriophage to expend the energy necessary to synthesize these relatively high-molecular-weight and complex molecules if they offered it no advantage. Frequently the toxicity of these substances is "unintentional" as far as the bacteria are concerned, considering that the primary goal of the microorganisms is to acquire nutrients and multiply rather than to harm the host.

### *Host-Mediated Pathogenesis*

The pathogenesis of many bacterial infections cannot be separated from the host immune response, for much of the tissue damage is caused by the host response rather than by bacterial factors. Classic examples of host response-mediated pathogenesis are seen in diseases such as Gram-negative bacterial sepsis, tuberculosis, and tuberculoid leprosy.

The tissue damage in these infections is caused by toxic factors released from the lymphocytes, macrophages, and polymorphonuclear neutrophils infiltrating the site of infection. Often the host response is so intense that host tissues are destroyed, allowing resistant bacteria to proliferate.

In lepromatous leprosy, in contrast, the absence of a cellular response to Mycobacterium leprae allows the bacteria to multiply to such large numbers in the skin that they become tightly packed and replace healthy tissue. The molecular basis for this specific immune anergy is poorly understood.

### *Intracellular Growth*

In general, bacteria that can enter and survive within eukaryotic cells are shielded from humoral antibodies and can be eliminated only by a cellular immune response. However, these bacteria must possess specialized mechanisms to protect them from the harsh effects of the lysosomal enzymes encountered within the cell.

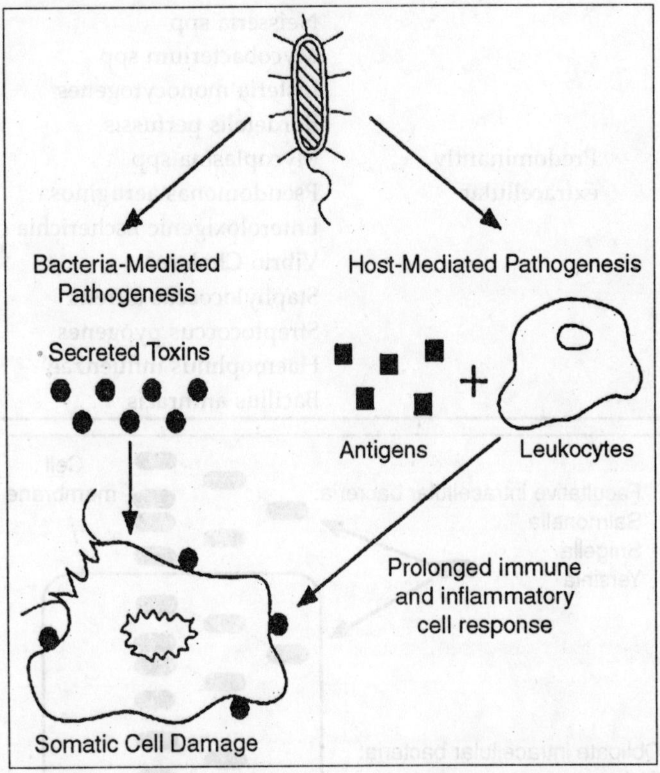


**Fig.** Generalized Mechanisms of Bacterial Pathogenesis: Bacteria-induced Toxicity or Host-mediated Damage.

Pathogenic bacteria can be grouped into three categories on the basis of their invasive properties for eukaryotic cells. Although some bacteria (e.g., *Rickettsia, Coxiella,* and *Chlamydia*) grow only inside host cells, others (e.g., *Salmonella, Shigella,* and *Yersinia*) are facultative intracellular pathogens, invading cells when it gives them a selective advantage in the host.

**Table. Intracellular or Extracellular Growth Preference Relative to Eukaryotic Cells.**

| *Category* | *Bacterial Pathogen* |
|---|---|
| Obligate intracellular | Rickettsia spp<br>Coxielfa burnetii<br>Chlamydia spp |

| | |
|---|---|
| Facultative Intracellular | Salmonella spp |
| | Shigella spp |
| | Legionella pneumophiia |
| | Invasive Escherichia coli |
| | Neisseria spp |
| | Mycobacterium spp |
| | Listeria monocytogenes |
| | Bordetelis perfussis |
| Predominantly extracellular | Mycoplasma spp |
| | Pseudomonas aeruginosa |
| | Enteroloxigenic Escherichia coli |
| | Vibrio Cholerae |
| | Staphylococcus aureus |
| | Streptococcus pyogenes |
| | Haemophilus influenzae |
| | Bacillus anthracis |

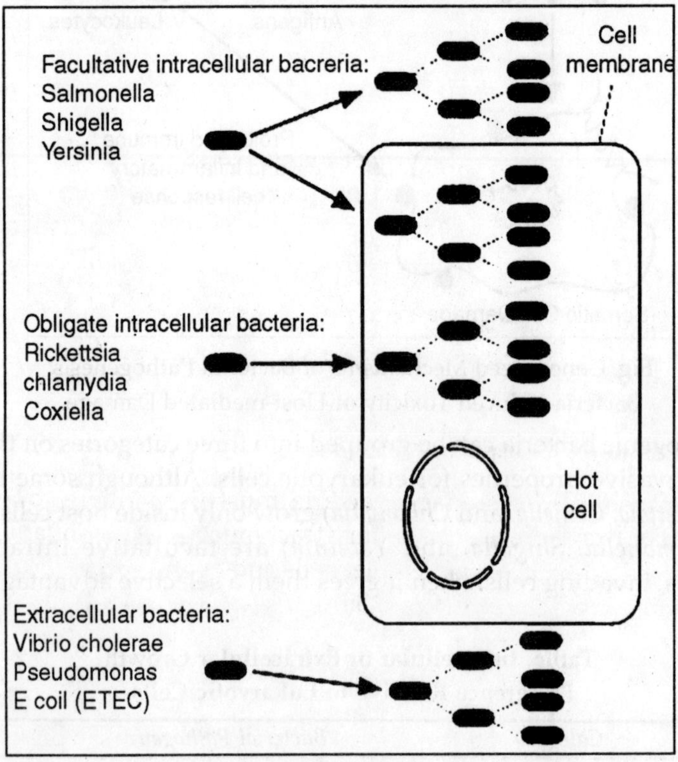


**Fig.** Examples of Pathogenic Bacteria, Indicating their Preferred Growth Phase Within the Host. (ETEC:enterotoxigenic *E Coli*)

Some bacteria survive the intracellular milieu by producing

phospholipases to dissolve the phagocytic vesicle surrounding them. This appears to be the case for R rickettsii, which destroys the phagosomal membrane with which the lysosomes fuse. Legionella pneumophila, which prefers the intracellular environment of macrophages for growth, appears to induce its own uptake and blocks lysosomal fusion by undefined mechanisms. Other bacteria have evolved to the point that they prefer the low-pH environment within the lysosomal granules, as may be the case for *Coxiella burnetii,* a highly resistant member of the rickettsial group. Salmonella and Mycobacterium species also appear to be very resistant to intracellular killing by phagocytic cells, but their mechanisms of resistance are not yet fully understood. Certainly, the capacity of bacteria to survive and multiply within host cells has great impact on the pathogenesis of the respective infections.

Most bacterial pathogens do not invade cells, proliferating instead in the extracellular environment enriched by body fluids. Some of these bacteria (e.g., V cholerae and Bordetella pertussis) do not even penetrate body tissues, but, rather, adhere to epithelial surfaces and cause disease by secreting potent protein toxins. Although bacteria such as *E coli* and *P aeruginosa* are termed noninvasive, they frequently spread rapidly to various tissues once they gain access to the body. All bacteria could at some point be considered intracellular once they become ingested by polymorphonuclear neutrophils and macrophages, but these organisms are not renowned for their capacity to survive the intracellular environment or to induce their own uptake by most host cells.

## SPECIFIC VIRULENCE FACTORS

The virulence factors of bacteria can be divided into a number of functional types. These are discussed in the following sections:

### Adherence and Colonization Factors

To cause infection, many bacteria must first adhere to a mucosal surface. For example, the alimentary tract mucosa is continually cleansed by the release of mucus from goblet cells and by the peristaltic flow of the gut contents over the epithelium. Similarly, ciliated cells in the respiratory tract sweep mucus and bacteria upward. In addition, the turnover of epithelial cells at these surfaces is fairly rapid. The intestinal epithelial cell monolayer is continually replenished, and the cells are pushed from the crypts to the villar tips in about 48 hours. To establish an infection at such a site, a bacterium must adhere to the epithelium and multiply before the mucus and extruded epithelial cells are swept away. To accomplish this, bacteria have evolved attachment mechanisms, such as pili (fimbriae), that recognize and attach the bacteria to cells. Colonization factors (as they are often called) are produced by numerous bacterial pathogens and constitute an important part of the pathogenic mechanism of these bacteria. Some examples of piliated, adherent bacterial pathogens are *V cholerae, E coli, Salmonella* spp, *N gonorrheae, N meningitidis,* and *Streptococcus pyogenes.*

### Invasion Factors

Mechanisms that enable a bacterium to invade eukaryotic cells facilitate entry at mucosal surfaces. Some of these invasive bacteria (such as Rickettsia and Chlamydia species) are obligate intracellular pathogens, but most are facultative intracellular pathogens. The specific bacterial surface factors that mediate invasion are not known in most instances, and often, multiple gene products are involved. Some *Shigella* invasion factors are encoded on a 140 megadalton plasmid, which, when conjugated into *E coli,* gives these noninvasive bacteria the capacity to invade cells. Other invasion genes have also recently been identified in *Salmonella* and *Yersinia* pseudotuberculosis. The mechanisms of invasion of *Rickettsia,* and *Chlamydia* species are not well known.

### Capsules and Other Surface Components

Bacteria have evolved numerous structural and metabolic virulence factors that enhance their survival rate in the host. Capsule formation has long been recognized as a protective mechanism for bacteria. Encapsulated strains of many bacteria (e.g., pneumococci) are more virulent and more resistant to phagocytosis and intracellular killing than are nonencapsulated strains.

Organisms that cause bacteremia (e.g., *Pseudomonas*) are less sensitive than many other bacteria to killing by fresh human serum containing complement components, and consequently are called serum resistant. Serum resistance may be related to the amount and composition of capsular antigens as well as to the structure of the lipopolysaccharide. The relationship between surface structure and virulence is important also in *Borrelia* infections. As the bacteria encounter an increasing specific immune response from the host, the bacterial surface antigens are altered by mutation, and the progeny, which are no longer recognized by the immune response, express renewed virulence. *Salmonella typhi* and some of the paratyphoid organisms carry a surface antigen, the *Vi antigen,* thought to enhance virulence. This antigen is composed of a polymer of galactosamine and uronic acid in 1,4-linkage. Its role in virulence has not been defined, but antibody to it is protective.

Some bacteria and parasites have the ability to survive and multiply inside phagocytic cells. A classic example is Mycobacterium tuberculosis, whose survival seems to depend on the structure and composition of its cell surface. The parasite *Toxoplasma gondii* has the remarkable ability to block the fusion of lysosomes with the phagocytic vacuole. The hydrolytic enzymes contained in the lysosomes are unable, therefore, to contribute to the destruction of the parasite. The mechanism(s) by which bacteria such as *Legionella pneumophila, Brucella abortus,* and *Listeria* monocytogenes remain unharmed inside phagocytes are not understood.

## Endotoxins

Endotoxin is comprised of toxic lipopolysaccharide components of the outer membrane of Gram-negative bacteria. Endotoxin exerts profound biologic effects on the host and may be lethal. Because it is omnipresent in the environment, endotoxin must be removed from all medical supplies destined for injection or use during surgical procedures. The term endotoxin was coined in 1893 by Pfeiffer to distinguish the class of toxic substances released after lysis of bacteria from the toxic substances (exotoxins) secreted by bacteria. Few, if any, other microbial products have been as extensively studied as bacterial endotoxins. Perhaps it is appropriate that a molecule with such important biologic effects on the host, and one produced by so many bacterial pathogens, should be the subject of intense investigation.

### *Structure of Endotoxin*

Figure illustrates the basic structure of endotoxin. Endotoxin is a molecular complex of lipid and polysaccharide; hence, the alternate name lipopolysaccharide. The complex is secured to the outer membrane by ionic and hydrophobic forces, and its strong negative charge is neutralized by $Ca^{2+}$ and $Mg^{2+}$ ions.

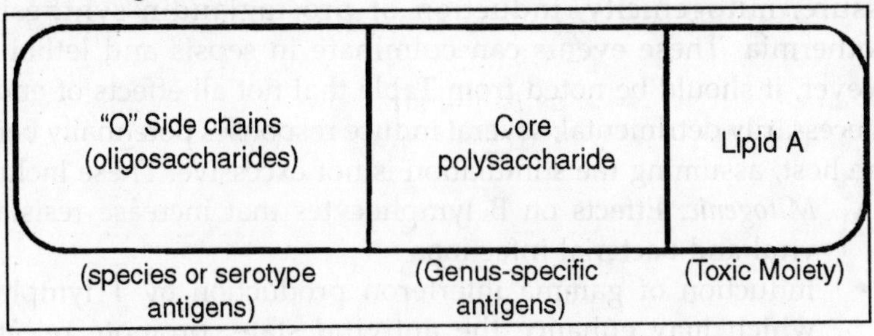


**Fig.** Basic Structure of Endotoxin (Lipopolysaccharide) from Gram-negative Bacteria.

The structure of endotoxin molecules from *Salmonella* spp and *E coli* is known in detail. Enough data on endotoxin from other Gram-negative organisms have been gathered to reveal a common pattern with genus and species diversity. Although all endotoxin molecules are similar in chemical structure and biologic activity, some diversity has evolved. Purified endotoxin appears as large aggregates.

The molecular complex can be divided into three regions: (1) the O-specific chains, which consist of a variety of repeating oligosaccharide residues, (2) the core polysaccharide that forms the backbone of the macromolecule, and (3) lipid A, composed usually of a glucosamine disaccharide with attached long-chain fatty acids and phosphate.

The polysaccharide portions are responsible for antigenic diversity, whereas the lipid A moiety confers toxicity. Dissociation of the complex has revealed that the polysaccharide is important in solubilizing the toxic lipid A component, and in the laboratory it can be replaced by carrier proteins (e.g., bovine serum albumin).

Members of the family Enterobacteriaceae exhibit O-specific chains of various lengths, whereas *N gonorrhoeae, N meningitidis,* and *B pertussis* contain only core polysaccharide and lipid A. Some investigators working on the latter forms of endotoxin prefer to call them lipooligosaccharides to emphasize the chemical difference from the endotoxin of the enteric bacilli. Nevertheless, the biologic activities of all endotoxin preparations are essentially the same, with some being more potent than others.

### *Biologic Activity of Endotoxin*

The biologic effects of endotoxin have been extensively studied. Purified lipid A (conjugated to bovine serum albumin) and endotoxin elicit the same biologic responses. Table lists some of the biologic effects of endotoxin.

The more pertinent toxic effects include pyrogenicity, leukopenia followed by leukocytosis, complement activation, depression in blood pressure, mitogenicity, induction of prostaglandin synthesis, and hypothermia. These events can culminate in sepsis and lethal shock. However, it should be noted from Table that not all effects of endotoxin are necessarily detrimental; several induce responses potentially beneficial to the host, assuming the stimulation is not excessive. These include:

- *Mitogenic:* Effects on B lymphocytes that increase resistance to viral and bacterial infections
- Induction of gamma interferon production by T lymphocytes, which may enhance the antiviral state, promote rejection of tumor cells, and activate macrophages and natural killer cells
- Activation of the complement cascade with the formation of C3a and C5a
- Induction of the formation of interleukin-1 by macrophages and interleukin-2 and other mediators by T lymphocytes.

**Table. Multiple Biologic Activities Exhibited by the Lipid A Component of Endotoxin**

| |
|---|
| Pyrogenicity |
| Leukopenia, leukocytosis |
| Complement activation |
| Depression of blood pressure |
| Hageman factor activation |
| Plafelei activation |
| Induction of plasminogen activator |
| Bone marrow necrosis |
| Hypothermia in mice |
| Lethal toxicity in mice |
| Shwartzman reaction |
| Induction of prostaglandin synthesis |
| Limulus lysate gelation |
| Induction of nonspecific resistance to infection |
| Induction of endotoxin tolerance |
| Adjuvant activity |
| Mitogenic activity for lymphocytes |
| Macrophage activation |
| Induction of interferon synthesis |
| Induction of tumor necrosis factor synthesis |

Potentially beneficial stimulatory effects of endotoxin in low doses. Current research focuses on exploiting some of the potential beneficial effects of "nontoxic" endotoxin derivatives and holds promise for development of future treatment regimens for stimulating the immune response.

For example, the toxicity of endotoxin is largely attributed to lipid A, attached to a polysaccharide carrier. The toxicity of lipid A is markedly reduced after hydrolysis of a phosphate group or deacylation of one or more fatty acids from the lipid A molecule.

Clinical trials are in progress to test a monophosphoryl lipid A for its potential of inducing low dose tolerance to endotoxin. Tolerance to endotoxin can be achieved by pretreatment of an animal with low doses of endotoxin or a detoxified lipid A derivative before challenge with high doses of endotoxin. Experimental studies have demonstrated that induction of tolerance to endotoxin reduces the dangerous effects of endotoxin. It is hoped that these relatively nontoxic lipid A derivatives may be useful in reducing the severity of bacterial sepsis in which bacterial endotoxin produces a life-threatening clinical course.

Endotoxin, which largely accumulates in the liver following injection of a sublethal dose by the intravenous route, can be devastating because of its ability to affect a variety of cell and host proteins. Kupffer cells, granulocytes, macrophages, platelets, and lymphocytes all have a cell receptor on their surface called CD14, which binds endotoxin. Endotoxin binding to the CD14 receptor on macrophages is enhanced by interaction with a host protein made in the liver (i.e., LPS-binding protein). The extent of involvement of each cell type probably depends on the level of endotoxin exposure.

The effects of endotoxin on such a wide variety of host cells result in a complex array of host responses that can culminate in the serious condition gram-negative sepsis, which often leads to shock and death. The effects of endotoxin on host cells are known to stimulate prostaglandin synthesis and to activate the kallikrein system, the kinin system, the complement cascade via the alternative pathway, the clotting system, and the fibrinolytic pathways.

When these normal host systems are activated and operate out of control, it is not surprising that endotoxin can be lethal. Although it is difficult to comprehend the mechanisms of all the cell responses and the myriad sequelae of the cell mediators released rather indiscriminately in the host following exposure to endotoxin, it does seem clear that the host cellular response to endotoxin, rather than a direct toxic effect of endotoxin, plays the major role in causing tissue damage.

### *Detection of Endotoxin in Medical Solutions*

Endotoxin is omnipresent in the environment. It is found in most deionized-water lines in hospitals and laboratories, for example, and affects virtually every biologic assay system ever examined. It tends to be a scapegoat for all biologic problems encountered in the laboratory, and, many times, this reputation is deserved. Because of its pyrogenic and destructive properties, extreme care must be taken to avoid exposing patients to medical solutions containing endotoxin.

Even though all supplies should be sterile, solutions for intravenous administration can become contaminated with endotoxin-containing bacteria after sterilization as a result of improper handling. Furthermore, water used in the preparation of such solutions must be filtered through ion exchange resins to remove endotoxin, because it is not removed by either autoclave sterilization or filtration through bacterial membrane filters. If endotoxin-containing solutions were used in such medical procedures as renal dialysis, heart bypass machines, blood transfusions, or surgical lavage, the patient would suffer immediate fever accompanied by a rapid and possibly lethal alterations in blood pressure.

Solutions for human or veterinary use are prepared under carefully controlled conditions to ensure sterility and to remove endotoxin. Representative samples of every manufacturing batch are checked for endotoxin by one of two procedures: the Limulus lysate test or the rabbit pyrogenicity test. The rabbit pyrogenicity test is based on the exquisite sensitivity of rabbits to the pyrogenic effects of endotoxin. A sample of the solution to be tested usually is injected intravenously into the ear veins of adult rabbits while the rectal temperature of the animal is monitored. Careful monitoring of the temperature responses provides a sensitive and reliable indicator of the presence of endotoxin and, importantly, one measure of the safety of the solution for use in patients.

The Limulus lysate test is more common and less expensive. This test, which is based on the ability of endotoxin to induce gelation of lysates of amebocyte cells from the horseshoe crab Limulus polyphemus, is simple, fast, and sensitive (about 1 ng/ml). It is so sensitive, however, that trace quantities of endotoxin in regular deionized water often obscure the results. It can be used for rapid detection of certain Gram-negative infections (e.g., of cerebrospinal fluid); however, blood contains inhibitors that prevent gelation. Test kits are commercially available. The amebocyte is the sole phagocytic immune cell of the horseshoe crab, and the gelation reaction is believed to be involved in sequestering invading Gram-negative bacteria.

### Exotoxins

Exotoxins, unlike the lipopolysaccharide endotoxin, are protein toxins released from viable bacteria. They form a class of poisons that is among the most potent, per unit weight, of all toxic substances. Most of the higher molecular-sized exotoxin proteins are heat labile; however, numerous low molecular-sized exotoxins are heat-stable peptides. Unlike endotoxin, which is a structural component of all Gram-negative cells, exotoxins are produced by some members of both Gram-positive and Gram-negative genera. The functions of these exotoxins for the bacteria are usually unknown, and the genes for most can be deleted with no noticeable effect on bacterial growth. In contrast to the extensive systemic and immune-system effects of endotoxin on the host, the site of action of most exotoxins is more localized and is confined to particular cell types or cell receptors. Tetanus toxin, for example, affects only internuncial neurons. In general, exotoxins are excellent antigens that elicit specific antibodies called antitoxins. Not all antibodies to exotoxins are protective, but some react with important binding sites or enzymatic sites on the exotoxin, resulting in complete inhibition of the toxic activity (i.e., neutralization).

Exotoxins can be grouped into several categories (e.g., neurotoxins, cytotoxins, and enterotoxins) based on their biologic effect on host cells.

Neurotoxins are best exemplified by the toxins produced by *Clostridium* spp, for example, the botulinum toxin formed by *C botulinum*. This potent neurotoxin acts on motor neurons by preventing the release of acetylcholine at the myoneural junctions, thereby preventing muscle excitation and producing flaccid paralysis. The cytotoxins constitute a larger, more heterogeneous grouping with a wide array of host cell specificities and toxic manifestations. One cytotoxin is diphtheria toxin, which is produced by *Corynebacterium diphtheriae*. This cytotoxin inhibits protein synthesis in many cell types by catalyzing the ADP-ribosylation of elongation factor II, which blocks elongation of the growing peptide chain.

Enterotoxins stimulate hypersecretion of water and electrolytes from the intestinal epithelium and thus produce watery diarrhea. Some enterotoxins are cytotoxic (e.g., shiga-like enterotoxin from *E coli*), while others perturb eukaryotic cell functions and are cytotonic (e.g., cholera toxin). Enterotoxins also can disturb normal smooth muscle contraction, causing abdominal cramping and decrease transit time for water absorption in the intestine. Enterotoxigenic *E coli* and *V cholerae* produce diarrhea after attaching to the intestinal mucosa, where they elaborate enterotoxins.

Neither pathogen invades the body in substantial numbers, except in the case of *E coli* species that have acquired an invasion plasmid. Importantly, cholera toxin and *E coli* heat-labile enterotoxins I and II cause ADP-ribosylation of cell proteins in a manner similar to diphtheria toxin, except that the primary target is the regulatory protein ($G_{s-}$) of adenylate cyclase, resulting in increased levels of cyclic 3',5'-adenosine monophosphate (cAMP). In contrast, the organisms responsible for shigellosis (*Shigella dysenteriae, S boydii, S flexneri,* and *S sonnei*) penetrate the mucosal surface of the colon and terminal ileum to proliferate and cause ulcerations that bleed into the intestinal lumen.

Despite causing extensive ulceration of the mucosa, the pathogens rarely enter the bloodstream. The Shiga enterotoxin produced by *Shigella* species and the Shiga-like enterotoxin elaborated by many isolates of *E coli* inhibit protein synthesis in eukaryotic cells. It is not clear how this cytotoxic enterotoxin causes hypersecretion of water and electrolytes from the intestinal epithelium. These enterotoxins differ from those secreted by *V cholerae* and *E coli* in that the Shiga toxins are cytotoxic and lethal, whereas the cholera toxin-like enterotoxins are not. The latter enterotoxins cause no structural damage to cells, and are described as cytotonic. The ensuing inflammatory response to the invading bacteria and/or their toxins appears to activate neurologic control mechanisms (e.g., prostaglandins, serotonin) that normally regulate water and electrolyte transport.

## Siderophores

Both animals and bacteria require iron for metabolism and growth, and the control of this limited resource is often used as a tactic in the conflict between pathogen and host. Animals have evolved mechanisms of "withholding" iron from tissue fluids in an attempt to limit the growth of invading bacteria.

Although blood is a rich source of iron, this iron is not readily available to bacteria since it is not free in solution. Most of the iron in blood is bound either to hemoglobin in erythrocytes or to transferrin in plasma. Similarly, the iron in milk and other secretions (e.g., tears, saliva, bronchial mucus, bile, and gastrointestinal fluid) is bound to lactoferrin. Some bacteria express receptors for eukoyotic iron-binding proteins (e.g., transferrin-binding outer membrane proteins on the surface of *Neisserira* spp). Via these specialized receptors iron acquisition is facilitated, providing the esssential element for bacterial growth.

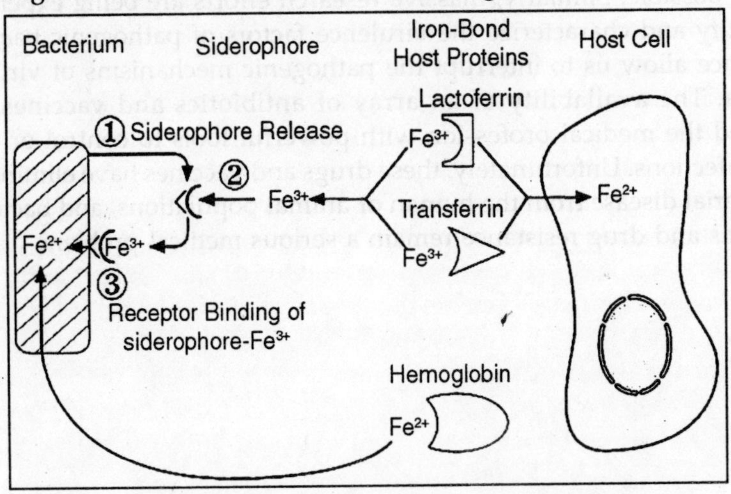


**Fig.** Competition between Host Cells and Bacterial Pathogens for Iron, Illustrating the Importance of Siderophores.

Other bacteria have evolved elaborate mechanisms to extract the iron from host proteins. Siderophores are substances produced by many bacteria (and some plants) to capture iron from the host. The absence of iron triggers transcription of the genes coding for the enzymes that synthesize siderophores, as well as for a set of surface protein receptors that recognize siderophores carrying bound iron. The binding constants of the siderophores for iron are so high that even iron bound to transferrin and lactoferrin is confiscated and taken up by the bacterial cells. An example of a bacterial siderophore is enterochelin, which is produced by *Escherichia* and *Salmonella* species. Classic experiments have demonstrated that Salmonella mutants that have lost the capacity to synthesize enterochelin lose virulence in an assay of lethality in mice. Injection of purified enterochelin along with the Salmonella mutants restores virulence to the bacteria. Therefore, siderophore production by many pathogenic bacteria is considered an important virulence mechanism.

Since free iron is scarce in tissue fluids and blood, bacterial siderophores compete effectively for $Fe^{3+}$ bound to lactoferrin and transferrin.

## Epilogue

Many factors determine the outcome of the bacterium-host relationship. The host must live in an environment filled with a diverse population of microorganisms. Because of the magnitude of the infectious-disease problem, we strive to understand the natural immune mechanisms of the host so that future improvements in resistance to bacterial infections may be possible. Similarly, massive research efforts are being expended to identify and characterize the virulence factors of pathogenic bacteria and hence allow us to interrupt the pathogenic mechanisms of virulent bacteria. The availability of an array of antibiotics and vaccines has provided the medical profession with powerful tools to control or cure many infections. Unfortunately, these drugs and vaccines have eliminated no bacterial disease from the human or animal populations, and bacterial infections and drug resistance remain a serious medical problem.

# Chapter 8

# Specific Acquired Immunity

## BASIS OF ACQUIRED RESISTANCE

Acquired resistance is mediated by antigen-specific immune mechanisms. This specificity may be acquired following a disease, by asymptomatic carriage of the pathogen, by harboring an organism with a similar structure (crossreacting,) or by vaccination.

Specific acquired immunity against infectious diseases may be mediated by antibodies and/or T lymphocytes. Immunity mediated by these two factors may be manifested by a direct effect upon a pathogen, such as (1) antibody-initiated complement-dependent bacteriolysis, (2) opsonophagocytosis and killing, as occurs for some bacteria, (3) neutralization of viruses so that these organisms cannot enter cells, or (4) by T lymphocytes which will kill a cell parasitized by a microorganism.

## PRIMARY VS OPPORTUNISTIC PATHOGENS

Among the almost infinite varieties of microorganisms, relatively few cause a disease in an otherwise normal or healthy individual. These highly virulent microorganisms are conveniently classified as primary pathogens. Opportunistic infectious disease may be caused by organisms that are ordinarily in contact with the host, such as bacteria or fungi in the colon or in the upper respiratory tract; following an injury, whether mechanical (such as a open fracture); or by a disease with immunosuppressive activity (such as measles or malaria, or one induced by cytotoxic chemotherapy). Organisms, which cause an infectious disease in a host with depressed resistance, are classified as opportunistic pathogens. Primary pathogens may also cause more explosive disease in a host with depressed resistance. Our knowledge of the protective antigens and specific acquired host immune factors is more complete for primary pathogens. Most primary pathogens are inhabitants of, and pathogens for, humans only. Opportunistic pathogens, in contrast, may cause disease in many species of mammals. Some exceptions are tetanus, anthrax and rabies which may inhabit and cause disease in many animal species including humans.

## PROTECTIVE ANTIGENS

Microorganisms adapt to cause disease by many mechanisms. Many bacteria, for example, produce macromolecules that cause (1) inflammation, (2) adherence to human tissues, or (3) are toxins that chemically alter host metabolism. But, in general, specific acquired immunity to human pathogens is directed to one (protective) antigen.

### Protein Antigens

Proteins, in most instances, are defined as T cell dependent antigens by two properties. First, proteins are hydrolyzed to peptides by intracellular proteases. Second, proteins may have multiple but unique specificities (epitopes). These two properties permit recognition by receptors (membrane immunoglobulin) of antibody-producing cells (plasma cells) and internalization and proteolysis by enzymes. Activation requires that the peptide fragment of the protein antigen interact with the plasma cell histocompatibility antigen to form a complex that attracts and activates T cells. The activated antigen-specific T cells release cytokines that cause plasma cells to divide and to increase their secretion of specific antibody.

The same mechanism serves to activate T cells by contact with host cells parasitized by intracellular microorganisms. Proteins secreted by intracellular organisms interact with histocompatibility antigens of the host cell (T cell epitopes) and provide a specific site for T cells that kill the parasitized host cell (cytotoxic T lymphocytes).

Some protective epitopes, especially those of viruses, are expressed only on the intact organism (conformation epitope). An example is the neutralizing epitope of polioviruses (D antigen) that requires the intact capsid to elicit neutralizing antibodies.

The protective protein antigens are highly specific and are unique to each pathogen. Acquisition of antibodies to protein protective antigens either follows infection with the pathogen or vaccination.

### Polysaccharide Antigens

Surface polysaccharides of pathogens may serve as protective antigens. These polysaccharides may be capsular polysaccharides, present on either Gram-negative or Gram-positive organisms or the outermost domain of the lipopolysaccharide of Gram-negative organisms. Polysaccharides have simple structures composed of identical repeating units so that each molecule will have relatively fewer epitopes than a protein. But in contrast to a protein, a polysaccharide is multivalent for each epitope. This multivalency explains why a polysaccharide can crosslink the receptors of plasma cells, resulting in activation and multiplication of plasma cells, both of which increase secretion of antibodies.

**Table. The Surface Polysaccharides of Primary Bacterial Pathogens Causing Systemic Infections.**

| *Name of pathogen* | *Polysaccharide designation* |
|---|---|
| | Capsular polysaccharides: |
| Streptococcus pneumoniae | Type |
| Haemophilus influenzae | Type |
| Neisseria meningitidis | Group |
| Group B streptococcus | Group |
| Escherichia coli | K antigen |
| Salmonella typhi | Vi |
| | Lipopolysaccharides: |
| Salmonelfa | Group |
| Shigella | Type |
| Enteroinvasive E coli (EIEC) | Type |
| Vibrio cholerae 01 | Serotype |

The protective epitopes of polysaccharides, in contrast to proteins, are widely shared in nature, and natural immunity: antibody synthesis in the absence of the homologous organism, occurs in almost every individual during development. Similarly, disease and often asymptomatic carriage will also stimulate serum polysaccharide antibodies. Lastly, because they do not interact with T cells, polysaccharides are designated as T cell independent.

Each bacterial species, such as pneumococci, may have many capsular polysaccharides but only a fraction of these will be associated with a disease. For example, there are now 89 reported types but most systemic pneumococcal infections are caused by about 23 types. In infants and children, most pneumococcal infections are caused by only 8 types. Similarly, of the six types of Haemophilus influenzae, almost all systemic infections, especially meningitis, are caused by type b.

It is important to understand that immunity may be directed towards the intact pathogens such as bacteria, viruses, protozoa, or fungi, or to individual extracellular antigens such as toxins (antitoxin).

## IMMUNE MECHANISMS

### Antibody and Secondary Biologic Activities

Serum antibodies are the signal and specific component of a complex inactivation system. Serum IgM and IgG antibodies exert their protective effect directly upon bacteria whose surface polysaccharides or proteins are protective antigens. This signal is generated by the configuration of the non-antibody combining site region of the heavy polypeptide chain after binding of antibody with an epitope. For some pathogens such as meningococci (Gram-negative), the antigen-antibody configuration will activate the serum complement protein cascade with deposition of a C8,9 peptide probe that drills itself through the outer membrane.

The resultant lesion causes release of intracellular components and lysis of the meningococci. In addition to bacteriolysis, other Gram-negatives, such as Haemophilus influenzae type b, also may be inactivated by serum antibody, affixed to the polysaccharide of this pathogen, that attracts and activates serum complement proteins to form C3 and C5 complexes.

The latter complexes attract and activate phagocytic cells that engulf and digest the pathogen (opsonophagocytosis). This antibody-initiated, complement-dependent phagocytosis and killing are required for Gram-positive bacteria whose cell wall is not susceptible to lysis by complement.

Serum antibodies may confer immunity by binding directly to viral pathogens. The effect of this simple interaction is neutralization or inactivation of the virus.

The exact protective mechanism of viral-specific antibodies is not known and may be unique for each pathogen. In many cases, antibody binding renders the viral pathogen incapable of infecting a host cell by preventing penetration of cells. As an example, antibodies to the fusion (F) protein of measles prevent the integration of virus with the cell membrane of the host. Experimental proof for this direct antiviral effect is that monovalent fragments of IgG, unable to activate complement, exert similar neutralization activity as the intact antibody. Some larger viral pathogens, coated with specific antibody, may be phagocytized and digested.

Antibodies that neutralize bacterial toxins (antitoxins), such as tetanus, diphtheria and pertussis toxins, are protective and therapeutic. The protective effects of antitoxins are varied. Antitoxin does not exert antibacterial action upon Clostridium tetani. Rather, antitoxin inactivates the functions of tetanus toxin that facilitates its migration up the neural sheath to the synapse, and antibodies to the enzymatic region inhibit its alteration of the synapse. The neutralizing activities of diphtheria and pertussis antitoxins, in contrast, exert secondary antibacterial actions upon their respective pathogens. Both toxins serve to condition the respiratory epithelium to permit colonization by Corynebacterium diphtheriae tox+ and Bordetella pertussis: the former by its cytotoxicity and the latter by its inactivation of the function of phagocytic cells. Antitoxins block these actions and facilitate the function of phagocytic cells.

## Cell-Mediated Immunity

Antigen-specific activation of T cells has been described (vide supra). The targets for activated T cells are parasitized host cells. The secondary or inactivation mechanisms invoked by activated T cell phagocytic host cell complexes are not clearly understood. One important mechanism is the release of nitrous oxide that results in killing of the host cell and of the pathogen. Cell-mediated immunity is largely, if not exclusively, a curative mechanism.

## Preventive Immunity

Immune resistance to an infectious disease requires a critical level of either antigen-specific antibodies and/or T cells when the host encounters the pathogen. Prevention of an infection requires immune mechanisms to kill or inactivate the inoculum of the pathogen.

This immunity may be expressed as a protective level of antibodies so that resistance to specific infections may be reliably predicted by a serologic assay such as the level of neutralizing antibodies to measles, mumps or Groups A, B, Y and W135 meningococci. These assays can predict resistance to a disease for individuals or can be used to assess the immune status of communities. Quantitation of antigen-specific T cells to predict immunity on a clinical basis is, as yet, an investigative tool.

## Disease-acquired Immunity

Convalescence from most infectious diseases confers immunity. This immunity, in most instances, may be transferred for a limited period to non-immune individuals by injection of IgG as FDA-licensed immunoglobulin or to the newborn by passively acquired maternal serum IgG. There is also evidence that secretory IgA acquired by breast feeding confers immunity to newborns.

These findings indicate that critical levels of antibodies are sufficient to prevent infectious diseases. Their preventive action is best explained by antibodies killing or inactivating the inoculum of the pathogen on epithelial surfaces. Herd immunity follows vaccination with Haemophilus type b conjugates, diphtheria toxoid and measles virus vaccines. The resulting immunity causes a decreased transmission of the pathogen. Since there is no animal vector for these pathogens, the incidence of the disease in the entire community is far below that percentage of the population that has been vaccinated (herd immunity). Antibody-mediated inactivation of the inoculum may also occur in the blood stream in the case of pathogens inoculated directly into the tissues or blood stream, such as hepatitis B or malaria. There is yet no evidence in humans that antigen-specific T lymphocytes can prevent infectious diseases.

**Table. U.S. Licensed Immunoglobulin For Passive Immunization**

| *Disease* | *Biologic* | *Indication* |
|---|---|---|
| Botulism | Specific equine lg | Treatment |
| CMV | Hyperimmune human IV lg | Prophytaxis |
| Diphtheria | Specific equine lg | Treatment |
| Hepatitis A, measles | Pooled human lg | Prophyiaxis |
| lg deficiency, ITP, Kawasaki disease | Pooled human lgG | Treatment |
| Hepatitis B lg | Immune human lg | Prophylaxis |
| Rabies (HRIG) | Immune human lg | Prophylaxis |
| Telanus lg (TIG) | Immune human lg | Treatment |
| Vaccinia | Immune human lg | Treatment |
| Varicella-zoster | Immune human lg | Prophylaxis |

The positive effect of passively administered immunoglobutin has been established for the immunodeficiency disease. X-linked hypogammaglobulinemia, in which bacterial diseases, such as otitis media, pneumonia, and meningitis are prevented. Prior to the advent of viral vaccines, passively administered immunoglobulin was routinely used for prevention of measles, rubella, mumps, poliomyelitis, and varicella. Human immune globutins are used for prevention of rabies and are being considered for respiratory syncytial virus infections.

## Natural Immunity

Acquisition of serum antibodies to surface polysaccharides of pathogens is age-related and often occurs without the individual encountering the homologous organism. An example is group A meningococci, the cause of epidemic meningitis. Despite the absence of this pathogen in the United States for about 50 years, either as a cause of meningitis or in asymptomatic carriers, most adults have antibodies to this capsular polysaccharide. Antigenic stimuli for Group A meningococcal antibodies are likely due to exposure to several Gram-positive and Gram-negative bacterial species in human stools. Another example is Shigella dysenteriae type 1, the cause of epidemic dysentery. Adults in Sweden and the United States have antibodies to the LPS of S dysenteriae type 1, despite the virtual absence of this pathogen in these countries during the past 50 years. The stimulus for these natural and protective antibodies is probably cross-reacting polysaccharides of the enteric and respiratory tract floras.

Natural serum antibodies to surface polysaccharides confer specific protection to adults and are transmitted to newborns. The highest attack rate and mortality occur during childhood when these maternally-acquired anti-polysaccharide antibodies have waned and adult levels have not been reached. Acquisition of natural antibodies is not uniform and many adults remain non-immune. Vaccination with polysaccharide-based vaccines increases the percentage of adults with protective levels to almost 100%. These principles are elegantly illustrated by the development of groups A and C meningococcal polysaccharide vaccines during the 1960's when outbreaks of meningitis caused by this pathogen occurred in armed forces recruits during the Vietnam conflict. Introduction of these polysaccharide vaccines rapidly eliminated these outbreaks.

## Vaccination-Induced Active and Passive Immunity

Vaccines are heterogeneous according to (1) the nature of the immunizing substance, whether they are inert or living, and (2) by the method of their administration. Vaccines include living attenuated strains of viruses (poliovirus) or bacteria (BCG), inactivated viruses (yellow fever) and bacteria (anthrax), purified polysaccharides (pneumococcal 23 valent) or polysaccharide-protein conjugates (Haemophilus type b conjugate). Inert and injected vaccines, such as tetanus toxoid, elicit mostly serum antibodies. Living vaccines, such as attenuated strains of viruses (poliovirus), elicit secretory antibodies and sensitized T cells.

To date, the FDA regulates vaccines and seroepidemiologic studies to assess the immune status of populations by measurement of biologically active antibodies only. Thus, the status of diphtheria immunity is evaluated by the percentage of the population with protective levels of neutralizing antibodies to diphtheria toxin (antitoxin). Similarly, the immune status to measles is evaluated by measurement of the percentage of the population with protective levels of neutralizing antibodies. For some vaccines, such as BCG, there is as yet no measure of immunity to assess their effectiveness.

Some investigational vaccines utilize recombinant DNA technology to mobilize genes governing the synthesis of protective antigens. These specific genes may be inserted into avirulent vectors. Administration of vectors is designed to stimulate the comprehensive immunity that follows disease with the pathogen itself. Naked DNA may be incorporated into plasmids that infect somatic cells and continually induce synthesis of protective antigens that stimulate antibodies and activated T cells.

## Curative Immunity

Infection with most pathogens does not result in death of the host and the offending organism is ultimately cleared after the symptoms of the disease have waned. The basis of the curative process of patients is not well understood, but it is likely mediated by expansion of both specific immune and effector mechanisms.

Quantitative increases of specific antibodies and activated T cells is accompanied by increases in serum complement levels and phagocytic cells. Also, infection increases the levels of cytokines and serum proteins known as acute phase reactants, such as C-reactive protein, alpha 1 trypsin inhibitor, and transferrin, that serve as scavengers or inhibitors of bacterial debris. Cure of infectious diseases is most likely the prolonged interaction of maximal levels of host specific and non-specific (effector) mechanisms with the pathogen.

## Chapter 9

# Epidemiology

## INTRODUCTION

This chapter reviews the general concepts of epidemiology, which is the study of the determinants, occurrence, distribution, and control of health and disease in a defined population. Epidemiology is a descriptive science and includes the determination of rates, that is, the quantification of disease occurrence within a specific population. The most commonly studied rate is the attack rate: the number of cases of the disease divided by the population among whom the cases have occurred. Epidemiology can accurately describe a disease and many factors concerning its occurrence before its cause is identified. For example, Snow described many aspects of the epidemiology of cholera in the late 1840s, fully 30 years before Koch described the bacillus and Semmelweis described puerperal fever in detail in 1861 and recommended appropriate control and prevention measures a number of years before the streptococcal agent was fully described. One goal of epidemiologic studies is to define the parameters of a disease, including risk factors, in order to develop the most effective measures for control. This chapter includes a discussion of the chain of infection, the three main epidemiologic methods, and how to investigate an epidemic.

**Table. Epidemiologic Methods and Investigation**

| *Descriptive* | *Analytic* | *Experiment* |
|---|---|---|
| Time | Case control | Manipulate cause |
| Secular | Cohort | and note effect |
| Periodic | | |
| Seasonal | | |
| Epidemic | | |
| Place | | |
| Person | | |

Proper interpretation of disease-specific epidemiologic data requires information concerning past as well as present occurrence of the disease.

An increase in the number of reported cases of a disease that is normal and expected, representing a seasonal pattern of change in host susceptibility, does not constitute an epidemic. Therefore, the regular collection, collation, analysis, and reporting of data concerning the occurrence of a disease is important to properly interpret short-term changes in occurrence.

A sensitive and specific surveillance programme is important for the proper interpretation of disease occurrence data. Almost every country has a national disease surveillance programme that regularly collects data on selected diseases. The quality of these programs varies, but, generally, useful data are collected that are important in developing control and prevention measures. There is an international agreement that the occurrence of three diseasescholera, plague, and yellow feverwill be reported to the World Health Organization in Geneva, Switzerland. In the United States, the Centers for Disease Control and Prevention (CDC), U.S. Public Health Service, and the state health officers of all 50 states have agreed to report the occurrence of 51 diseases weekly and of another 10 diseases annually from the states to the CDC. Many states have regulations or laws that mandate reporting of these diseases and often of other diseases of specific interest to the state health department.

The methods of case reporting vary within each state. Passive reporting is one of the main methods. In such a case, physicians or personnel in clinics or hospitals report occurrences of relevant diseases by telephone, postcard, or a reporting form, usually at weekly intervals. In some instances, the report may be initiated by the public health or clinical laboratory where the etiologic agent is identified. Some diseases, such as human rabies, must be reported by telephone as soon as diagnosed. In an active surveillance programme, the health authority regularly initiates the request for reporting. The local health department may call all or some health care providers at regular intervals to inquire about the occurrence of a disease or diseases. The active system may be used during an epidemic or if accurate data concerning all cases of a disease are desired.

The health care provider usually makes the initial passive report to a local authority, such as a city or county health department. This unit collates its data and sends a report to the next highest health department level, usually the state health department.

The number of cases of each reportable disease are presented weekly, via computer linkage, by the state health department to the CDC. Data are analyzed at each level to develop needed information to assist public health authorities in disease control and prevention. For some diseases, such as hepatitis, the CDC requests preparation of a separate case reporting form containing more specific details.

In addition, the CDC prepares and distributes routine reports summarizing and interpreting the analyses and providing information on epidemics and other appropriate public health matters. Most states and some county health departments also prepare and distribute their own surveillance reports. The CDC publishes Morbidity and Mortality Weekly Report, which is available for a small fee from the Massachusetts Medical Society. The CDC also prepares more detailed surveillance reports for specific diseases, as well as an annual summary report, all of which can also be obtained through the Massachusetts Medical Society.

Infection is the replication of organisms in the tissue of a host; when defined in terms of infection, disease is overt clinical manifestation. In an inapparent (subclinical) infection, an immune response can occur without overt clinical disease.

A carrier (colonized individual) is a person in whom organisms are present and may be multiplying, but who shows no clinical response to their presence. The carrier state may be permanent, with the organism always present; intermittent, with the organism present for various periods; or temporary, with carriage for only a brief period. Dissemination is the movement of an infectious agent from a source directly into the environment; when infection results from dissemination, the source, if an individual, is referred to as a dangerous disseminator.

Infectiousness is the transmission of organisms from a source, or reservoir, to a susceptible individual. A human may be infective during the preclinical, clinical, postclinical, or recovery phase of an illness. The incubation period is the interval in the preclinical period between the time at which the causative agent first infects the host and the onset of clinical symptoms; during this time the agent is replicating. Transmission is most likely during the incubation period for some diseases such as measles; in other diseases such as shigellosis, transmission occurs during the clinical period. The individual may be infective during the convalescent phase, as in diphtheria, or may become an asymptomatic carrier and remain infective for a prolonged period, as do approximately 5% of persons with typhoid fever.

The spectrum of occurrence of disease in a defined population includes sporadic (occasional occurrence); endemic (regular, continuing occurrence); epidemic (significantly increased occurrence); and pandemic (epidemic occurrence in multiple countries).

## CHAIN OF INFECTION

The chain of infection includes the three factors that lead to infection: the etiologic agent, the method of transmission, and the host. These links should be characterized before control and prevention measures are proposed. Environmental factors that may influence disease occurrence must be evaluated.

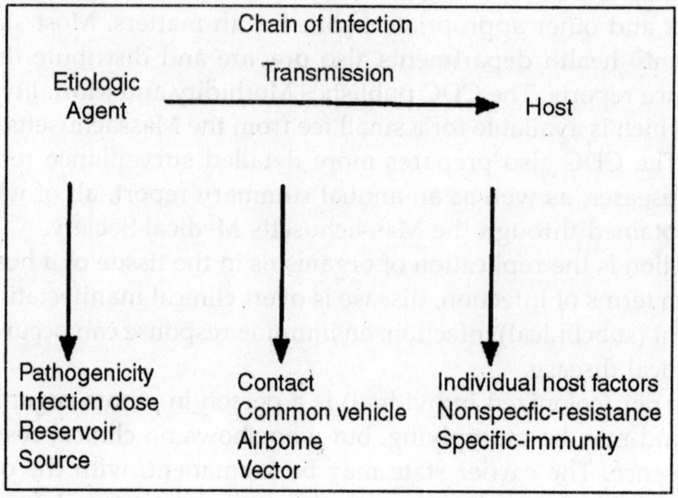


**Fig.** Summary of Important Aspects Involved in the Chain of any Infection.

## Etiologic Agent

The etiologic agent may be any microorganism that can cause infection. The pathogenicity of an agent is its ability to cause disease; pathogenicity is further characterized by describing the organism's virulence and invasiveness. Virulence refers to the severity of infection, which can be expressed by describing the morbidity (incidence of disease) and mortality (death rate) of the infection. An example of a highly virulent organism is *Yersinia pestis,* the agent of plague, which almost always causes severe disease in the susceptible host.

The invasiveness of an organism refers to its ability to invade tissue. *Vibrio cholerae* organisms are noninvasive, causing symptoms by releasing into the intestinal canal an exotoxin that acts on the tissues. In contrast, *Shigella* organisms in the intestinal canal are invasive and migrate into the tissue.

No microorganism is assuredly avirulent. An organism may have very low virulence, but if the host is highly susceptible, as when therapeutically immunosuppressed, infection with that organism may cause disease. For example, the poliomyelitis virus used in oral polio vaccine is highly attenuated and thus has low virulence, but in some highly susceptible individuals it may cause paralytic disease.

Other factors should be considered in describing the agent. The infecting dose (the number of organisms necessary to cause disease) varies according to the organism, method of transmission, site of entrance of the organism into the host, host defenses, and host species. Another agent factor is specificity; some agents (for example, *Salmonella typhimurium)* can infect a broad range of hosts; others have a narrow range of hosts. *S typhi,* for example, infects only humans. Other agent factors include antigenic composition, which can vary within a species (as in influenza virus or *Streptococcus* species); antibiotic sensitivity; resistance transfer plasmids; and enzyme production.

The reservoir of an organism is the site where it resides, metabolizes, and multiplies. The source of the organism is the site from which it is transmitted to a susceptible host, either directly or indirectly through an intermediary object. The reservoir and source can be different; for example, the reservoir for *S typhi* could be the gallbladder of an infected individual, but the source for transmission might be food contaminated by the carrier. The reservoir and source can also be the same, as in an individual who is a permanent nasal carrier of *S aureus* and who disseminates organisms from this site. The distinction can be important when considering where to apply control measures

## Method of Transmission

The method of transmission is the means by which the agent goes from the source to the host. The four major methods of transmission are by contact, by common vehicle, by air or via a vector.

In contact transmission the agent is spread directly, indirectly, or by airborne droplets. Direct contact transmission takes place when organisms are transmitted directly from the source to the susceptible host without involving an intermediate object; this is also referred to as person-to-person transmission. An example is the transmission of hepatitis A virus from one individual to another by hand contact. Indirect transmission occurs when the organisms are transmitted from a source, either animate or inanimate, to a host by means of an inanimate object. An example is transmission of *Pseudomonas* organisms from one individual to another by means of a shaving brush. Droplet spread refers to organisms that travel through the air very short distances, that is, less than 3 feet from a source to a host. Therefore, the organisms are not airborne in the true sense. An example of a disease that may be spread by droplets is measles.

Common-vehicle transmission refers to agents transmitted by a common inanimate vehicle, with multiple cases resulting from such exposure. This category includes diseases in which food or water as well as drugs and parenteral fluids are the vehicles of infection. Examples include food-borne salmonellosis, waterborne shigellosis, and bacteremia resulting from use of intravenous fluids contaminated with a gram-negative organism.

The third method of transmission, airborne transmission, refers to infection spread by droplet nuclei or dust. To be truly airborne, the particles should travel more than 3 feet through the air from the source to the host. Droplet nuclei are the residue from the evaporation of fluid from droplets, are light enough to be transmitted more than 3 feet from the source, and may remain airborne for prolonged periods. Tuberculosis is primarily an airborne disease; the source may be a coughing patient who creates aerosols of droplet nuclei that contain tubercle bacilli. Infectious agents may be contained in dust particles, which may become resuspended and transmitted to hosts. An example occurred in an outbreak of salmonellosis in a newborn nursery in which *Salmonella*-contaminated dust in a vacuum cleaner bag was resuspended when the equipment was used repeatedly, resulting in infections among the newborns.

The fourth method of transmission is vector borne transmission, in which arthropods are the vectors. Vector transmission may be external or internal. External, or mechanical, transmission occurs when organisms are carried mechanically on the vector (for example, *Salmonella* organisms that contaminate the legs of flies). Internal transmission occurs when the organisms are carried within the vector. If the pathogen is not changed by its carriage within the vector, the carriage is called harborage (as when a flea ingests plague bacilli from an infected individual or animal and contaminates a susceptible host when it feeds again; the organism is not changed while in the flea). The other form of internal transmission is called biologic. In this form, the organism is changed biologically during its passage through the vector (for example, malaria parasites in the mosquito vector).

An infectious agent may be transmitted by more than one route. For example, *Salmonella* may be transmitted by a common vehicle (food) or by contact spread (human carrier). *Francisella tularensis* may be transmitted by any of the four routes.

## Host

The third link in the chain of infection is the host. The organism may enter the host through the skin, mucous membranes, lungs, gastrointestinal tract, or genitourinary tract, and it may enter fetuses through the placenta. The resulting disease often reflects the point of entrance, but not always: meningococci that enter the host through the mucous membranes may nonetheless cause meningitis. Development of disease in a host reflects agent characteristics and is influenced by host defense mechanisms, which may be nonspecific or specific.

Nonspecific defense mechanisms include the skin, mucous membranes, secretions, excretions, enzymes, the inflammatory response, genetic factors, hormones, nutrition, behavioral patterns, and the presence of other diseases. Specific defense mechanisms or immunity may be natural, resulting from exposure to the infectious agent, or artificial, resulting from active or passive immunization.

The environment can affect any link in the chain of infection. Temperature can assist or inhibit multiplication of organisms at their reservoir; air velocity can assist the airborne movement of droplet nuclei; low humidity can damage mucous membranes; and ultraviolet radiation can kill the microorganisms. In any investigation of disease, it is important to evaluate the effect of environmental factors. At times, environmental control measures are instituted more on emotional grounds than on the basis of epidemiologic fact. It should be apparent that the occurrence of disease results from the interaction of many factors. Some of these factors are outlined here.

**Table. General Factors That Influence the Occurrence of Infectious Disease.**

| *Pathogenic Agent* | *Host* |
|---|---|
| Growth charcteristics | Incubation period |
| Stability | Nonspecific defense meachanisms |
| Ability to form spores | Age |
| Possession of antibiotic resistance plasmids | Sex |
| Expression of antigens | Skin |
| Enzyme production | Secretions |
| Pathogenicity–ability to induce disease | Cough |
| Virulence–influencing disease severity, morbidity, and mortality | Cilry function |
| | Peristalsis |
| Invasiveness | Inflammation |
| Dose | Nutrition |
| Reservoir | Genetic factors |
| Source | Hormones |
| Mode of dissemination | Personal education |
| Host specificity | Behavior patterns |
| | Chronic disease |
| Disease Transmission | |
| Contact | Specific defense mechanisms (immunity) |
| Direct | Natural |
| Indirect | Active–apparent, inapparent |
| Droplets | Passive–trans- |

| | placental antibody |
|---|---|
| Common vehicle | Artificial |
| Food | Active–vaccine, |
| | toxoid |
| Water | Passive–immune |
| | serum globulin |
| Medication | |
| Solution | Environment |
| Airbome | Temperature |
| Droplet nuclei | Rainfall |
| Dust | Humidity |
| Skin squames | Radiation |
| Vectorbome–arthropods | Air currents |
| External | |
| Internal–harborage, true biologic transmission | |

## EPIDEMIOLOGIC METHODS

The three major epidemiologic techniques are descriptive, analytic, and experimental. Although all three can be used in investigating the occurrence of disease, the method used most is descriptive epidemiology. Once the basic epidemiology of a disease has been described, specific analytic methods can be used to study the disease further, and a specific experimental approach can be developed to test a hypothesis.

### Descriptive Epidemiology

In descriptive epidemiology, data that describe the occurrence of the disease are collected by various methods from all relevant sources. The data are then collated by time, place, and person.

Four time trends are considered in describing the epidemiologic data. The secular trend describes the occurrence of disease over a prolonged period, usually years; it is influenced by the degree of immunity in the population and possibly nonspecific measures such as improved socioeconomic and nutritional levels among the population. For example, the secular trend of tetanus in the United States since 1920 shows a gradual and steady decline.

The second time trend is the periodic trend. A temporary modification in the overall secular trend, the periodic trend may indicate a change in the antigenic characteristics of the disease agent. For example, the change in antigenic structure of the prevalent influenza A virus every 2 to 3 years results in periodic increases in the occurrence of clinical influenza caused by lack of natural immunity among the population. Additionally, a lowering of the overall immunity of a population or a segment thereof (known as herd immunity) can result in an increase in the occurrence of the disease. This can be seen with some immunizable diseases when periodic decreases occur in the level of immunization in a defined population. This may then result in an increase in the number of cases, with a subsequent rise in the overall level of herd immunity. The number of new cases then decreases until the herd's immunity is low enough to allow transmission to occur again and new cases then appear.

The third time trend is the seasonal trend. This trend reflects seasonal changes in disease occurrence following changes in environmental conditions that enhance the ability of the agent to replicate or be transmitted. For example, food-borne disease outbreaks occur more frequently in the summer, when temperatures favour multiplication of bacteria. This trend becomes evident when the occurrence of salmonellosis is examined on a monthly basis.

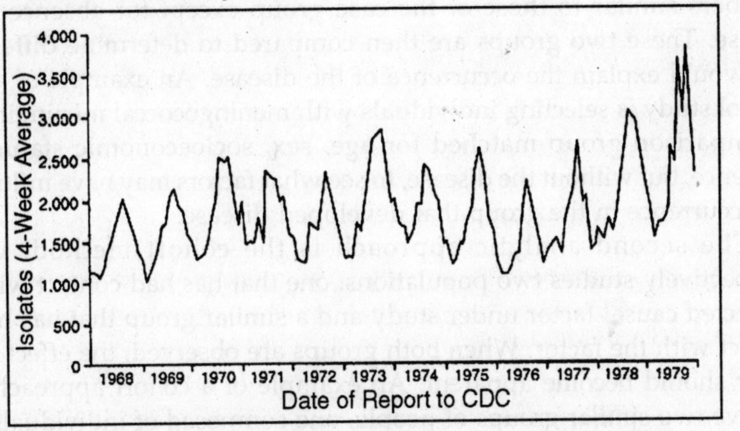


Fig. An Example of a Disease Showing a Seasonal Trend.

Reported human *Salmonella* isolations, by 4-week average, in the United States from 1968 to 1980.

The fourth time trend is the epidemic occurrence of disease. An epidemic is a sudden increase in occurrence due to prevalent factors that support transmission.

A description of epidemiologic data by place must consider three different sites: where the individual was when disease occurred; where the individual was when he or she became infected from the source; and where the source became infected with the etiologic agent. Therefore, in an outbreak of food poisoning, the host may become clinically ill at home from food eaten in a restaurant. The vehicle may have been undercooked chicken, which became infected on a poultry farm. These differences are important to consider in attempting to prevent additional cases.

The third focus of descriptive epidemiology is the infected person.

All pertinent characteristics should be noted: age, sex, occupation, personal habits, socioeconomic status, immunization history, presence of underlying disease, and other data.

Once the descriptive epidemiologic data have been analyzed, the features of the epidemic should be clear enough that additional areas for investigation are apparent.

## Analytic Epidemiology

The second epidemiologic method is analytic epidemiology, which analyzes disease determinants for possible causal relations. The two main analytic methods are the case-control (or case-comparison) method and the cohort method. The case-control method starts with the effect (disease) and retrospectively investigates the cause that led to the effect. The case group consists of individuals with the disease; a comparison group has members similar to those of the case group except for absence of the disease. These two groups are then compared to determine differences that would explain the occurrence of the disease. An example of a case-control study is selecting individuals with meningococcal meningitis and a comparison group matched for age, sex, socioeconomic status, and residence, but without the disease, to see what factors may have influenced the occurrence in the group that developed disease.

The second analytic approach is the cohort method, which prospectively studies two populations: one that has had contact with the suspected causal factor under study and a similar group that has had no contact with the factor. When both groups are observed, the effect of the factor should become apparent. An example of a cohort approach is to observe two similar groups of people, one composed of individuals who received blood transfusions and the other of persons who did not. The occurrence of hepatitis prospectively in both groups permits one to make an association between blood transfusions and hepatitis; that is, if the transfused blood was contaminated with hepatitis B virus, the recipient cohort should have a higher incidence of hepatitis than the nontransfused cohort.

The case-control approach is relatively easy to conduct, can be completed in a shorter period than the cohort approach, and is inexpensive and reproducible; however, bias may be introduced in selecting the two groups, it may be difficult to exclude subclinical cases from the comparison group, and a patient's recall of past events may be faulty. The advantages of a cohort study are the accuracy of collected data and the ability to make a direct estimate of the disease risk resulting from factor contact; however, cohort studies take longer and are more expensive to conduct.

Another analytic method is the cross-sectional study, in which a population is surveyed over a limited period to determine the relationship between a disease and variables present at the same time that may influence its occurrence.

## Experimental Epidemiology

The third epidemiologic method is the experimental approach. A hypothesis is developed and an experimental model is constructed in which one or more selected factors are manipulated. The effect of the manipulation will either confirm or disprove the hypothesis. An example is the evaluation of the effect of a new drug on a disease. A group of people with the disease is identified, and some members are randomly selected to receive the drug. If the only difference between the two is use of the drug, the clinical differences between the groups should reflect the effectiveness of the drug.

## Epidemic Investigation

An epidemic investigation describes the factors relevant to an outbreak of disease; once the circumstances related to the occurrence of disease are defined, appropriate control and prevention measures can be identified. In an epidemic investigation, data are collected, collated according to time, place, and person, and analyzed and inferences are drawn.

In the investigation, the first action should be to confirm the existence of the epidemic by noting from past surveillance data the number of cases suspected and comparing this with the number of cases initially reported. Additionally, the investigator should discuss the occurrence of the disease with physicians or others who have seen or reported cases after examining patients and reviewing laboratory and hospital records. These diagnoses should then be verified. A case definition should be developed to differentiate patients who represent actual cases, those who represent suspected or presumptive cases, and those who should be omitted from further study. Additional cases may be sought or additional patient data obtained, and a rough case count made.

This initial phase consists basically of collecting data, which then must be organized according to time, place, and person. The population at risk should be identified and a hypothesis developed concerning the occurrence of the disease. If appropriate, specimens should be collected and transported to the laboratory. More specific studies may be indicated. Additional data from these studies should be analyzed and the hypothesis confirmed or altered. After analysis, control and prevention measures should be developed and, as far as possible, implemented. A report containing this information should be prepared and distributed to those involved in investigating the outbreak and in implementing control and/ or prevention measures. Continued surveillance activities may be appropriate to evaluate the effectiveness of the control and prevention measures.

In the United States, the CDC assists state health departments by providing epidemiologic and laboratory support services on request. Its assistance supports disease investigations and diagnostic laboratory activities and includes various training programs conducted in the states and at the CDC.

A close working relationship exists between the CDC and state health departments. Additionally, physicians frequently consult with CDC personnel on a variety of health-related problems and attend public health training programs.

The use of epidemiology to characterize a disease before its etiology has been identified is exemplified by the initial studies of acquired immune deficiency syndrome (AIDS). The first cases came to the attention of the CDC late in 1981 when an increase was observed in requests for pentamidine for treatment of *Pneumocystis carinii* pneumonia. This initiated specific surveillance activities and epidemiologic studies that provided important information about this newly diagnosed disease.

Initial symptoms include fever, loss of appetite, weight loss, extreme fatigue, and enlargement of lymph nodes. A severe immune deficiency then develops, which appears to be associated with opportunistic infections.

These infections include *P carinii* pneumonia, diagnosed in 52 per cent of cases; Kaposi sarcoma in 26 per cent of cases; and both *P carinii* pneumonia and Kaposi sarcoma in 7 per cent of cases. The remaining 15 per cent of AIDS patients have other parasitic, fungal, bacterial, or viral infections associated with immunodeficiencies.

Among the first 2,640 cases reported to the CDC, there were 1,092 deaths, a case-fatality rate of 41 per cent. Approximately 95 per cent of the cases were male; 70 per cent were 20 to 49 years of age at the time of diagnosis.

Approximately 40 per cent of the cases were reported from New York City, 12 per cent from San Francisco, 8 per cent from Los Angeles, and the remainder from 32 other states. Cases were reported from at least 16 other countries. Among the 90 per cent of patients who were categorized according to possible risk factors, those at highest risk were homosexuals or bisexuals (70 per cent), intravenous drug abusers (17 per cent), Haitian entrants into the United States (9.5 per cent), and persons with hemophilia (1 per cent).

Analysis of these initial data, collected before the etiologic agent of

AIDS was identified, supported the hypothesis that transmission occurred primarily by sexual contact, receipt of contaminated blood or blood products, or contact with contaminated intravenous needles.

Spread through casual contact did not seem likely. The epidemiologic data indicated that AIDS was an infectious disease. It has now been determined that AIDS results from infection with a retrovirus of the human T cell leukemia/lymphoma virus family, which has been designated human immunodeficiency virus type I (HIV-l). The initial hypotheses have been proven as shown by analysis of data subsequently collected.

## Chapter 10

# Principles of Diagnosis

Some infectious diseases are distinctive enough to be identified clinically. Most pathogens, however, can cause a wide spectrum of clinical syndromes in humans. Conversely, a single clinical syndrome may result from infection with any one of many pathogens. Influenza virus infection, for example, causes a wide variety of respiratory syndromes that cannot be distinguished clinically from those caused by streptococci, mycoplasmas, or more than 100 other viruses.

Most often, therefore, it is necessary to use microbiologic laboratory methods to identify a specific etiologic agent. Diagnostic medical microbiology is the discipline that identifies etiologic agents of disease. The job of the clinical microbiology laboratory is to test specimens from patients for microorganisms that are, or may be, a cause of the illness and to provide information (when appropriate) about the in vitro activity of antimicrobial drugs against the microorganisms identified.

The staff of a clinical microbiology laboratory should be qualified to advise the physician as well as process specimens. The physician should supply salient information about the patient, such as age and sex, tentative diagnosis or details of the clinical syndrome, date of onset, significant exposures, prior antibiotic therapy, immunologic status, and underlying conditions.

The clinical microbiologist participates in decisions regarding the microbiologic diagnostic studies to be performed, the type and timing of specimens to be collected, and the conditions for their transportation and storage. Above all, the clinical microbiology laboratory, whenever appropriate, should provide an interpretation of laboratory results.

### MANIFESTATIONS OF INFECTION

The manifestations of an infection depend on many factors, including the site of acquisition or entry of the microorganism; organ or system tropisms of the microorganism; microbial virulence; the age, sex, and immunologic status of the patient; underlying diseases or conditions; and the presence of implanted prosthetic devices or materials. The signs and symptoms of infection may be localized, or they may be systemic, with fever, chills, and hypotension. In some instances the manifestations of an infection are sufficiently characteristic to suggest the diagnosis; however, they are often nonspecific.

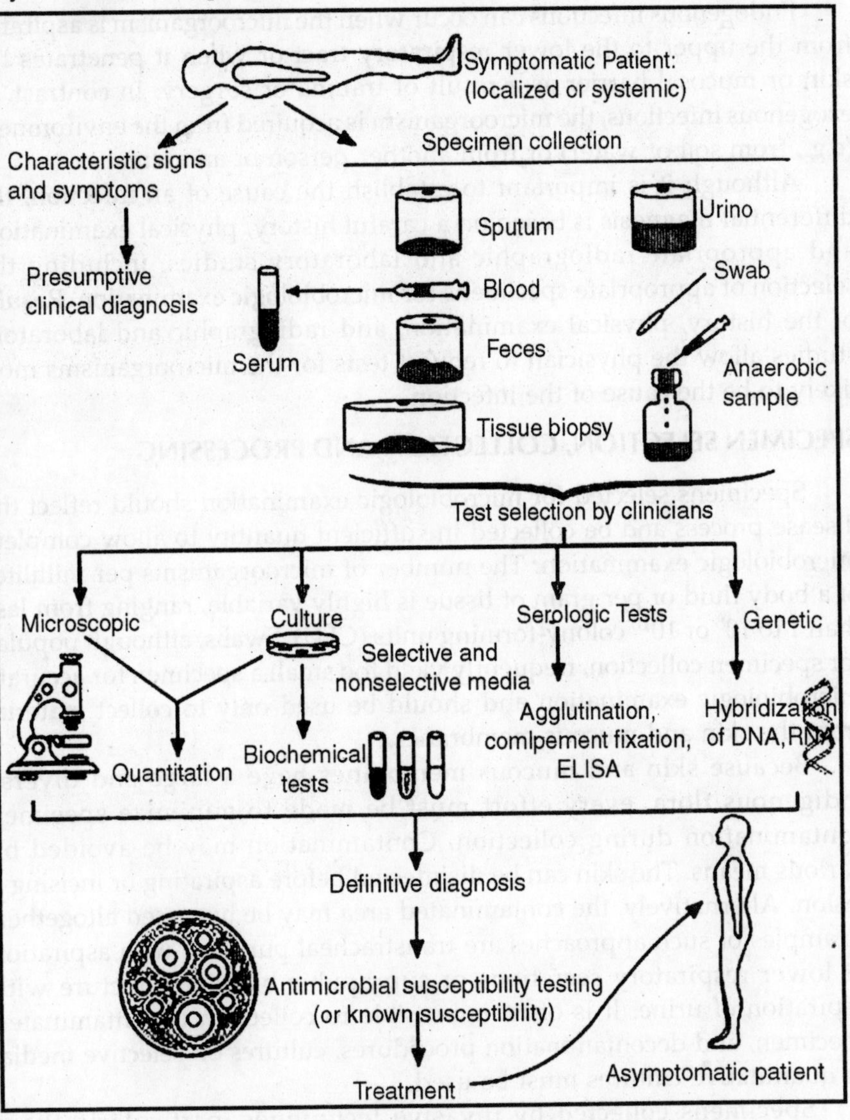


**Fig.** Laboratory Procedures Used in Confirming a Clinical Diagnosis of Infectious Disease with a Bacterial Etiology.

## MICROBIAL CAUSES OF INFECTION

Infections may be caused by bacteria (including mycobacteria, chlamydiae, mycoplasmas, and rickettsiae), viruses, fungi, or parasites. Infection may be endogenous or exogenous. In endogenous infections, the microorganism (usually a bacterium) is a component of the patient's indigenous flora.

Endogenous infections can occur when the microorganism is aspirated from the upper to the lower respiratory tract or when it penetrates the skin or mucosal barrier as a result of trauma or surgery. In contrast, in exogenous infections, the microorganism is acquired from the environment (e.g., from soil or water) or from another person or an animal.

Although it is important to establish the cause of an infection, the differential diagnosis is based on a careful history, physical examination, and appropriate radiographic and laboratory studies, including the selection of appropriate specimens for microbiologic examination. Results of the history, physical examination, and radiographic and laboratory studies allow the physician to request tests for the microorganisms most likely to be the cause of the infection.

## SPECIMEN SELECTION, COLLECTION AND PROCESSING

Specimens selected for microbiologic examination should reflect the disease process and be collected in sufficient quantity to allow complete microbiologic examination. The number of microorganisms per milliliter of a body fluid or per gram of tissue is highly variable, ranging from less than 1 to $10^8$ or $10^{10}$ colony-forming units (CFU). Swabs, although popular for specimen collection, frequently yield too small a specimen for accurate microbiologic examination and should be used only to collect material from the skin and mucous membranes.

Because skin and mucous membranes have a large and diverse indigenous flora, every effort must be made to minimize specimen contamination during collection. Contamination may be avoided by various means. The skin can be disinfected before aspirating or incising a lesion. Alternatively, the contaminated area may be bypassed altogether. Examples of such approaches are transtracheal puncture with aspiration of lower respiratory secretions or suprapubic bladder puncture with aspiration of urine. It is often impossible to collect an uncontaminated specimen, and decontamination procedures, cultures on selective media, or quantitative cultures must be used.

Specimens collected by invasive techniques, particularly those obtained intraoperatively, require special attention. Enough tissue must be obtained for both histopathologic and microbiologic examination. Histopathologic examination is used to distinguish neoplastic from inflammatory lesions and acute from chronic inflammations. The type of inflammation present can guide the type of microbiologic examination performed. If, for example, a caseous granuloma is observed histopathologically, microbiologic examination should include cultures for mycobacteria and fungi. The surgeon should obtain several samples for examination from a single large lesion or from each of several smaller lesions. If an abscess is found, the surgeon should collect several milliliters of pus, as well as a portion of the wall of the abscess, for microbiologic examination. Swabs should be kept out of the operating room.

If possible, specimens should be collected before the administration of antibiotics. Above all, close communication between the clinician and the microbiologist is essential to ensure that appropriate specimens are selected and collected and that they are appropriately examined.

## MICROBIOLOGIC EXAMINATION

### Direct Examination

Direct examination of specimens frequently provides the most rapid indication of microbial infection. A variety of microscopic, immunologic, and hybridization techniques have been developed for rapid diagnosis.

**Table. Rapid Tests Commonly Used to Detect Microorganisms in Specimens**

| *Specimen* | *Test* | *Application* |
|---|---|---|
| Blood | Giemsa | Plasmodia, microfilarie |
| | EIA | Hepatitis A and B virus, human immunodeficiency virus |
| Cerebrospinal fluid | Gram stain | Bacteria |
| | LA; COA | Haemophilus influenzae, Neisseria meningitidis |
| | India ink wet-mount or LA | Streptococcus pneumoniae Cryptococcus neoformans |
| Wound exudates, pus | Gram stain | Bacteria |
| Respiratory secretions | Gram stain | Bacteria |
| | Acid-fast stain | Mycobacteria, nocardiae |
| | IFA or genetic probe | Legionelfa species, Streptococcus pyogenes |
| | KOH wet mount | Fungi |
| | Gomori methenamine | Fungi, Pneumocystis carinil |
| | silver stain FA, EIA | Respiratory syncytial virus |
| Urine | Gram stain | Bacteria |
| Urethral or cervical | Gram stain, EIA, IFA, | Neisseria gonorrhoeae |
| | EIA, or genetic probe | Chlamydia trachomatis, |

| | | |
|---|---|---|
| | | papillomaviruses |
| Genital ulcer | FA, EIA, or genetic probe | Herpes simplex virus |
| Feces | Methylene blue stain | Leukocyles |
| | Eosin wet mount, trichrome stain | Parasites |
| | EM, LA, EIA | Rotaviruses |
| | EIA | Adenoviruses, Clostridium difficile |

## Sensitivity and Specificity

The sensitivity of a technique usually depends on the number of microorganisms in the specimen. Its specificity depends on how morphologically unique a specific microorganism appears microscopically or how specific the antibody or genetic probe is for that genus or species.

For example, the sensitivity of Cram stains is such that the observation of two bacteria per oil immersion field (X 1,000) of a Gram-stained smear of uncentrifuged urine is equivalent to the presence of > $10^5$ CFU/ml of urine.

The sensitivity of the Gram-stained smear for detecting Gram-negative coccobacilli in cerebrospinal fluid from children with *Haemophilus influenzae* meningitis is approximately 75 per cent because in some patients the number of colony-forming units per milliliter of cerebrospinal fluid is less than $10^4$. At least $10^4$ CFU of tubercle bacilli per milliliter of sputum must be present to be detected by an acid-fast smear of decontaminated and concentrated sputum.

An increase in the sensitivity of a test is often accompanied by a decrease in specificity. For example, examination of a Gram-stained smear of sputum from a patient with pneumococcal pneumonia is highly sensitive but also highly nonspecific if the criterion for defining a positive test is the presence of any Gram-positive cocci. If, however, a positive test is defined as the presence of a preponderance of Gram-positive, lancet-shaped diplococci, the test becomes highly specific but has a sensitivity of only about 50 per cent. Similar problems related to the number of microorganisms present affect the sensitivity of immunoassays and genetic probes for bacteria, chlamydiae, fungi and viruses. In some instances, the sensitivity of direct examination tests can be improved by collecting a better specimen. For example, the sensitivity of fluorescent antibody stain for Chlamydia trachomatis is higher when endocervical cells are obtained with a cytobrush than with a swab. The sensitivity may also be affected by the stage of the disease at which the specimen is collected. For example, the detection of herpes simplex virus by immunofluorescence, immunoassay, or culture is highest when specimens from lesions in the vesicular stage of infection are examined. Finally, sensitivity may be improved through the use of an enrichment or enhancement step in which microbial or genetic replication occurs to the point at which a detection method can be applied.

## Techniques

For microscopic examination it is sufficient to have a compound binocular microscope equipped with low-power (1OX), high-power (40X), and oil immersion (1OOX) achromatic objectives, 10X wide-field oculars, a mechanical stage, a substage condenser, and a good light source. For examination of wet-mount preparations, a darkfield condenser or condenser and objectives for phase contrast increases image contrast. An exciter barrier filter, darkfield condenser, and ultraviolet light source are required for fluorescence microscopy.

For immunologic detection of microbial antigens, latex particle agglutination, coagglutination, and enzyme-linked immunosorbent assay (ELISA) are the most frequently used techniques in the clinical laboratory. Antibody to a specific antigen is bound to latex particles or to a heat-killed and treated protein A-rich strain of Staphylococcus aureus to produce agglutination.

There are several approaches to ELISA; the one most frequently used for the detection of microbial antigens uses an antigen-specific antibody that is fixed to a solid phase, which may be a latex or metal bead or the inside surface of a well in a plastic tray. Antigen present in the specimen binds to the antibody as in Fig.The test is then completed by adding a second antigen-specific antibody bound to an enzyme that can react with a substrate to produce a colored product. The initial antigen antibody complex forms in a manner similar to that shown in Figure. When the enzyme-conjugated antibody is added, it binds to previously unbound antigenic sites, and the antigen is, in effect, sandwiched between the solid phase and the enzyme-conjugated antibody. The reaction is completed by adding the enzyme substrate.

Genetic probes are based on the detection of unique nucleotide sequences with the DNA or RNA of a microorganism. Once such a unique nucleotide sequence, which may represent a portion of a virulence gene or of chromosomal DNA, is found, it is isolated and inserted into a cloning vector (plasmid), which is then transformed into Escherichia coli to produce multiple copies of the probe. The sequence is then reisolated from plasmids and labeled with an isotope or substrate for diagnostic use. Hybridization of the sequence with a complementary sequence of DNA or RNA follows cleavage of the double-stranded DNA of the microorganism in the specimen.

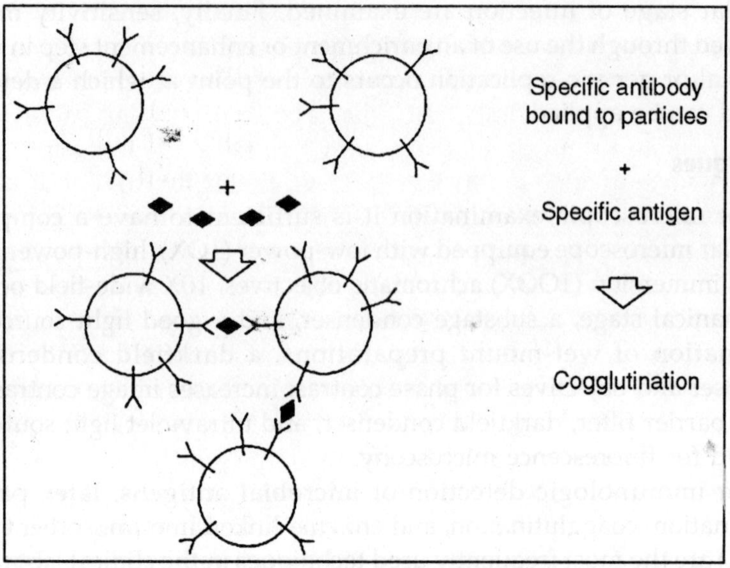


**Fig.** Agglutination test in which inert particles (latex beads or heat-killed *S aureus* Cowan 1 strain with protein A) are coated with antibody to any of a variety of antigens and then used to detect the antigen in specimens or in isolated bacteria.

The use of molecular technology in the diagnoses of infectious diseases has been further enhanced by the introduction of gene amplication techniques, such as the polymerase chain reaction (PCR) in which DNA polymerase is able to copy a strand of DNA by elongating complementary strands of DNA that have been initiated from a pair of closely spaced oligonucleotide primers. This approach has had major applications in the detection of infections due to microorganisms that are difficult to culture (e.g. the human immunodeficiency virus) or that have not as yet been successfully cultured (e.g. the Whipple's disease bacillus).

## Culture

In many instances, the cause of an infection is confirmed by isolating and culturing microorganism either in artificial media or in a living host. Bacteria (including mycobacteria and mycoplasmas) and fungi are cultured in either liquid (broth) or on solid (agar) artificial media. Liquid media provide greater sensitivity for the isolation of small numbers of microorganisms; however, identification of mixed cultures growing in liquid media requires subculture onto solid media so that isolated colonies can be processed separately for identification. Growth in liquid media also cannot ordinarily be quantitated. Solid media, although somewhat less sensitive than liquid media, provide isolated colonies that can be quantified if necessary and identified. Some genera and species can be recognized on the basis of their colony morphologies.

In some instances one can take advantage of differential carbohydrate fermentation capabilities of microorganisms by incorporating one or more carbohydrates in the medium along with a suitable pH indicator. Such media are called differential media (e.g., eosin methylene blue or MacConkey agar) and are commonly used to isolate enteric bacilli. Different genera of the Enterobacteriaceae can then be presumptively identified by the colour as well as the morphology of colonies.

Culture media can also be made selective by incorporating compounds such as antimicrobial agents that inhibit the indigenous flora while permitting growth of specific microorganisms resistant to these inhibitors.

One such example is Thayer-Martin medium, which is used to isolate *Neisseria gonorrhoeae*. This medium contains vancomycin to inhibit Gram-positive bacteria, colistin to inhibit most Gram-negative bacilli, trimethoprim-sulfamethoxazole to inhibit *Proteus* species and other species that are not inhibited by colistin and anisomycin to inhibit fungi. The pathogenic *Neisseria* species, *N gonorrhoeae* and *N meningitidis*, are ordinarily resistant to the concentrations of these antimicrobial agents in the medium.

The number of bacteria in specimens may be used to define the presence of infection. For example, there may be small numbers ($< 10^3$ CFU/ml) of bacteria in clean-catch, midstream urine specimens from normal, healthy women; with a few exceptions, these represent bacteria that are indigenous to the urethra and periurethral region. Infection of the bladder (cystitis) or kidney (pyelone-phritis) is usually accompanied by bacteriuria of about $> 10^4$ CFU/ml.

For this reason, quantitative cultures of urine must always be performed. For most other specimens a semiquantitative streak method over the agar surface is sufficient. For quantitative cultures, a specific volume of specimen is spread over the agar surface and the number of colonies per milliliter is estimated.

For semiquantitative cultures, an unquantitated amount of specimen is applied to the agar and diluted by being streaked out from the inoculation site with a sterile bacteriologic loop.

The amount of growth on the agar is then reported semiquantitatively as many, moderate, or few (or 3+, 2+, or 1+), depending on how far out from the inoculum site colonies appear. An organism that grows in all streaked areas would be reported as 3+.

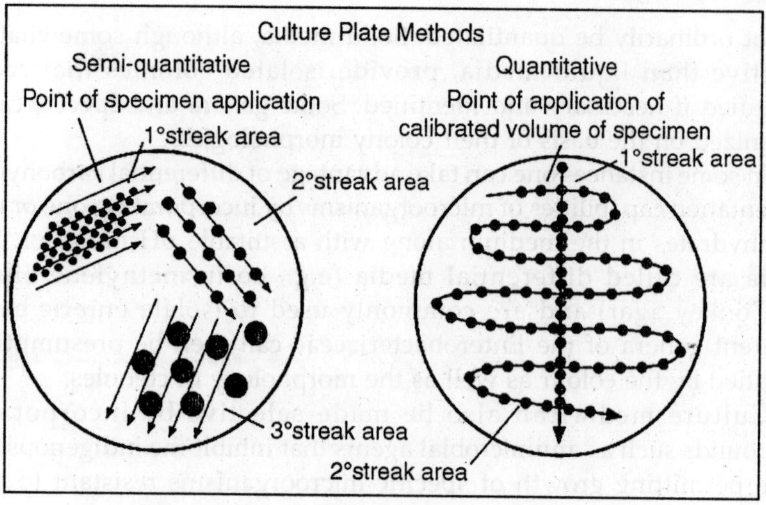


**Fig.** Quantitative Versus Semiquantitative Culture, Revealing the Number of Bacteria in Specimens.

Chlamydiae and viruses are cultured in cell culture systems, but virus isolation occasionally requires inoculation into animals, such as suckling mice, rabbits, guinea pigs, hamsters, or primates. Rickettsiae may be isolated with some difficulty and at some hazard to laboratory workers in animals or embryonated eggs. For this reason, rickettsial infection is usually diagnosed serologically. Some viruses, such as the hepatitis viruses, cannot be isolated in cell culture systems, so that diagnosis of hepatitis virus infection is based on the detection of hepatitis virus antigens or antibodies.

Cultures are generally incubated at 35 to 37°C in an atmosphere consisting of air, air supplemented with carbon dioxide (3 to 10 per cent), reduced oxygen (microaerophilic conditions), or no oxygen (anaerobic conditions), depending upon requirements of the microorganism. Since clinical specimens from bacterial infections often contain aerobic, facultative anaerobic, and anaerobic bacteria, such specimens are usually inoculated into a variety of general purpose, differential, and selective media, which are then incubated under aerobic and anaerobic conditions.

The duration of incubation of cultures also varies with the growth characteristics of the microorganism. Most aerobic and anaerobic bacteria will grow overnight, whereas some mycobacteria require as many as 6 to 8 weeks.

## MICROBIAL IDENTIFICATION

Microbial growth in cultures is demonstrated by the appearance of turbidity, gas formation, or discrete colonies in broth; colonies on agar;

cytopathic effects or inclusions in cell cultures; or detection of genus- or species-specific antigens or nucleotide sequences in the specimen, culture medium, or cell culture system.

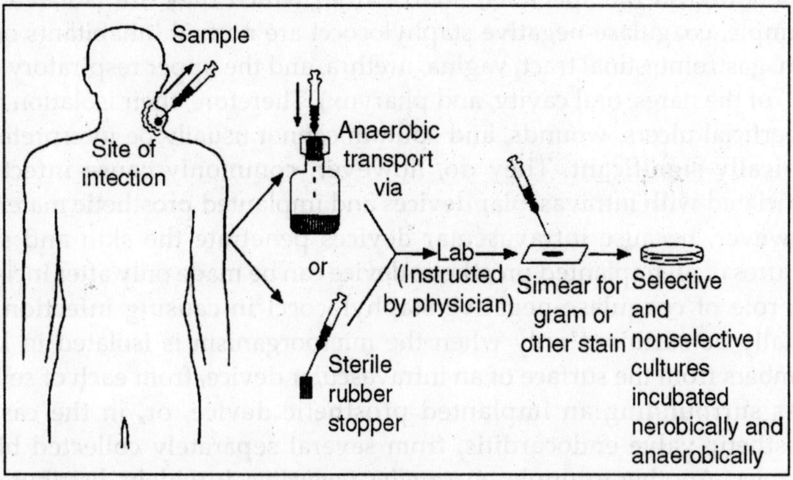


**Fig.** General Procedure for Collecting and Processing Specimens for Aerobic and/or Anaerobic Bacterial Culture.

Identification of bacteria (including mycobacteria) is based on growth characteristics (such as the time required for growth to appear or the atmosphere in which growth occurs), colony and microscopic morphology, and biochemical, physiologic, and, in some instances, antigenic or nucleotide sequence characteristics. The selection and number of tests for bacterial identification depend upon the category of bacteria present (aerobic versus anaerobic, Gram-positive versus Gram-negative, cocci versus bacilli) and the expertise of the microbiologist examining the culture.

Gram-positive cocci that grow in air with or without added $CO_2$ may be identified by a relatively small number of tests. The identification of most Gram-negative bacilli is far more complex and often requires panels of 20 tests for determining biochemical and physiologic characteristics. The identification of filamentous fungi is based almost entirely on growth characteristics and colony and microscopic morphology. Identification of viruses is usually based on characteristic cytopathic effects in different cell cultures or on the detection of virus- or species-specific antigens or nucleotide sequences.

## Interpretation of Culture Results

Some microorganisms, such as *Shigella dysenteriae, Mycobacterium tuberculosis, Coccidioides immitis,* and influenza virus, are always considered clinically significant. Others that ordinarily are harmless components of the indigenous flora of the skin and mucous membranes or that are common in the environment may or may not be clinically significant, depending on the specimen source from which they are isolated. For example, coagulase-negative staphylococci are normal inhabitants of the skin, gastrointestinal tract, vagina, urethra, and the upper respiratory tract (i.e., of the nares, oral cavity, and pharynx). Therefore, their isolation from superficial ulcers, wounds, and sputum cannot usually be interpreted as clinically significant. They do, however, commonly cause infections associated with intravascular devices and implanted prosthetic materials. However, because intravascular devices penetrate the skin and since cultures of an implanted prosthetic device can be made only after incision, the role of coagulase-negative staphylococci in causing infection can usually be surmised only when the microorganism is isolated in large numbers from the surface of an intravascular device, from each of several sites surrounding an implanted prosthetic device, or, in the case of prosthetic valve endocarditis, from several separately collected blood samples. Another example, *Aspergillus fumigatus,* is widely distributed in nature, the hospital environment, and upper respiratory tract of healthy people but may cause fatal pulmonary infections in leukemia patients or in those who have undergone bone marrow transplantation. The isolation of *A fumigatus* from respiratory secretions is a nonspecific finding, and a definitive diagnosis of invasive aspergillosis requires histologic evidence of tissue invasion.

Physicians must also consider that the composition of microbial species on the skin and mucous membranes may be altered by disease, administration of antibiotics, endotracheal or gastric intubation, and the hospital environment. For example, potentially pathogenic bacteria can often be cultured from the pharynx of seriously ill, debilitated patients in the intensive care unit, but may not cause infection.

## Serodiagnosis

Infection may be diagnosed by an antibody response to the infecting microorganism. This approach is especially useful when the suspected microbial agent either cannot be isolated in culture by any known method or can be isolated in culture only with great difficulty. The diagnosis of hepatitis virus and Epstein-Barr virus infections can be made only serologically, since neither can be isolated in any known cell culture system. Although human immunodeficiency virus type 1 (HIV-1) can be isolated in cell cultures, the technique is demanding and requires special containment facilities. HIV-1 infection is usually diagnosed by detection of antibodies to the virus.

The disadvantage of serology as a diagnostic tool is that there is usually a lag between the onset of infection and the development of antibodies to the infecting microorganism. Although IgM antibodies may appear relatively rapidly, it is usually necessary to obtain acute- and convalescent-phase serum samples to look for a rising titer of IgG antibodies to the suspected pathogen. In some instances the presence of a high antibody titer when the patient is initially seen is diagnostic; often, however, the high titer may reflect a past infection, and the current infection may have an entirely different cause. Another limitation on the use of serology as a diagnostic tool is that immunosuppressed patients may be unable to mount an antibody response.

## Antimicrobial Susceptibility

The responsibility of the microbiology laboratory includes not only microbial detection and isolation but also the determination of microbial susceptibility to antimicrobial agents. Many bacteria, in particular, have unpredictable susceptibilities to antimicrobial agents, and their susceptibilities can be measured in vitro to help guide the selection of the most appropriate antimicrobial agent.

Antimicrobial susceptibility tests are performed by either disk diffusion or a dilution method. In the former, a standardized suspension of a particular microorganism is inoculated onto an agar surface to which paper disks containing various antimicrobial agents are applied. Following overnight incubation, any zone diameters of inhibition about the disks are measured and the results are reported as indicating susceptibility or resistance of the microorganism to each antimicrobial agent tested. An alternative method is to dilute on a log2 scale each antimicrobial agent in broth to provide a range of concentrations and to inoculate each tube or, if a microplate is used, each well containing the antimicrobial agent in broth with a standardized suspension of the microorganism to be tested. The lowest concentration of antimicrobial agent that inhibits the growth of the microorganism is the minimal inhibitory concentration (MIC). The MIC and the zone diameter of inhibition are inversely correlated. In other words, the more susceptible the microorganism is to the antimicrobial agent, the lower the MIC and the larger the zone of inhibition. Conversely, the more resistant the microorganism, the higher the MIC and the smaller the zone of inhibition.

The term susceptible means that the microorganism is inhibited by a concentration of antimicrobial agent that can be attained in blood with the normally recommended dose of the antimicrobial agent and implies that an infection caused by this microorganism may be appropriately treated with the antimicrobial agent. The term resistant indicates that the microorganism is resistant to concentrations of the antimicrobial agent that can be attained with normal doses and implies that an infection caused by this microorganism could not be successfully treated with this antimicrobial agent.

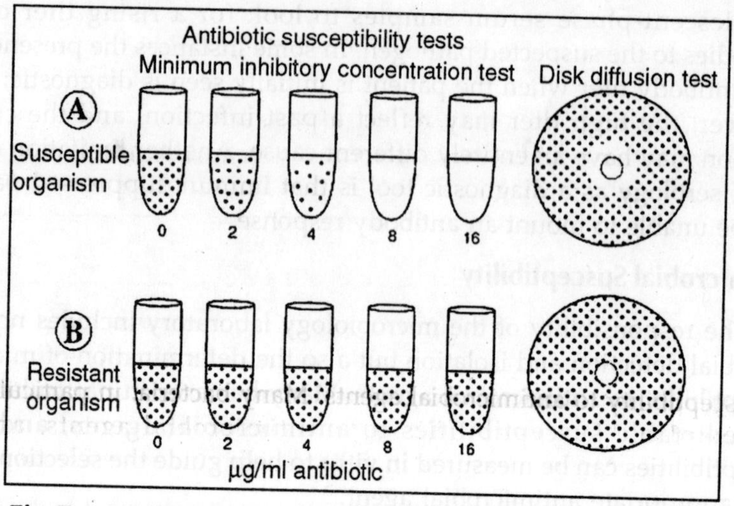


**Fig.** Two Methods for Performing Antibiotic Susceptibility Tests.

(A) Disk diffusion method.

(B) Minimum inhibitory concentration (MIC) method. In the example shown, two different microorganisms are tested by both methods against the same antibiotic. The MIC of the antibiotic for the susceptible microorganism is 8 μg/ml. The corresponding disk diffusion test shows a zone of inhibition surrounding the disk. In the second sample, a resistant microorganism is not inhibited by the highest antibiotic concentration tested (MIC > 16 μg/ml) and there is no zone of inhibition surrounding the disk. The diameter of the zone of inhibition is inversely related to the MIC.

# Chapter 11

# Antimicrobial Chemotherapy

The earliest evidence of successful chemotherapy is from ancient Peru, where the Indians used bark from the cinchona tree to treat malaria. Other substances were used in ancient China, and we now know that many of the poultices used by primitive peoples contained antibacterial and antifungal substances. Modern chemotherapy has been dated to the work of Paul Ehrlich in Germany, who sought systematically to discover effective agents to treat trypanosomiasis and syphilis. He discovered p-rosaniline, which has antitrypanosomal effects, and arsphenamine, which is effective against syphilis.

Ehrlich postulated that it would be possible to find chemicals that were selectively toxic for parasites but not toxic to humans. This idea has been called the "magic bullet" concept. It had little success until the 1930s, when Gerhard Domagk discovered the protective effects of prontosil, the forerunner of sulfonamide. Ironically, penicillin G was discovered fortuitously in 1929 by Fleming, who did not initially appreciate the magnitude of his discovery.

In 1939 Florey and colleagues at Oxford University again isolated penicillin. In 1944 Waksman isolated streptomycin and subsequently found agents such as chloramphenicol, tetracyclines, and erythromycin in soil samples. By the 1960s, improvements in fermentation techniques and advances in medicinal chemistry permitted the synthesis of many new chemotherapeutic agents by molecular modification of existing compounds.

Progress in the development of novel antibacterial agents has been great, but the development of effective, nontoxic antifungal and antiviral agents has been slow. Amphotericin B, isolated in the 1950s, remains an effective antifungal agent, although newer agents such as fluconazole are now widely used. Nucleoside analogs such as acyclovir have proved effective in the chemotherapy of selected viral infections.

## BIOCHEMICAL BASIS OF ANTIMICROBIAL ACTION

Bacterial cells grow and divide, replicating repeatedly to reach the large numbers present during an infection or on the surfaces of the body. To grow and divide, organisms must synthesize or take up many types of biomolecules.

Antimicrobial agents interfere with specific processes that are essential for growth and/or division. They can be separated into groups such as inhibitors of bacterial and fungal cell walls, inhibitors of cytoplasmic membranes, inhibitors of nucleic acid synthesis, and inhibitors of ribosome function.

Antimicrobial agents may be either bactericidal, killing the target bacterium or fungus, or bacteriostatic, inhibiting its growth. Bactericidal agents are more effective, but bacteriostatic agents can be extremely beneficial since they permit the normal defenses of the host to destroy the microorganisms.

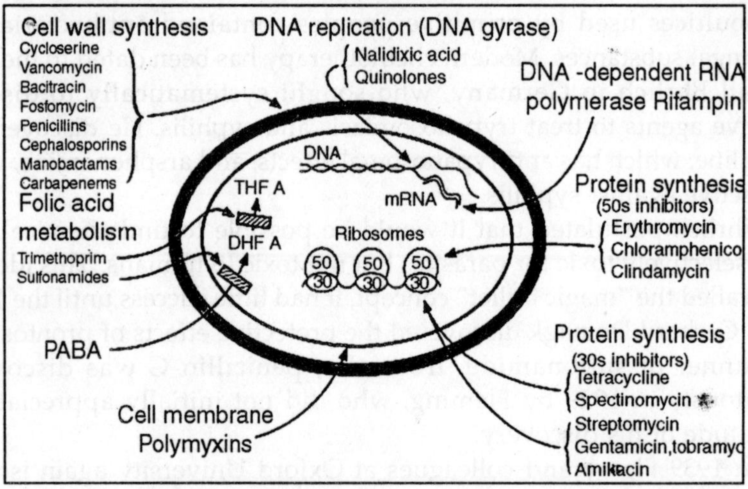


**Fig.** Sites of Action of Different Antimicrobial Agents.

PABA, paraminobenzoic acid; DHFA, dihydrofolic acid; THFA, tetrahydrofolic acid.

**Table. Mechanisms of Action of Antimicrobial Agents.**

| Inhibitors of Bacterial Cell Wall Synthesis | Inhibitors of Ribosome Function |
|---|---|
| Drugs that inhibit biosynthetic enzymes | Inhibitors of 30S units |
| Fosfomycin | Streptomycin |
| Cycloserine | Kanamycin, gentamicin, amikacin |
| Drugs that combine with carrier molecules | Spectinomycin |
| Bacitracin | Tetracyclines |
| Drugs that combine with cell wall substrates | Inhibitors of 50S units |
| Vancomycin | Chloramphenicol |

Drugs that inhibit polymerization and attachment of new pepidoglycan to cell wall

Penicilins

Cephalosporins

Carbapenems

Monobactams

Inhibitors of Cytoplasmic Membranes

Drugs that disorgnize the cytoplasmic membrane

Tyrocidins

Polymyxins

Drugs that produce pores in membranes

Gramicidins

Drugs that alter structure of fungi

Polyenes (amphotericin)

Imidazoles (ketoconazole, fluconazole)

Inhibitors of Nucleic Acid Synthesis

Inhibitors of nucleotide metabolism

Adenosine arabinoside (viruses)

Acyclovir (viruses)

Flucytosine (fungi)

Agents that impair DNA template function

Intercalating agents

Chloroquine (parasites)

Inhibitors of DNA replication

Quinolones

Nitroimidazoles

Inhibitors of RNA polymerase

Rifampin

Clindamycin

Erythromycin

Fusidic acid

Inhibitors of Folate Metabolism

Inhibitor or pteroic acid synthelase

Sulfonamides

Inhibitor of dihydrotolate reduclase

Trimephoprim

## INHIBITION OF BACTERIAL CELL WALL SYNTHESIS

As noted in earlier chapters, bacteria are classified as Gram-positive and Gram-negative organisms on the basis of staining characteristics. Gram-positive bacterial cell walls contain peptidoglycan and teichoic or teichuronic acid, and the bacterium may or may not be surrounded by a protein or polysaccharide envelope. Gram-negative bacterial cell walls contain peptidoglycan, lipopolysaccharide, lipoprotein, phospholipid, and protein. The critical attack site of anti-cell-wall agents is the peptidoglycan layer. This layer is essential for the survival of bacteria in hypotonic environments; loss or damage of this layer destroys the rigidity of the bacterial cell wall, resulting in death.

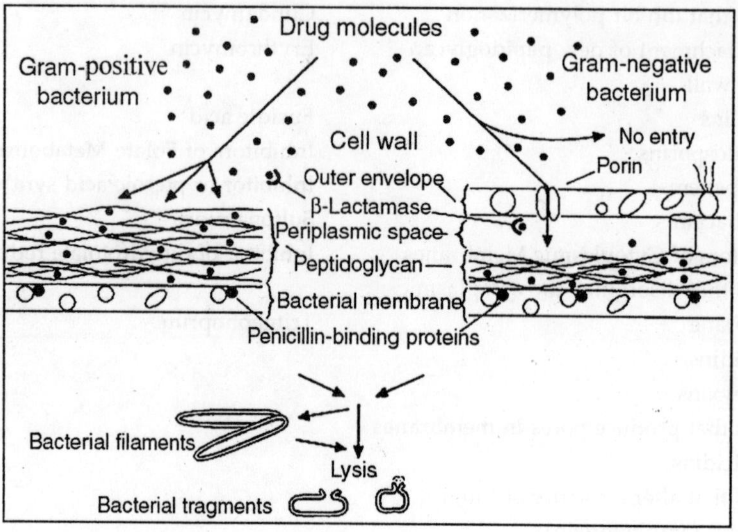


**Fig.** Outer wall of Gram-positive and Gram-negative Species and Detail of Porin Channels of Gram-negative Bacteria.

Antimicrobial agents diffuse easily through the loose outer wall of Gram-positive bacteria, but must go through the narrow channels of the Gram-negative species.

Peptidoglycan synthesis occurs in three stages. The first stage takes place in the cytoplasm, where the low-molecular-weight precursors UDP-GlcAc and UDP-MurNAc-L-Ala-D-Glu-meso-Dap-D-Ala-D-Ala are synthesized. A number of antimicrobial agents interfere with these early steps in cell wall biosynthesis. UTP and N-acetylglucosamine alpha-1-P are converted to UDP-N-acetylglucosamine, which is subsequently converted by the enzyme phosphoenolpyruvate: UDP-GlcNAc-3-enol-pyruvyltransferase.

Fosfomycins block this transfer by a direct nucleophilic attack on the enzyme. Because mammalian enzymes such as enolase, pyruvate kinase, carboxykinases, and the shikimate enlaces are not inhibited by these compounds, fosfomycins have no effect on the host. Three amino acids are added to the muramyl peptide to yield a tripeptide to which two more amino acids will be linked. The dipeptide D-alanyl-D-alanine is synthesized from two molecules of D-alanine by the enzyme D-alanyl-D-alanine synthetase. D-Alanine is produced from L-alanine by an alanine racemase. Cycloserine inhibits both alanine racemase and D-alanyl-D-alanine synthetase owing to the structural similarity of cycloserine and D-alanine and to the fact that cycloserine actually binds to the enzymes better than the D-alanine.

The second stage of cell wall synthesis is catalyzed by membrane-bound enzymes. The nonnucleotide portion of the precursor molecules previously made are transferred sequentially to a carrier in the cytoplasmic membrane.

This carrier is a phosphorylated undecaprenyl alcohol. The lipid carrier functions as a point of attachment to the membrane for the precursors and allows for transport of the subunits across the hydrophobic interior of the cytoplasmic membrane to the outside surface. Bacitracin is a peptide antibiotic that specifically interacts with the pyrophosphate derivate of the undecaprenyl alcohol, preventing further transfer of the muramylpentapeptide from the precursor nucleotide to the nascent peptidoglycan.

The third stage of cell wall synthesis involves polymerization of the subunits and the attachment of nascent peptidoglycan to the cell wall. Polymerization occurs by transfer of the new peptidoglycan chain from its carrier in the membrane to the nonreducing N-acetylglucosamine of the new saccharide-peptide that is attached to the membrane. The new peptidoglycan is attached to preexisting cell wall peptidoglycan by a transpeptidase reaction that involves peptide chains in both polymers, one of which must possess a D-alanyl-D-alanine terminus. It is believed that the transpeptidase enzyme cleaves the peptide bond between two D-alanyl residues in the pentapeptide and become acylated via the carbonyl group of the penultimate D-alanine residue.

This final reaction is inhibited by β-lactam antibiotics. These antibiotics contain a critical four-membered ring, which undergoes an acylation reaction with the transpeptidases that cross-link the polymers mentioned above. The β-lactam antibiotics are the penicillins (penams), cephalosporins (including oxacephems and cephamycins), penems, thienamycins (carbapenems), and aztreonam (monobactams). The enzymes involved in this final process of cell wall formation are called penicillin-binding proteins since they were discovered by labeling with radioactive penicillin G.

The enzymes are different in Gram-positive and Gram-negative bacteria and in anaerobic species. Differences in the penicillin-binding proteins explain, to some extent, differences in antibacterial activity of the β-lactam antibiotics. The penicillin-binding protein, to which a particular β-lactam antibiotic binds, affects the morphologic response of the bacterium to the agent. For example, some antibiotics bind to a penicillin-binding protein that is involved in forming the septum between dividing cells; as a result, the bacteria continue to grow into long filaments, which eventually die. Binding to another penicillin-binding protein results in rapid lysis of a bacterium because the wall bulges and the bacterium bursts. β-Lactams such as mecillinam (an amidino penicillin) do not bind to the penicillin-binding proteins of Gram-positive bacteria and therefore do not affect these bacteria. Aztreonam binds only to Gram-negative penicillin-binding proteins and does not inhibit Gram-positive or anaerobic species.

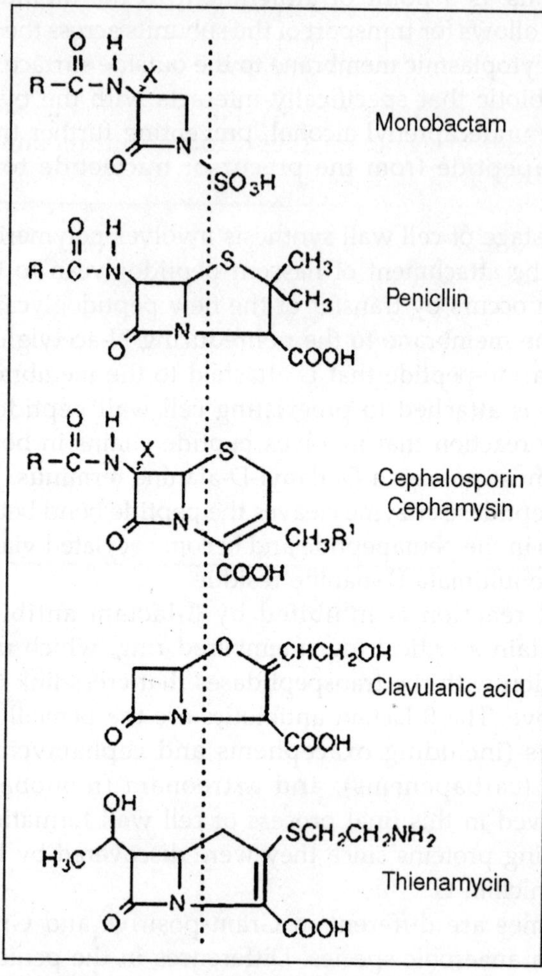


**Fig.** Basic Structures of β-lactam Antibiotics.

Penicillins and cephalosporins/cephamycins are widely used to inhibit both Gram-positive and Gram-negative bacilli. Monobactams inhibit only aerobic Gram-negative bacilli, clavulanic acid acts as a β-lactamase inhibitor, and thienamycin inhibits a wide range of aerobic and anaerobic species. R and R′ represent various carbon groups. X can be either hydrogen or a mehoxy group.

Vancomycin interrupts cell wall synthesis by forming a complex with the C-terminal D-alanine residues of peptidoglycan precursors. Complex formation at the outer surface of the cytoplasmic membrane prevents the transfer of the precursors from a lipid carrier to the growing peptidoglycan wall by transglycosidases. Biochemical reactions in the cell wall catalyzed by transpeptidases and D, D-carboxypeptidases are also inhibited by vancomycin and other glycopeptide antimicrobials.

Because of its large size and complex structure, vancomycin does not penetrate the outer membrane of gram-negative organisms. With resistance to beta-lactams increasing in frequency among staphylococci and enterococci, glycopeptides such as vancomycin remain important therapeutic agents against such bacteria.

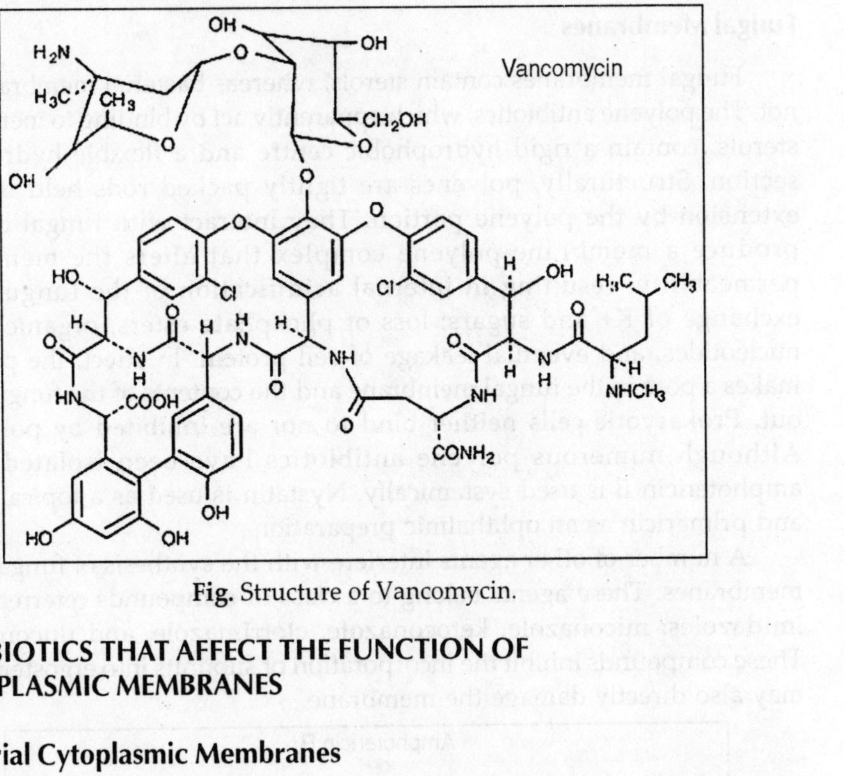


**Fig.** Structure of Vancomycin.

## ANTIBIOTICS THAT AFFECT THE FUNCTION OF CYTOPLASMIC MEMBRANES

### Bacterial Cytoplasmic Membranes

Biologic membranes are composed basically of lipid, protein, and lipoprotein. The cytoplasmic membrane acts as a diffusion barrier for water, ions, nutrients, and transport systems. Most workers now believe that membranes are a lipid matrix with globular proteins randomly distributed to penetrate through the lipid bilayer. A number of antimicrobial agents can cause disorganization of the membrane. These agents can be divided into cationic, anionic, and neutral agents. The best-known compounds are polymyxin B and colistemethate (polymyxin E). These high-molecular-weight octapeptides inhibit Gram-negative bacteria that have negatively charged lipids at the surface. Since the activity of the polymyxins is antagonized by $Mg^{2+}$ and $Ca^{2+}$, they probably competitively displace $Mg^{2+}$ or $Ca^{2+}$ from the negatively charged phosphate groups on membrane lipids. Basically, polymyxins disorganize membrane permeability so that nucleic acids and cations leak out and the cell dies. The polymyxins are of virtually no use as systemic agents since they bind to various ligands in body tissues and are potent toxins for the kidney and nervous system. Gramicidins are also membrane-active antibiotics that appear to act by producing aqueous pores in the membranes. They also are used only topically.

## Fungal Membranes

Fungal membranes contain sterols, whereas bacterial membranes do not. The polyene antibiotics, which apparently act by binding to membrane sterols, contain a rigid hydrophobic centre and a flexible hydrophilic section. Structurally, polyenes are tightly packed rods held in rigid extension by the polyene portion. They interact with fungal cells to produce a membrane-polyene complex that alters the membrane permeability, resulting in internal acidification of the fungus with exchange of K+ and sugars; loss of phosphate esters, organic acids, nucleotides; and eventual leakage of cell protein. In effect, the polyene makes a pore in the fungal membrane and the contents of the fungus leak out. Prokaryotic cells neither bind to nor are inhibited by polyenes. Although numerous polyene antibiotics have been isolated, only amphotericin B is used systemically. Nystatin is used as a topical agent and primaricin as an ophthalmic preparation.

A number of other agents interfere with the synthesis of fungal lipid membranes. These agents belong to a class of compounds referred to as imidazoles: miconazole, ketoconazole, clotrimazole, and fluconazole. These compounds inhibit the incorporation of subunits into ergosterol and may also directly damage the membrane.

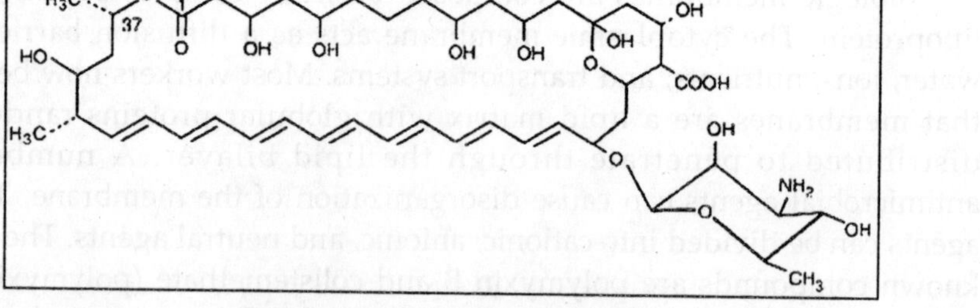


**Fig.** Structure of Amphotericin B.

## ANTIBIOTICS THAT INHIBIT NUCLEIC ACID SYNTHESIS

Antimicrobial agents can interfere with nucleic acid synthesis at several different levels. They can inhibit nucleotide synthesis or interconversion; they can prevent DNA from functioning as a proper template; and they can interfere with the polymerases involved in the replication and transcription of DNA.

### Interference with Nucleotide Synthesis

A large number of agents interfere with purine and pyrimidine synthesis or with the interconversion or utilization of nucleotides. Other agents act as nucleotide analogs that are incorporated into polynucleotides.

Flucytosine (5-fluorocytosine) is an antifungal agent that inhibits yeast species. It is converted in the fungal cell to 5-fluorouracfl, which inhibits thymidylate synthetase resulting in a deficit of thymine nucleotides and impaired DNA synthesis. Adenosine arabinoside inhibits viruses. It is phosphorylated in virus-infected cells and acts as a competitive analog of DATP, inhibiting the incorporation of DATP into DNA.

Acyclovir is a nucleoside analog that, after being converted to a triphosphate, inhibits the thymidine kinase and DNA polymerase of herpes viruses. Zidovudine (AZT) inhibits human immunodeficiency virus (HM replication by interfering with viral RNA-dependent DNA polymerase (reverse transcriptase).

### Agents That Impair the Template Function of DNA

A number of substances bind to DNA by intercalation. None of them is useful as an antibacterial agent; however, chloroquine and miracil D (lucanthone) inhibit plasmodia and schistosomes, respectively. These agents are thought to intercalate into the DNA and thereby to inhibit further nucleic acid synthesis. Acridine dyes such as proflavine act by this intercalation mechanism, but because they are toxic and carcinogenic in mammals they cannot be used as antibacterial agents.

### Inhibition of DNA-Directed DNA Polymerase

Rifamycins are a class of antibiotics that inhibit DNA-directed RNA polymerase. Polypeptide chains in RNA polymerase attach to a factor that confers specificity for the recognition of promoter sites that initiate transcription of the DNA. Rifampin binds noncovalently but strongly to a subunit of RNA polymerase and interferes specifically with the initiation process. However, it has no effect once polymerization has begun.

## INHIBITION OF DNA REPLICATION

DNA gyrase and topoisomerase I act in concert to maintain an optimum supercoiling state of DNA in the cell. In this capacity, DNA gyrase is essential for relieving torsional strain during replication of circular chromosomes in bacteria. The enzyme is a tetrameric protein composed of two A and two B subunits.

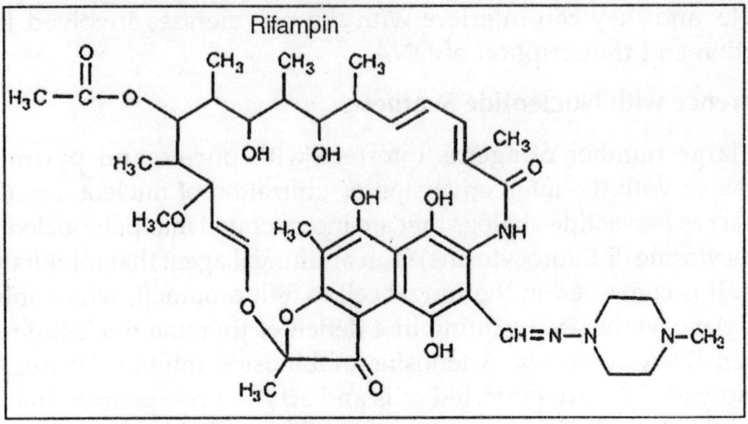


**Fig.** Structure of Rifamphin, which Inhibits the DNA-directed RNA Polymerase.

A transient, covalent bond between the A subunit and DNA occurs during the double strand passage reaction catalyzed by gyrase. Quinolones such as nalidixic acid , bind to the cleavage complex composed of DNA and gyrase during this strand passage.

This interaction of quinolone acts to stabilize the cleavage intermediate which has a detrimental effect on the normal DNA replication process. The effects of this inhibition result in the death of the bacterial cell. The newer fluoroquinolones such as ciprofloxacin, norfloxacin, and ofloxacin also interact with DNA gyrase and possess a broad spectrum of antimicrobial activity.

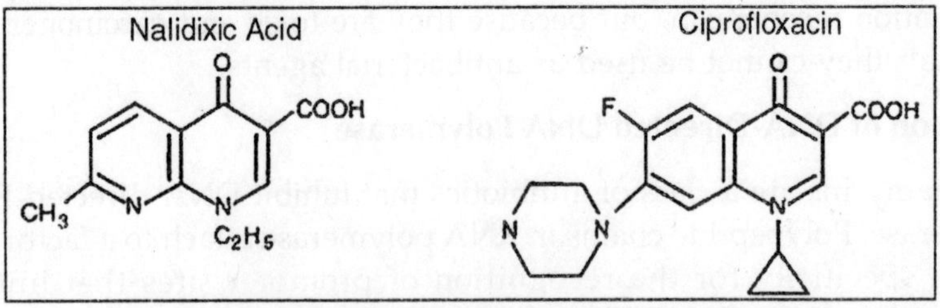


**Fig.** Structure of Quinolone Antibiotica.

Nalidixic inhibits only aerobic Gram-negative species. In ciprofloxicin, the flourine provides Gram-positive activity, the piperazine group increases activity against members of the *Enterovacteriaceae,* and the piperazine and cylopropyl groups give activity against *Pseudomonas* species. Nitroimidazoles such as metronidazole inhibit anaerobic bacteria and protozoa. The nitro group of the nitrosohydroxyl amino moiety is reduced by an electron transport protein in anaerobic bacteria. The reduced drug causes strand breaks in the DNA. Mammalian cells are unharmed because they lack enzymes to reduce the nitro group of these agents.

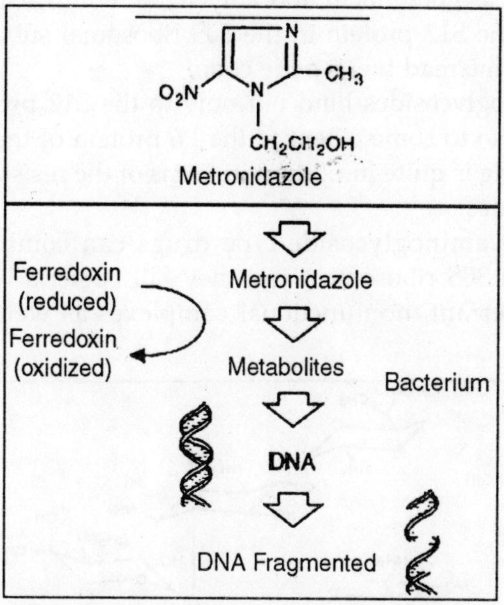


**Fig.** Structure of Metronidazole and Its Mechanism of Action.

Metronidazole enters an aerobic bacterium where, via the electron transport protein ferrodoxin, it is reduced. The drug then binds to DNA, and DNA breakage occurs.

## Antimicrobial Inhibitors of Ribosome Function

The basic structure and function of ribosomes are presented in Fig.. A number of antibacterial agents act by inhibiting ribosome function. Bacterial ribosomes contain two subunits, the 50S and 30S subunits, and it is possible to localize the action of antibiotics to one or both subunits. It is also possible to isolate the specific ribosomal proteins to which an agent binds and to isolate bacterial mutants that lack a specific ribosomal protein and therefore show resistance to a particular agent.

Aminoglycosides act by binding to specific ribosomal subunits. Aminoglycosides are complex sugars connected in glycosidic linkage. They differ both in the molecular nucleus, which can be streptidine or 2-deoxystreptidine, and in the aminohexoses linked to the nucleus. Essential to the activity of these agents are free NH, and OH groups by which aminoglycosides bind to specific ribosomal proteins. Streptomycin, the first aminoglycoside studied, was a useful too] in elucidating protein synthesis.

However, it is rarely used clinically today except to treat tuberculosis, and its mode of action differs to some extent from that of the other clinically useful aminoglycosides, which are 2-deoxystreptidine derivatives such as gentamicin, tobramycin, and amikacin. Streptomycin binds to a specific S12 protein in the 30S ribosomal subunit and causes the ribosome to misread the genetic code.

Other aminoglycosides bind not only to the S12 protein of the 30S ribosome, but also to some extent to the L6 protein of the 50S ribosome. This latter binding is quite important in terms of the resistance of bacteria to aminoglycosides.

Indeed, the aminoglycoside-type drugs can combine with other binding sites on 30S ribosomes, and they kill bacteria by inducing the formation of aberrant, nonfunctional complexes as well as by causing misreading.

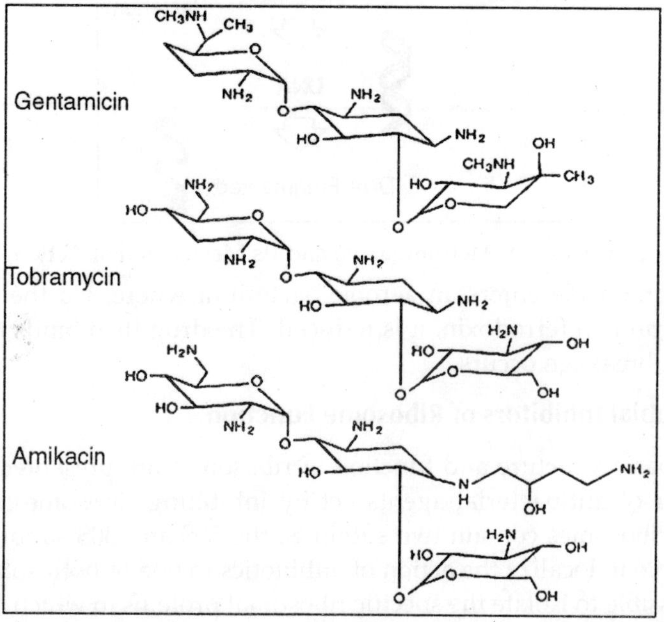


**Fig.** Structures of Three Aminoglycoside Antibiotics Used Clinically.

Critical aspectos of the molecules are the amino and hydroxy groups that bind to proteins in the ribosomes. Spectinomycin is an aminocylitol antibiotic that is closely related to the aminoglycosides. It binds to a different protein in the ribosome and is bacteriostatic but not bactericidal. It is used to treat penicillin-resistant gonorrhea.

Other agents that bind to 30S ribosomes are the tetracyclines. These agents appear to inhibit the binding of aminoacyl-tRNA into the A site of the bacteria] ribosome.

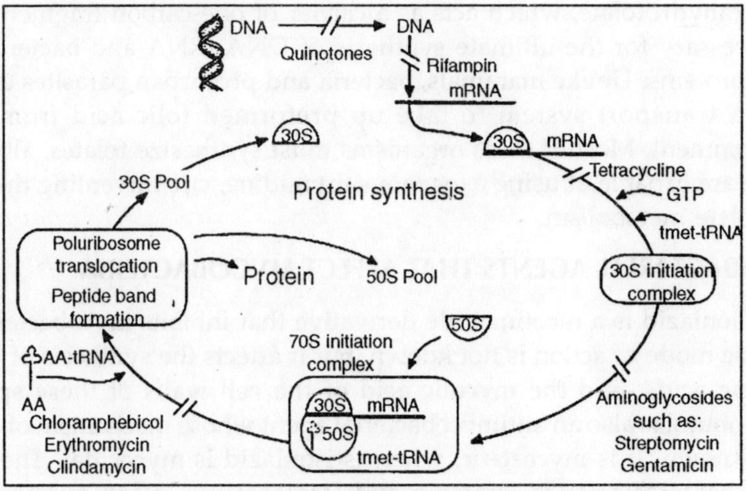


**Fig.** Diagrammatic Representation of Inhibition Sites of Protein Biosynthesis by Various Antibiotics that Bind to the 30S and 50S Ribosomes.

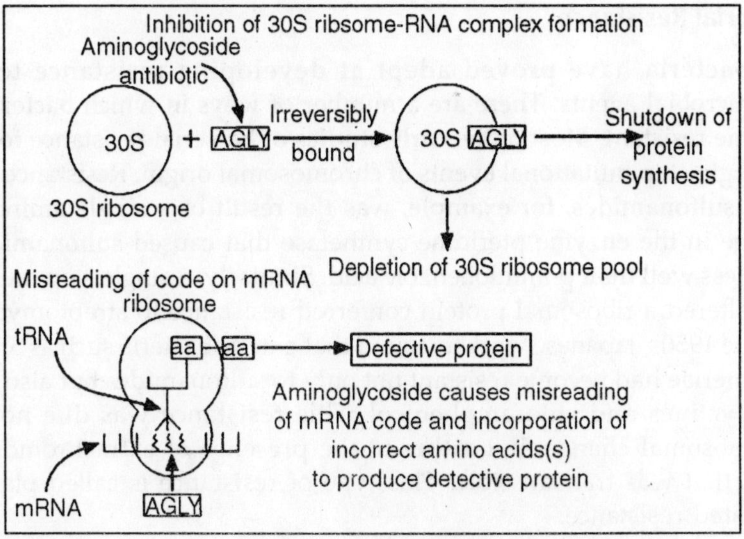


**Fig.** Inhibition of Protein Biosynthesis by Aminoglycosides.

Tetracycline binding is transient, so these agents are bacteriostatic. Nonetheless, they inhibit a wide variety of bacteria, chlamydias, and mycoplasmas and are extremely useful antibiotics.

## DRUGS THAT INHIBIT OTHER BIOCHEMICAL TARGETS

Both trimethoprim and the sulfonamides interfere with folate metabolism in the bacterial cell by competitively blocking the biosynthesis of tetrahydrofolate, which acts as a carrier of one-carbon fragments and is necessary for the ultimate synthesis of DNA, RNA and bacterial cell wall proteins. Unlike mammals, bacteria and protozoan parasites usually lack a transport system to take up preformed folic acid from their environment. Most of these organisms must synthesize folates, although some are capable of using exogenous thymidine, circumventing the need for folate metabolism.

## ANTIBACTERIAL AGENTS THAT AFFECT MVCOBACTERIA

Isoniazid is a nicotinamide derivative that inhibits mycobacteria. Its precise mode of action is not known, but it affects the synthesis of lipids, nucleic acids, and the mycolic acid of the cell walls of these species. Ethambutol is also an antimycobacterial agent whose mechanism of action is unknown. It is mycostatic, whereas isoniazid is mycocidal. The other antituberculosis drugs, rifampin and streptomycin, affect mycobacteria in the same manner that they inhibit bacteria. Pyrazinamide is a synthetic analog of nicotinamide. It is bactericidal, but its exact mechanism is unknown

### Bacterial Resistance

Bacteria have proved adept at developing resistance to new antimicrobial agents. There are a number of ways in which bacteria can become resistant. Most of the early studies of bacterial resistance focused on single-step mutational events of chromosomal origin. Resistance to the early sulfonamides, for example, was the result of a single amino acid change in the enzyme pteridine synthetase that caused sulfonamides to bind less well than p-aminobenzoic acid. Similarly, a single step mutation that altered a ribosomal protein conferred resistance to streptomycin. In the late 1950s, Japanese workers found that enteric bacteria such as Shigella dysenteriae had become resistant not only to sulfonamides but also to the tetracyclines and chloramphenicol. This resistance was due not to a chromosomal change, but rather to the presence of extrachromosomal DNA that was transmissible. This type of resistance is called plasmid-mediated resistance.

**Table. Mechanisms of Resistance**

| |
|---|
| Alteration of target |
| Modification to insensitivity to inhibitor |
| Reduction in physiologic importance of target |

Synthesis of new target enzyme that duplicates function of inhibited target
Prevention of access to target
Efflux of more drug than enters cell
Failure of modified drug to enter cell
Inactivation of agent
Destruction of the agent
Modification of the agent so it fails to bind to target
Failure to convet an inactive precursor agent to its active form

Resistance-conferring plasmids are present in virtually all bacteria. For example, resistance to ampicillin appeared in *Haemophilus influenzae* in 1974 and in Neisseria gonorrhoeae in 1976. In the last several years, organisms such as enterococci have been shown to contain plasmids that confer resistance to drugs such as ampicillin and aminoglycosides.

## Mechanisms of Resistance

The basic mechanisms by which a microorganism can resist an antimicrobial agent are (1) to alter the receptor for the drug (the molecule on which it exerts its effect); (2) to decrease the amount of drug that reaches the receptor by altering entry or increasing removal of the drug; (3) to destroy or inactivate the drug; and (4) to develop resistant metabolic pathways. Bacteria can possess one or all of these mechanisms simultaneously.

## Resistance Due to Altered Receptors

*β-Lactam Resistance:* The ability to analyse changes in receptors for β-lactams by competition experiments in which [ $^{14}$C]penicillin is inhibited from binding to penicillin-binding proteins has explained a number of cases of bacterial resistance to penicillins and cephalosporins. In 1977, *Streptococcus pneumoniae* strains resistant to penicillin G were encountered in South Africa.

Plasmids were not the cause of the resistance. Penicillin-resistant *S pneumoniae* cells have altered penicillin-binding proteins, which bind penicillin less well. Resistance of *S pneumoniae* to penicillin has been increasing, and there are now relatively resistant isolates (minimal inhibitory concentration [MIC], 0.1 to 1 mg/ml) in many parts of the world.

Altered penicillin-binding proteins also explain the resistance of some *Staphylococcus aureus* strains to β-lactamase-stable penicillins (the so-called methicillin-resistant strains). The β-lactams induce synthesis of a new penicillin-binding protein, PBP2a, which does not bind any β-lactam. The β-lactam resistance of coagulase-negative staphylococci is also the result of altered penicillin-binding proteins. Staphylococcal organisms resistant to methicillin are resistant to all penicillins, cephalosporins, and carbapenems.

The resistance of group D streptococci to β-lactam antibiotics appears to be the result of lower affinity of the penicillin-binding proteins for the penicillins. Enterococci are resistant to all cephalosporins because of failure to bind to the penicillin-binding proteins. One Gram-negative species for which resistance to β-lactam antibiotics can be correlated with diminished affinity of the target enzymes is *N gonorrhoeae*.

*Vancomycin Resistance:* Certain transposable genetic elements encode special cell wall-synthesizing enzymes which change the structure of the normal D-Ala-D-Ala side chain in the peptidoglycan assembly pathway. The altered side chain (D-Ala-D-Lac) does not bind vancomycin and allows normal peptidoglycan polymerization to occur in the presence of the drug. Depending upon the nature of the vancomycin resistance gene, high-level resistance can occur to glycopeptides. Thus far, this type of resistance has been found in enterococci but not in multi-resistant isolates of *Staphylococcus aureus*.

*Macrolide-lincomycin Resistance:* Macrolide-lincomycin resistance in clinical isolates of staphylococci and streptococci has been recognized for several decades. The resistance is due to methylation of two adenine nucleotides in the 23S component of 50S RNA. This resistance is plasmid mediated, and the resistance is encoded on transposons. Resistance results from induction of an enzyme that is normally repressed. The methylated RNA binds macrolidelincomycin-type drugs less well than unmethylated RNA does. Induction of resistance varies by species, and in most Gram-positive species erythromycin is a more effective inducer of resistance than is clindamycin. The plasmids that mediate macrolide-lincomycin resistance in streptococci and staphylococci have extensive structural similarity, indicating that these plasmids readily pass between these species.

*Rifampin Resistance:* The resistance of bacteria to rifampin is caused by an alternation of one amino acid in DNA-directed RNA polymerase, which results in reduced binding of rifampin. The degree of resistance is related to the degree to which the enzyme is changed, but does not correlate strictly with enzyme inhibition. This form of resistance occurs at a low level in any population of bacteria so that resistance develops by natural selection during a course of therapy. Naturally resistant organisms are more common among members of the Enterobacteriaceae, explaining why agents of urinary tract infections rapidly became resistant to rifampin. The resistance of *Neisseria meningitides* to rifampin appeared in closed military settings in which rifampin has been used for prophylaxis.

*Sulfonamide-trimethoprim Resistance:* Sulfonamide can be rendered ineffective by altered or new dihydropteroic synthetase that has poor affinity for sulfonamides and preferentially binds p-aminobenzoic acid. Sulfonamide resistance of this type can result from a point mutation or from acquisition of a plasmid that causes synthesis of the new enzyme. A most serious resistance problem is an increase in resistance to trimethoprim. This plasmid- and transposon-mediated resistance is due to production of an altered dihydrofolate reductase that has markedly reduced affinity for trimethoprim.

*Quinolone Resistance:* Resistance to quinolones can be caused by mutations in DNA gyrase subunits A or B, reduced outer membrane permeability in gram-negative cells, or to active efflux transporters found in many bacteria. The highest level of resistance to the newer fluoroquinolones is most frequently associated with chromosomal mutations, causing amino acid substitutions in a highly conserved region in the A subunit of DNA gyrase. Multiple-mechanisms of resistance can occur in a single isolate of bacteria, leading to a higher level of resistance to many fluoroauinolones.

## RESISTANCE DUE TO DECREASED ENTRY OF A DRUG

*Tetracycline Resistance:* The uptake of tetracycline by members of the Enterobacteriaceae is biphasic. In an initial energy-independent rapid phase, tetracycline binds to cell surface layers and passes by diffusion through the outer layers of the cell. In the second, energy-dependent phase, tetracycline crosses the cytoplasmic membrane, probably by means of a proton-motive force. The precise transport system has not been identified.

Tetracycline resistance is common in both Gram-positive and Gram-negative bacteria. In most cases it is plasmid encoded and inducible; however, chromosomal, constitutive resistance is found in some organisms such as *Proteus* species. Many plasmid-encoded specified tetracycline resistance determinants have been found in enteric bacteria. The most common of these determinants, TetB, is also present in *H influenzae*. Tetracycline resistance in *Staphylococcus aureus* is due primarily to small multicopy plasmids; chromosomal resistance is rare. Tetracycline resistance is found on nonconjugative plasmids in *Streptococcus faecalis* and on the chromosome of *S pneumoniae, S agalactiae* (group B streptococci), and oral streptococci. *Clostridium* species such as *C difficile* harbor chromosomal genes for tetracycline resistance.

Basically, tetracycline resistance is due to a decrease in the levels of drug accumulation. Decreased uptake and increased efflux both probably participate. Resistant bacteria bind less tetracycline, and the tetracycline they do accumulate is lost by an energy-dependent process when they are in a drug-free milieu.

Plasmid-mediated resistance to tetracyclines can be partially overcome in Gram-positive species by modifying the tetracycline nucleus. Hence, achievable concentrations of minocycline and doxycycline, in particular, will inhibit some tetracycline-resistant streptococci such as *S pneumoniae*, and some *S aureus* strains. Molecular modification has not been successful in overcoming the tetracycline resistance of members of the Enterobacteriaceae or *Pseudomonas* or most *Bacteroides* species.

Tetracycline resistance is a major concern because it is located on plasmids near insertion sites, and these plasmids readily acquire other genetic information to enlarge the spectrum of resistance.

The widespread use of tetracycline in animal feeds may be a factor in the extensive, worldwide resistance of members of the Enterobacteriaceae, particularly enteric species such as *Salmonella,* to tetracyclines and subsequently to many other drugs. Not only can tetracycline resistance move among members of the Enterobacteriaceae on plasmids, but plasmids mediating tetracycline resistance have moved between *S aureus, S epidermidis, S pyogenes, S pneumoniae,* and *S faecalis.*

*Fosfomycin Resistance:* Fosfomycin and fosmidomycin, which inhibit cell wall synthesis, enter bacteria by means of a glycerol-phosphate or glucose-6-phosphate transport system. Gram-positive bacteria in which the glucose-6-phosphate transport system is poorly developed do not take up these drugs in concentrations adequate to inhibit the cell wall synthesis. This resistance usually is chromosomal. The resistance of Gram-negative bacteria to these agents is related primarily to the presence in the population of some bacteria that can function without the transport system. Plasmids and transposons that transfer resistance to fosfomycin have been found in bacteria such as *Serratia marcescens.*

*Aminoglycoside Resistance:* In the most important form of aminoglycoside resistance, the compound is modified outside the cell and resistance is due partly to poor uptake of the altered compound. Also, all aminoglycosides have free amino and hydroxy groups that are essential for binding to ribosomal proteins. A number of enzymes can acetylate the amino groups and phosphorylate or adenylate the hydroxyl groups. Other forms of resistance, such as altered binding site on 30S ribosomes, are much less common.

In members of the Enterobacteriaceae and in *Pseudomonas* species, the aminoglycosides pass through the cell wall via channels designed to admit cationic molecules to the periplasmic space. These channels, called porin channels, are lined by the porin protein. Aminoglycosides are then translocated across the cell membrane by an energy-dependent proton-motive force and, in the cytoplasm, bind to ribosomes located just below the membrane. Aminoglycosides bind only to ribosomes actively engaged in protein synthesis. Binding to the ribosomes induces a protein involved in the uptake of the aminoglycosides.

Bacteria may contain in the periplasmic space enzymes that acetylate, phosphorylate, or adenylate aminoglycosides to various degrees. It is not clear whether the enzymes are free in the periplasmic space or bound to the cytoplasmic membrane. The modified aminoglycosides do not bind well to ribosomes, and hence uptake is poor or absent.

## RESISTANCE DUE TO DESTRUCTION OR INACTIVATION OF A DRUG

*Chloramphenicol Resistance:* Many Gram-positive and Gram-negative bacteria, including some recently discovered *H influenzae* strains, are resistant to chloramphenicol because they possess the enzyme chloramphenicol transacetylate, which acetylates hydroxyl groups on the chloramphenicol structure. This enzyme, unlike the aminoglycoside-inactivating enzymes and β-lactamases, is an intracellular enzyme of higher molecular weight and subunit structure. Acetylated chloramphenicol binds less well to the 50S ribosome.

*β-Lactam Resistance:* The best-known mechanism of bacterial resistance is the resistance to β-lactams, which is mediated by penicillinase enzymes. Resistance of *E coli* to penicillin was recognized in 1940, before sufficient penicillin was made to be clinically useful. In the 1940s, resistance of staphylococci was shown to be due to a penicillinase. As these enzymes also attack other β-lactam compounds such as cephalosporins, carbapenems, and monobactams, they would be more appropriately designated β-lactamases. The most important activity of these enzymes is alteration of the β-lactam nucleus. β-lactamases are widely distributed in nature and are usually classified on the basis of the principal compounds they destroy (e.g., as penicillinases or cephalosporinases). β-Lactamases may be chromosomally or plasmid mediated, and they may be constitutive or inducible.

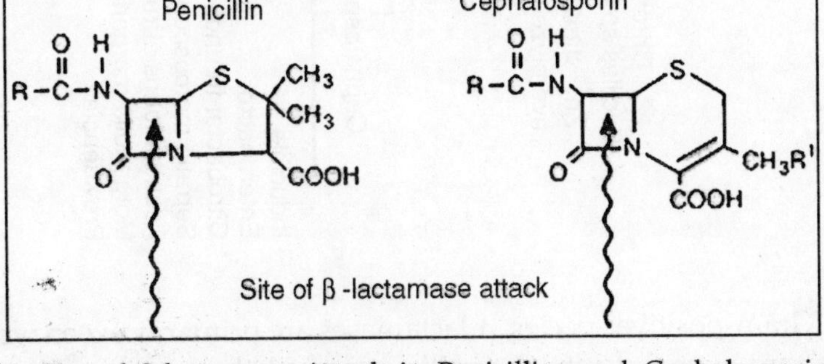


**Fig.** Site of β-lactamase Attack in Penicillins and Cephalosporins.

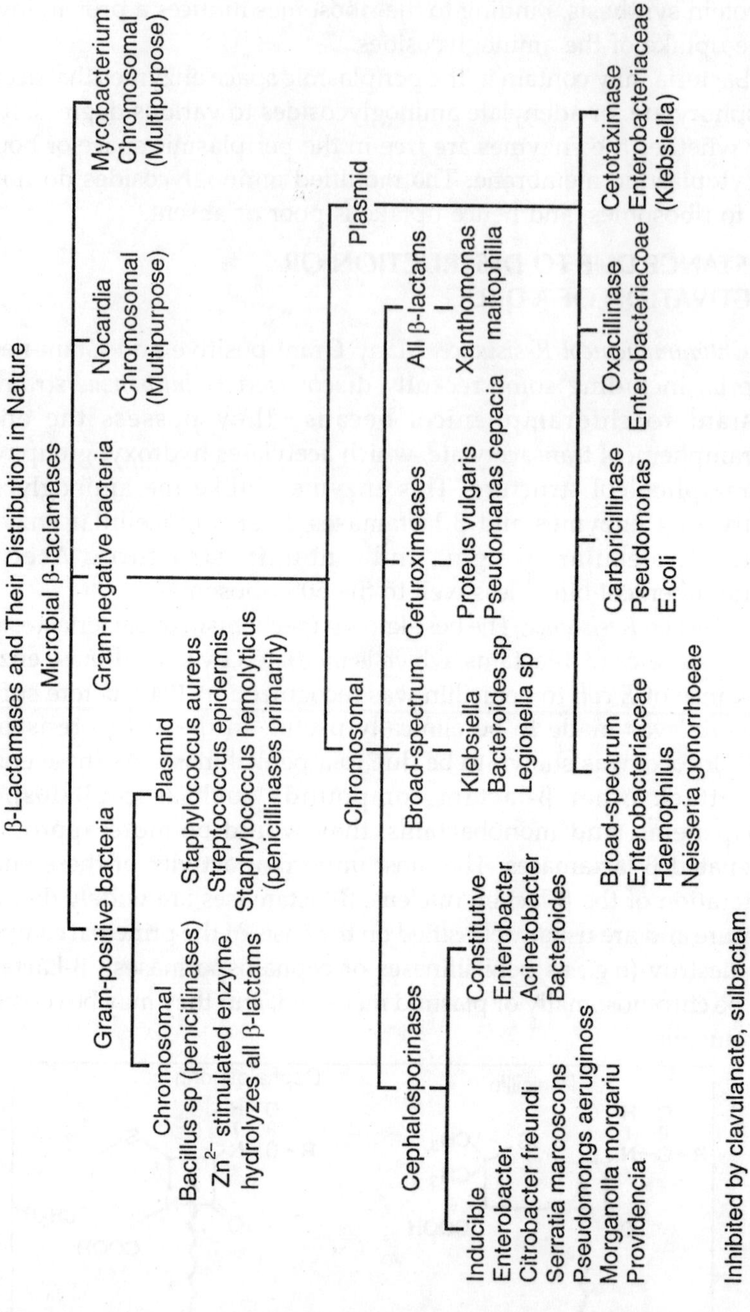


**Fig.** β-Lactamase Found in Bacteria and Their Classification and Synthesis, Whether Chromosomally or Plasmid Mediated.

In Gram-positive species, β-lactamases are primarily exoenzymes; that is, they are excreted into the milieu around the bacteria. Virtually all hospital isolates of staphylococci, both *S aureus* and *S epidermidis*, have β-lactamases, and 50 to 80 per cent of community-acquired staphylococcal isolates produce β-lactamases. In Gram-negative species, both aerobic and anaerobic, β-lactamases are contained in the periplasmic space, thus effectively protecting the penicillin-binding proteins.

Resistance of staphylococci to β-lactams was soon overcome with the antistaphylococcal penicillins and the cephalosporins. Some strains of *S aureus* produce more β-lactamase constitutively and can destroy some of the cephalosporins.

In 1974, *H influenzae* was shown to possess a plasmid mediated β-lactamase. At present 10 to 35 per cent of *H influenzae* strains in the United States produce β-lactamases.

The TnA transposon has become more widespread, and the resistance of *Haemophilus* species to penicillin G and ampicillin seems to be increasing yearly. The *Haemophilus* β-lactamase is the same structurally as the enzyme found in *E coli, Salmonella, Shigella,* and *N gonorrhoeae*. The enzyme has generally been called the TEM enzyme after the initials of the Greek girl from whom an *E coli* strain containing a plasmid β-lactamase was first isolated.

These enzymes are also called Richmond-Sykes class IIIa enzymes from a classification proposed by Richmond and Sykes in 1973. By far the most common plasmid β-lactamase found in nature is TEM-1, which accounts for 75 to 80 per cent of plasmid-mediated β-lactamase resistance worldwide.

Recently new β-lactamases have been found that hydrolyze compounds such as inomethoxy cephalosporin, which were not destroyed by other plasmid-encoded β-lactamases.

The new β-lactamases have an altered amino acid composition, which permits binding to the cephalosporin and subsequent hydrolysis. How common these new enzymes will become is unknown.

Chromosomally mediated β-lactamases are present in many *Enterobacter, Citrobacter, Proteus-Providencia,* and *Pseudomonas* species. All *Klebsiella* species possess a β-lactamase, which acts primarily as a penicillinase and is chromosomally mediated. Constitutively produced β-lactamases are also present in many anaerobic species.

Table lists the major β-lactamases of clinical importance. β-Lactamases vary in their ability to destroy penicillins and cephalosporins.

β-Lactamase activity is only one component of the β-lactam resistance of Gram-negative bacteria, since resistance to β-lactams is a combination of decreased entry, β-lactamase stability and affinity of the compounds for penicillin-binding proteins.

**Table. R-Plasmid-Mediated Resistance.**

| *Antibiotic* | *Mechanism* | *Organisms* |
|---|---|---|
| Penicillin, ampicillin, carbenicillin, etc. | β-Lactamase hydrolysis | Staphylococci, enterococci (rare) Enterobacteriaceae, pseudomonads, bacteroides |
| Oxacillin, methicillin, etc. | β-Lactamase hydrolysis | Enterobacteriaceae, pseudomonads |
| Cephalosporins | β-Lactamase | Staphylococci, Enterobacteriaceae, pseudomonads, bacteroides |
| Chloramphenicol | Acetylation | Staphylococci, enterococci, streptococci, Enterobacteriaceae, pseudomonads |
| Tetracyclines | Permeability block | Staphylococci, enterococci, streptococci, Enterobacteriaceae, pseudomonads, bacteroides |
| Aminoglycosides | | |
| Streptomycin | Acetylation | Staphylococci, enterococci, |
| Neomycin | Phosphorylation | Enterobacteriaceae, |
| Kanamycin | Adenylation (alters binding to ribosomes and uptake of drug) | pseudomonads |
| Tobramycin | | |
| Amikacin | | |
| Macrolides-Lincinoids Erythromycin | Altered 23S RNA | Staphylococci, enterococci, streptococci, bacteroides |
| Clindamycin | | |
| Trimethoprim | Altered dihydrofolate | Staphylococci, Enterobacteriaceae |
| Sulfonamides | Altered tetraphydropteroic synthetase | Staphylococci, enterococci, streplococci, Enterobacteriaceae, pseudomonads |
| Fosfomycin | Altered glucose | Staphylococci, Enterobacteriaceae |
| Vancomycin | New protein | Enterococci |

## SYNTHESIS OF RESISTANT METABOLIC PATHWAY

No synthesis of a new type of cell wall resistant to β-lactams has occurred, but some bacteria, particularly some streptococci, lack the hydrolytic enzymes necessary for forming a new cell wall, and so β-lactams do not lyse these bacteria. An altered hydrolytic system thus converts a bactericidal antibiotic into a bacteriostatic agent. Whether such resistance occurs in Gram-negative species is not clear. Some thymidine-requiring streptococci are not inhibited by trimethoprim and sulfonamides and so are not killed by these agents. These organisms are a rare cause of urinary tract infections. Other bacteria produce adequate deoxyribosylthymine 5'-monophosphate (DTMP) by alternative methods and, as a result. survive exposure to these folate inhibitors.

Certain *Candida* or *Cryptococcus* yeasts are resistant to flucytosine because they cannot convert it to its active component, fluorouracil. Other fungi can resist the polyenes and imidazoles because they synthesize membrane components by different metabolic mechanisms.

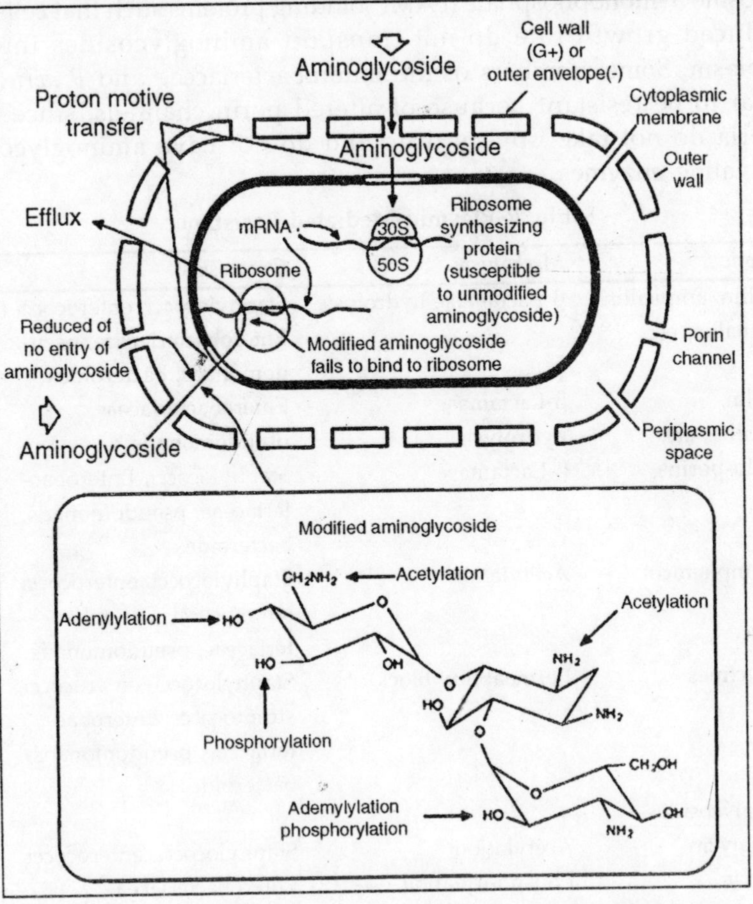


Fig. Diagrammatic Representation of Transfer and Transfer Reduction of Aminoglycoside Across the Bacterial Cell Wall.

If it is modified by acetylation, adenylation, or phosphorylation, the drug will not bind to ribosomes and will leave the bacterial cell.

Aminoglycoside-modifying enzymes have been found in Gram-positive species such as *S aureus, S faecalis, S pyogenes,* and enzymes are particularly prevalent in members of the Enterobacteriaceae and *P aeruginosa. S pneumoniae.* These Many of the genes for aminoglycoside-modifying enzymes are carried on transposons

Anaerobic organisms such as *Bacteroides* species are resistant to aminoglycosides because they lack an oxygen-dependent transport system to move the drugs across the cytoplasmic membrane. Although most resistance of *S aureus* to aminoglycosides is due to aminoglycoside-modifying enzymes, small-colony variants of staphylococci also show resistance, which may be due to a defect in adenylate cyclase or in cyclic adenosine 5'-monophosphate (cAMP)binding proteins such that cells with a reduced growth rate do not transport aminoglycosides into the cytoplasm. Some members of the Enterobacteriaceae and *P aeruginosa* appear to be resistant because of altered porin channels, since these bacteria do not take up any drug and do not have aminoglycoside-inactivating enzymes.

**Table. R-Plasmid-Mediated Resistance.**

| *Antibiotic* | *Mechanism* | *Organisms* |
|---|---|---|
| Penicillin, ampicillin, carbenicillin, etc. | β-Lactamase hydrolysis | Staphylococci, enterococci (rare) Enterobacteriaceae, pseudomonads, bacteroides |
| Oxacillin, methicillin, etc. | β-Lactamase hydrolysis | Enterobacteriaceae, pseudomonads |
| Cephalosporins | β-Lactamase | Staphylococci, Enterobacteriaceae, pseudomonads, bacteroides |
| Chloramphenicol | Acetylation | Staphylococci, enterococci, streptococci, Enterobacteriaceae, pseudomonads |
| Tetracyclines | Permeability block | Staphylococci, enterococci, streptococci, Enterobacteriaceae, pseudomonads, bacteroides |
| Aminoglycosides | | |
| Streptomycin | Acetylation | Staphylococci, enterococci, |
| Neomycin | Phosphorylation | Enterobacteriaceae, |
| Kanamycin | Adenylation (alters binding to ribosomes and uptake of drug) | pseudomonads |

| | | |
|---|---|---|
| Tobramycin | | |
| Amikacin | | |
| Macrolides-Lincinoids Erythromycin | Altered 23S RNA | Staphylococci, enterococci, streptococci, bacteroides |
| Clindamycin | | |
| Trimethoprim | Altered dihydrofolate | Staphylococci, Enterobacteriaceae |
| Sulfonamides | Altered tetraphydropteroic synthetase | Staphylococci, enterococci, streplococci, Enterobacteriaceae, pseudomonads |
| Fosfomycin | Altered glucose | Staphylococci, Enterobacteriaceae |
| Vancomycin | New protein | Enterococci |

Bacteria also contain transposons, which can insert into plasmids and also into the chromosome. Transposon-mediated resistance to most of the major antibiotics has been found in the past few years.

Antimicrobial agents exert a strong selective pressure on the development of both chromosomal and plasmid-mediated resistance, as discussed below. Administration of an antibiotic destroys the susceptible bacteria in a population, but may permit resistant ones to proliferate. From an epidemiologic viewpoint, plasmid-mediated resistance is the most important type, since it is transmissible, is usually highly stable, confers resistance to many different classes of antibiotics simultaneously, and often is associated with other characteristics that enable a microorganism to colonize and invade a susceptible host.

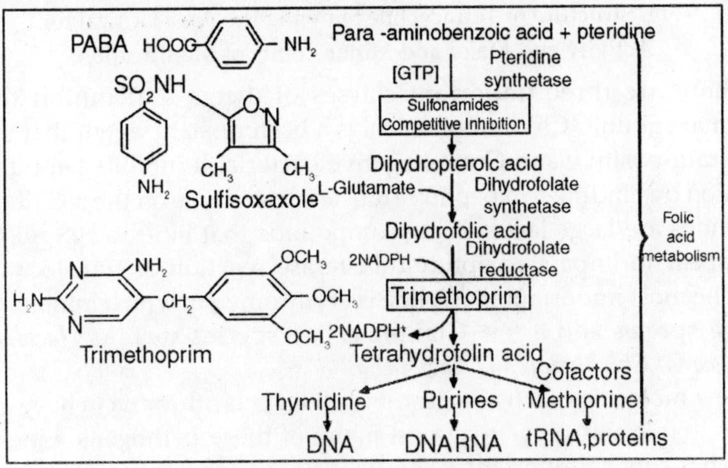


**Fig.** Structure of Sulfonamide and Trimethoprim with Sites of Inhibition of Folic Metabolism

Sulfonamides competitively block the conversion of pteridine and p-aminobenzoic acid (PABA) to dihydrofolic acid by the enzyme pteridine synthetase. Sulfonamides have a greater affinity than p-aminobenzoic acid for pteridine synthetase. Trimethoprim has a tremendous affinity for bacterial dihydrofolate reductase (10,000 to 100,000 times higher than for the mammalian enzyme); when bound to this enzyme, it inhibits the synthesis of tetrahydrofolate.

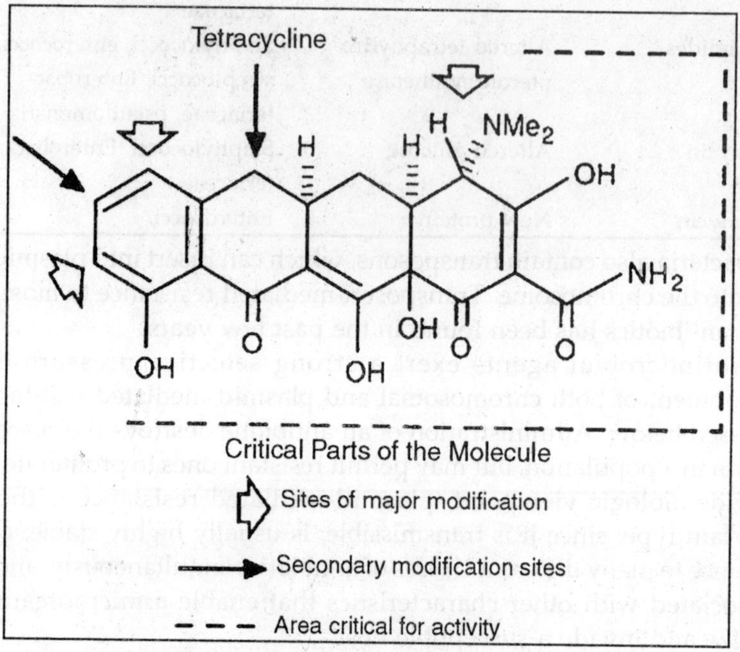


**Fig.** Structure of Tetracycline Showing the Area Critical for Activity and Major and Minor Points of Modification.

There are three important classes of drugs that inhibit the 50S ribosomal subunit. Chloramphenicol is a bacteriostatic agent that inhibits both Gram-positive and Gram-negative bacteria. It inhibits peptide bond formation by binding to a peptidyltransferase enzyme on the 50S ribosome. Macrolides are large lactone ring compounds that bind to 50S ribosomes and appear to impair a peptidyltransferase reaction or translocation, or both. The most important macrolide is erythromycin, which inhibits Gram-positive species and a few Gram-negative species such as *Haemophilus, Mycoplasma, Chlamydia, and Legionella*.

New molecules such as azithromycin and clarithromycin have greater activity than erythromycin against many of these pathogens. Lincinoids, of which the most important is clindamycin, have a similar site of activity. Both macrolides and lincinoids are generally bacteriostatic. inhibiting only the formation of new peptide chains.

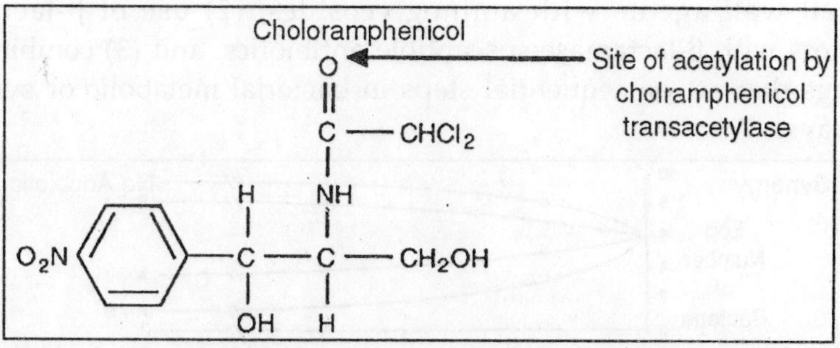


**Fig.** Structure of Chloranphenicol.

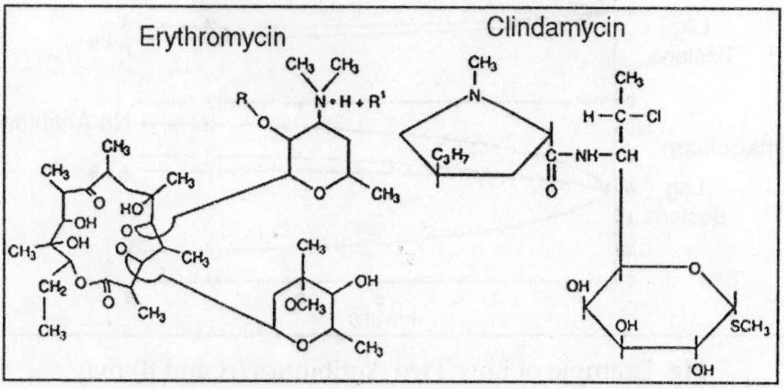


**Fig.** Structure of Erythromycin (Prototype or Macrolide) and Clindamycin.

Although extremely different in structure, both compounds inhibit protein synthesis by binding to 50S ribosome.

## COMBINATIONS OF ANTIMICROBIAL AGENTS

Antibiotics are frequently used in combination for the following reasons: (1) to treat a life-threatening infection; (2) to prevent emergence of bacterial resistance; (3) to treat mixed infections of aerobic and anaerobic bacteria; (4) to enhance antibacterial activity (synergy); and (5) to use lower doses of a toxic drug. Combined treatment is reasonable when the precise agents of a serious infection are unknown.

Use of two or more drugs to prevent the emergence of resistance is effective for tuberculosis and for therapy of some chronic infections. The use of combinations to achieve synergy is more complicated.

Synergy occurs when a combination of two drugs causes inhibition or killing when used at a fourfold-lower concentration than that of either component drug used separately. However, indifference or antagonism may occur instead. Indifference means that the combined action is the same as with either component; antagonism refers to a reduction in the activity of one or both components in the presence of the other.

Important examples of bacterial synergy include (1) combinations of anti-cell wall agents with aminoglycosides, (2) use of β-lactamase inhibitors with β-lactamase-susceptible antibiotics, and (3) combinations of drugs that act on sequential steps in bacterial metabolic or synthetic pathways.

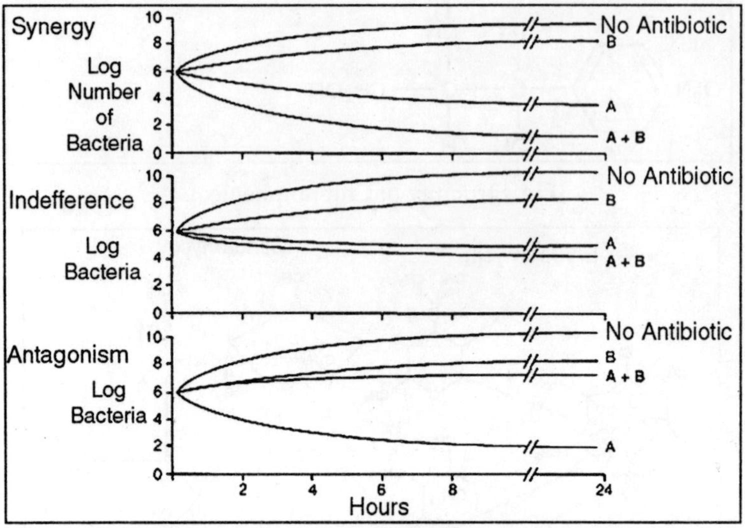


**Fig.** Example of how Two Antibiotics (A and B) may Interact with Synergy, Indifference, or Antagonism.

Examples of synergy that have proved clinically important are the combination of penicillin and streptomycin to treat *Enterococcus faecalis* endocarditis and the combination of carbenicillin and gentamicin to treat P aeruginosa infections. Recently the β-lactamase inhibitor clavulanic acid or sulbactam has been combined with amino penicillins to inhibit *S aureus, Klebsiella pneumoniae, H influenzae,* and anaerobic organisms such as *Bacteroides* species, all of which are resistant to amoxicillin or ampicillin when they contain β-lactamases. The combination of sulfamethoxazole and trimethoprim attacks two parts of the folic acid cycle and synergistically inhibits many bacteria. Finally, combinations of two penicillins that affect different stages of cell wall synthesis in Gram-negative bacteria are synergistic. This is true for the combination of mecillinam (amdinocillin), which binds to PBP2 of bacteria, and penicillins or cephalosporins, which bind to PBP lb or 3.

Antagonism can occur when a bacteriostatic agent is combined with a bactericidal agent. The classic example has been the combination of chlortetracycline and penicillin in treatment of pneumococcal meningitis. This effect has not been explained from a molecular standpoint, and tetracyclines and penicillins or cephalosporins are used to treat mixed infections such as pelvic inflammatory disease due to *N gonorrhoeae* and

*Chlamydia*. Some β-lactam antibiotics can induce β-lactamases that inactivate other β-lactams, and antagonism can be shown in the test tube, but the relevance for clinical infections is not established.

## TOXICOLOGY OF ANTIMICROBIAL AGENTS

Antimicrobial agents can be directly toxic, can interact with other drugs to increase their toxicity, or can alter microbial flora to cause infection by organisms that are normally saprophytic. Allergic reactions can be caused by an agent, but penicillins can produce either immediate, IgE-mediated, or delayed hypersensitivity reactions. Cutaneous reactions have been reported with every class of antimicrobial agent. Hematologic reactions can range from the life-threatening blood dyscrasia that occurs in 1 in 60,000 individuals who receive chloramphenicol to hemolytic anemia due to sulfonamides in individuals who lack the enzyme glucose-6-phosphate dehydrogenase. Depression of blood platelet activity has occurred with many agents. By altering the gastrointestinal flora, almost all antibiotics can cause overgrowth of *Clostridium difficile*, which produces a toxin that causes diarrhea and even pseudomembranous colitis. Alteration of intestinal flora by antibiotics can also result in overgrowth of *Candida* in the mouth, vagina, or gastrointestinal tract. Since a number of antibiotics are metabolized in the liver, damage to the liver can occur. This has been of particular concern with isoniazid, which is used to treat tuberculosis. Damage to the kidneys can follow the use of aminoglycosides. Neurologic toxicity is fortunately fairly uncommon, but the aminoglycosides can damage the auditory or vestibular apparatus if the dosage is not closely monitored.

Bacteria continue to evolve new mechanisms of resistance to old and to new antimicrobial agents. Some bacteria such as *P aeruginosa* are particularly adept at utilizing a number of different mechanisms simultaneously to become resistant to agents in virtually every class and those with such diverse sites of action as cell wall, protein biosynthesis, or DNA and RNA synthesis. Progress in medicine will keep patients alive who have nosocomial infections with resistant pathogens.

## MECHANISM TO REDUCE BACTERIAL RESISTANCE

Proper selection of new antibiotics will be a major force in slowing the development of antimicrobial resistance. Proper hygiene practices will reduce plasmid transfer and the establishment of multiple drug-resistant bacteria in the hospital and will delay the appearance of such species in the community. Table lists a number of mechanisms to prevent bacterial resistance. The health care provider must be continually alert to the appearance of antibiotic resistance within the hospital and community.

**Table. Mechanisms to Reduce Antibiotic Resistance**

- Control, reduce, or cycle antibiotic usage
- Improve hygiene in hospitals and among hospital personnel and reduce movement of patients to eliminate the dissemination of resistant organisms within hospitals
- Discover or develop new antibiotics
- Modify existing antibiolics chemically to produce compounds nert to known mechanisms of resistance
- Develop inhibitors of antibiotic-modifving enxymes
- Define agents that would "cure" resistance plasmids

# Chapter 12

# Nutrition and Cultivation of Bacteria

The survival of microorganisms in the laboratory, as well as in nature, depends on their ability to grow under certain chemical and physical conditions. An understanding of these conditions enables us to characterize isolates and differentiate between different types of bacteria. Such knowledge can also be applied to control the growth of microorganisms in practical situations.

Media used in the laboratory for the cultivation of bacteria must supply all of the necessary nutrients required for cellular growth and maintenance of the organisms. A wide variety of culture media is employed by the bacteriologist for the isolation, growth and maintenance of pure cultures and also for the identification of bacteria according to their biochemical and physiological properties.

A culture medium must supply suitable carbon and energy sources and other nutrients, sometimes including "growth factors". It is important to note that no one medium will support the growth of all microorganisms. Accordingly, the elements required for the maintenance, growth and reproduction of all organisms will be used by different organisms in different ways.

When one prepares a medium for the cultivation of microorganisms, one dissolves various organic and/or inorganic compounds sequentially in pure, distilled water. The importance of water cannot be overestimated. Water is the universal solvent in which all nutrients must be dissolved and all chemical reactions will take place. It can supply some hydrogen and oxygen in certain chemical reactions. Water containing a significant amount of solutes may not be osmotically compatible or "available" for use by microorganisms, so the concept of water availability needs to be considered.

## NUTRITIONAL CLASSIFICATION OF MICROORGANISMS (BASED ON ENERGY AND CARBON REQUIREMENTS)

The various life forms may be categorized as being either chemotrophs or phototrophs regarding the source of the energy which is used in ATP generation. Chemotrophs obtain their energy purely from the oxidation of chemical compounds. Phototrophs use light as the ultimate source of energy. Phototrophs include plants, algae, cyanobacteria, and the purple and green anoxygenic bacteria.

| *Type of Organism* | *General Process and Major Features* | |
|---|---|---|
| Chemotroph | Respiration | • Derive energy by oxidative phosphorylation.<br>• Most respirers use oxygen; this is aerobic respiration. Some respirers may also use nitrate or some other "oxygen substitute" in the process of anaerobic respiration.<br>• Certain organisms can only perform anaerobic respiration – for example, the methane producers and many sulfate reducers. |
| | Fermentation | • Derive energy by substrate-level phosphorylation. |
| Phototroph | Phototrophy | • Derive energy by photophosphorylation.<br>• Phototrophs may be oxygenic (oxygen-evolving) or anoxygenic (not oxygen-evolving). |

When it comes to describing any specific organism regarding its growth characteristics and means of obtaining energy, the terms "strictly aerobic," "facultatively anaerobic" and "strictly anaerobic" can be incomplete and misleading – as discussed here. In place of these terms – or in addition to them – one or more of the more-clearly-definable methods of energy generation of which the organism is capable should be indicated:

- Aerobic respiration
- Anaerobic respiration
- Fermentation
- Anoxygenic phototrophy
- Oxygenic phototrophy

The common laboratory test for "oxygen relationships" determines whether an organism (able to grow under the conditions given) can respire (aerobically) and/or ferment and is generally used to describe and compare chemoheterotrophic bacteria.

Another method of classifying organisms nutritionally is by the source of reducing power utilized. All organisms need reducing power in the form of electrons for biosynthesis. Organisms that oxidize organic compounds are called organotrophs and those that oxidize inorganic compounds are called lithotrophs (which literally means "stone eaters").

Some authors apply these terms exclusively to chemotrophs, using the terms "chemoorganotrophs" and "chemolithotrophs." In the generation of reducing power and energy, phototrophs may also oxidize either organic or inorganic compounds, and we thus have the categories of "photoorganotrophs" and "photolithotrophs." Examples of inorganic substances oxidized by lithotrophs include ammonium, nitrites, sulfur, sulfides, thiosulfates and hydrogen gas.

As carbon is a major and essential element in all living things, organisms may also be classified according to the nature of their source of carbon. Organisms which assimilate organic compounds for their carbon needs are termed heterotrophs. Those which utilize carbon dioxide are called autotrophs.

Considering the various requirements for carbon and energy with the above-defined terms, nearly all living things can be placed in one of the descriptive categories listed below. A very general overview of catabolism is given here.

- *Chemoheterotrophs:* As these organisms are generally organotrophic, they may also be termed chemoorganotrophs. These organisms may use a variety of organic compounds as both carbon and energy sources. A common sugar so used is glucose. ATP is generated by either substrate-level or oxidative phosphorylation.
- *Chemoautotrophs:* As these organisms are generally lithotrophic, they may also be termed chemolithotrophs. ATP is usually generated by oxidative phosphorylation.
- Myxotrophs do not follow the correlations noted for the above two groups. These organisms are actually "chemolithotrophic heterotrophs" and include the genus *Beggiatoa.*
- *Phototrophs:* Traditionally, these organisms are thought of as being autotrophs and lithotrophs, taking in carbon dioxide and generating reducing power by the oxidation of water with the release of oxygen. The classic "photosynthetic equation" we grew up with and applied to trees and other plants is based on this. In place of water, the purple and green sulfur bacteria substitute hydrogen sulfide ($H_2S$) or hydrogen ($H_2$). In Bacteriology 102, we study the purple non-sulfur bacteria, all of which are heterotrophs and organotrophs – obtaining carbon and reducing power at the expense of organic compounds. (Autotrophic/lithotrophic growth similar to that of the purple sulfur bacteria is an alternative for some bacteria in this group.) For phototrophs in general, ATP is generated by photophosphorylation.

The supply of carbon and energy for a particular organism may be relatively simple such as (1) providing light and an atmosphere containing carbon dioxide for photoautotrophs, or (2) providing glucose for the majority of the chemoheterotrophs.

## OTHER NUTRITIONAL AND PHYSICAL REQUIREMENTS

Besides carbon, other required major elements include hydrogen, oxygen, nitrogen, sulfur, phosphorus, potassium, and – to a lesser extent – magnesium, iron, calcium, chlorine and sodium. Other elements, generally required at relatively very low levels, include manganese, cobalt, zinc, molybdenum and copper.

(Attempting to group elements according to importance is somewhat arbitrary.) Certain organisms may use one or more of the first four elements in this listing (H, O, N, S) in their simplest, pure molecular forms. Otherwise, elements are always taken in as part of compounds with other elements. For example, organisms which are termed aerobic and facultatively anaerobic regularly use molecular oxygen ($O_2$) in respiration; see our oxygen relationships page. Also, nitrogen-fixing bacteria can obtain their nitrogen from the reduction of atmospheric nitrogen ($N_2$) to ammonium; the nitrogen becomes incorporated into amino acids and ultimately proteins.

Many of the latter elements in the above listing are required in such small amounts that one can depend on their compounds to be present as inorganic chemical contaminants in the various ingredients used to make media. Such elements not individually added are termed trace elements.

To a greater or lesser degree, various organisms may require preformed organic compounds which these organisms are incapable of synthesizing. Depending on a particular organism's capabilities of producing the essential organic compounds it needs for structure or metabolism, certain amino acids, fatty acids, nucleic acids, vitamins or other compounds may have to be supplied to that organism. A growth factor is therefore defined as a specific organic compound that is required – generally in a very small amount – by a particular organism as it cannot be synthesized by that organism. Organisms termed fastidious tend to require a variety of growth factors.

Each organism has its range of growth and its optimum pH value. Organisms themselves may change the pH of their immediate environment. For example, the pH of a medium tends to decrease when microbial fermentations take place, producing acidic products. Buffers, such as phosphates and calcium carbonate, are often utilized to help stabilize the pH during the growth of the cultures studied.

Incubation conditions must be appropriate for the organism under study. Considerations include the provision of a suitable atmosphere, a suitable temperature, and anything else which may be required such as a light source for the cultivation of phototrophs.

## PUTTING TOGETHER A CULTURE MEDIUM

The ingredients in culture media range from pure chemical compounds to complex materials such as extracts or digests of plant and animal tissues. If all the ingredients of a culture medium are known, both qualitatively and quantitatively, the medium is called a chemically-defined medium. These media are of great value in studying the nutritional requirements of microorganisms or in studying a great variety of their metabolic activities. In a complex medium, the exact chemical composition is not known, and such a medium is often prepared from very complex materials, e.g., body fluids, tissue extracts and infusions, and peptones. A peptone is a commercially-available digest of a particular plant or animal protein, made available to organisms as peptides and amino acids to help satisfy requirements for nitrogen, sulfur, carbon and energy. Peptones also contain small amounts of various organic and inorganic compounds, and a peptone solution can serve as a complete medium for many organisms including *E. coli*. Complex media often contain all nutrients, known and unknown, which may be required for optimal growth of a wide assortment of bacteria. Here are a couple examples of media, each formulated for a purpose:

- A broth (i.e., liquid) medium which is designed to detect acid produced by an organism from glucose fermentation (i.e., Glucose Fermentation Broth) may include the following ingredients:
  - Distilled water
  - *Glucose:* Supplied for carbon and energy. Without glucose, a "facultative anaerobe" would not be able to grow anaerobically, and an "aerotolerant anaerobe" would probably not be able to grow at all.
  - *Peptone:*
  - *pH indicator:* Supplied to detect the change to an acidic condition.

An organism which is not fermenting glucose may still be able to grow in the medium by respiring the glucose and/or one or more of the amino acids in the peptone. In any event, an alkaline reaction will occur at the top of the medium if an organism deaminates amino acids aerobically, producing ammonium. A cautionary note: The alkaline reaction from this ammonium can overneutralize acid which permeates throughout the medium from glucose fermentation (an anaerobic process), and acid may not be detectable at all if the peptone concentration is too high. So, one is careful regarding the addition of the peptone and usually any acid from fermentation is detectable at least in the lower part of the tube. With Glucose O/F Medium, the formulation elevates the amount of glucose and decreases that of peptone such that even the very small amount of acid associated with glucose respiration is detectable for organisms which do not ferment. Glucose Fermentation Broth and O/F Medium are discussed more fully here. How competing acid and alkaline reactions in a differential medium can be used to advantage in bacterial identification is discussed here.

- Here we have an example of a medium which supplies the basic needs for prototrophic strains (i.e., strains typical of their species regarding their biosynthetic capabilities and requirements) of a common intestinal bacterium, *Escherichia coli*. Such a medium formulated with nutritional requirements of a given species in mind is called a "minimal medium" as discussed below. Any required element not seen in this list of ingredients is still assumed to be part of the actual medium – having come into the medium as a trace element in one or more of the individual ingredients.
  - Distilled water
  - *Glucose:* Supplied for carbon and energy.
  - *Ammonium sulfate* ($(NH_4)_2SO_4$): Supplied as the nitrogen source.
  - Magnesium sulfate ($MgSO_4$)
  - *Di- and monopotassium phosphate* ($K_2HPO_4$ *and* $KH_2PO_4$): In a given combination of amounts, these can provide a certain pH and, as "pH buffers," assist in preventing the pH from varying widely due to an organism's formation of metabolic products, such as acids from the fermentation of the glucose.

If one is studying an auxotrophic strain of *E. coli* – i.e., one which cannot produce (from the constituents of the *E. coli* minimal medium) a compound essential for its metabolic needs which prototrophic (typical) strains can so produce – that compound will have to be added specifically to the medium in which case it is then termed a growth factor.

One may ask the question as to whether this example is a chemically defined or complex medium. Given that trace elements may be present as chemical contaminants of the listed ingredients, which (furthermore) are not indicated as being provided in specific amounts, one would have to call this medium complex. Chemically-defined media – as strictly defined – are very exceptional, utilizing ingredients of extreme purity and including a long list of additional compounds to compensate for the lack of trace elements in those pure ingredients.

## SOLID MEDIA

Agar is the major solidifying agent used in bacteriological media. It is an impure polysaccharide gum obtained from certain marine algae. It is added as a powder at a more or less standard concentration (1.5% for plates and slanted media, 0.5% or less for "semisolid" media), usually after the other medium components have been added and dissolved in the water. Agar dissolves at approximately 100°C, and an agar-containing medium thus heated will not solidify until the temperature is brought down to about 43°C.

Once solidified, the medium will not melt until brought back up to about 100°C. Among the advantages of this interesting temperature-related property are the following: (1) The medium can be inoculated while in a liquid state at a low enough temperature (approx. 43-50°C) such that the cells will not die off, and (2) the medium, once solidified, will stay solid over a wide range of incubation conditions.

Two additional attributes of agar are its resistance to degradation by nearly all organisms and its relative clarity, permitting easy viewing of growth on or in the medium. One drawback to agar is the fact that it is very difficult, if not impossible, to purify it fully of trace impurities. Thus, when agar is added to a chemically-defined liquid medium, the medium must be considered complex. If an absolutely chemically-defined solid medium is required, silicon-based solidifying agents can be employed.

Previous to agar, potato slices and gelatin were utilized to form solid substrates upon which microbial colonies could be grown and studied. These materials were unacceptable for general use due to their ability to be broken down by a wide variety of microorganisms. Furthermore, gelatin liquefies in a warm room, and potato slices are opaque. In 1881, Fanny Eilshemius Hesse, a technician in the laboratory of Robert Koch in Germany, introduced the concept of agar to bacteriology, having used it for many years in the preparation of homemade jellies.

*Classification of Culture Media*

A classification of media based on their respective uses follows. Note that these categories can overlap. Furthermore, by now you should be using these terms correctly: Medium is always the singular form of the word, and media is always (and only!) the plural form.

- A MINIMAL MEDIUM is one which supplies only the minimal nutritional requirements of a particular organism. As an example, a typical, prototrophic strain of *E. coli* is able to synthesize all of its cell components from a simple solution containing several "mineral salts" plus glucose as the source of carbon and energy – such as the medium given on the previous page. Minimal media vary in composition according to the minimal nutritional requirements of the particular species under study. A minimal medium for a "fastidious" organism such as *Lactobacillus* may contain many growth factors such as vitamins and amino acids.

- Ar ALL-PURPOSE MEDIUM is rich in a wide variety of nutrients (including many growth factors) and will, therefore, support the growth of a wide range of bacteria. All-purpose media include Nutrient Agar, APT Agar, Plate Count Agar, Heart Infusion Agar, Brain Heart Infusion Agar and Penassay Agar.
- A SELECTIVE MEDIUM supports the growth of desired organisms while inhibiting the growth of many or most of the unwanted ones – either by purposely adding one or more selective agents which "poison" certain types of organisms or by including or deleting certain nutrients such that the desired organisms and few others are able to grow. Examples on how these things may be accomplished are as follows:
  - *MacConkey Agar:* This is an example of a medium where selective agents are added which directly suppress the growth of undesired organisms as much as possible. The particular selective agents chosen for this medium – bile salts and crystal violet – inhibit gram-positive bacteria, allowing the near-exclusive growth of gram-negative bacteria.
  - *Nitrogen-Free Broth:* Here the medium is made selective by the deletion of a required element; no nitrogen compounds are present. Therefore, the only organisms which can grow after inoculation into this medium are those which can utilize gaseous nitrogen ($N_2$) which diffuses in from the atmosphere. These are the nitrogen-fixing bacteria. While this medium does not utilize selective agents, it is still restrictive to an extensive number of various organisms.
  - *Succinate Broth:* In this example, a particular nutrient utilized by the desired organism – and few others – is included as the only carbon source. This medium is used for the enrichment of the purple non-sulfur photosynthetic bacteria; most other organisms tend not to metabolize succinate under the anaerobic conditions utilized. This is another example of a restrictive medium which does not contain selective agents.
- A DIFFERENTIAL MEDIUM is one which allows two or more different types of organisms to grow, but it contains dyes and/or other components upon which different organisms act in various ways to produce a variety of end products or effects,

often detected by variations in colour. These differences are often very apparent among colonies of a mixed culture growing in a petri dish. Pure cultures, growing in separate tubes of the same differential medium, may also be characterized and differentiated from one another according to a particular biochemical characteristic. Examples of differential media include the following:

- *MacConkey Agar*: This medium is used in plates. Organisms which ferment the lactose in the medium will lower the pH due to the production of acids. The pH indicator (neutral red) will turn red, and the colonies will consequently have a reddish appearance. Other colonies on the same plate which do not contain lactose-fermenting cells should appear whitish. (As this medium also appears in the above category, it is termed a selective-differential medium.)
- *Glucose Fermentation Broth:* This medium is used in tubes, usually with Durham tubes. As discussed, organisms which ferment the sugar (glucose) will cause the pH indicator to change colour upon production of acidic products. Additionally, if insoluble gas ($H_2$) is produced during fermentation, a bubble will be seen in the inverted Durham tube.
- Examples of differential media which do not involve pH-related reactions include Motility Medium (exploiting a morphological characteristic – i.e., production of flagella), Nutrient Gelatin, Starch Agar, Kligler Iron Agar (for $H_2S$ production) and Blood Agar.

## COMMONLY-USED CONSTITUENTS IN MICROBIOLOGICAL MEDIA

- *AGAR:* Agar is used as a solidifying agent in media. It is an impure polysaccharide gum obtained from certain marine algae. Agar dissolves and melts around 100°C and solidifies around 43°C. Generally agar itself is not used as a nutrient by microorganisms.
- *BODY FLUIDS:* Whole or defibrinated blood, plasma, serum or other body fluids are frequently added to culture media for the isolation and cultivation of many pathogens. Body fluids contribute many growth factors and/or substances which detoxify certain inhibitors.
- *BUFFERS:* These compounds are incorporated to maintain the optimum pH range of the organism. Substances like sodium and potassium phosphates and calcium carbonate prevent marked

changes in pH which otherwise would result from microbial production of organic acids or bases. Crude organic preparations such as peptones also act as buffers.

- *EXTRACTS:* Eucaryotic tissues (yeast, beef muscle, liver, brain, heart, etc.) are extracted by boiling and then concentrated to a paste or dried to a powder. These extracts are frequently used as a source of amino acids, vitamins and coenzymes, including many needed as growth factors by fastidious organisms. Trace elements and other minerals and usually some sugar are also present. (The term infusion refers to the aqueous extracts of these materials used for these purposes without being dried or otherwise concentrated, although "infusion" is sometimes used synonymously with "extract.")
- *PEPTONES:* These complex mixtures of organic and inorganic compounds are obtained by digestion of protein-containing tissues of animals and plants such as meat scraps, beef muscle, gelatin, milk protein (casein) and soybean meal. These materials are then dried down to a powder and made commercially available to microbiology laboratories. Peptones primarily contain peptides and single amino acids. Being crude digests of complex materials, they contain a great variety of other organic and inorganic materials, but they may be deficient in certain minerals and vitamins. Three examples of brand names of peptones are Tryptone (or Trypticase; a pancreatic digest of casein), Phytone (or Soytone; a papaic digest of soybean meal) and simply Peptone (a digest of beef muscle). Peptones are used frequently in conjunction with extracts for the cultivation of fastidious organisms, and a simple peptone solution will support the growth of many organisms. Peptone in a concentration of 0.1% is often used as a diluent.
- *pH INDICATORS:* An acid-base indicator is often added to differential media to detect changes in hydrogen ion concentration during the growth of an organism such as in Carbohydrate Fermentation Broth, Kligler Iron Agar, Simmons Citrate Agar, MacConkey Agar and Glucose O/F Medium. Brom-cresol purple, brom-thymol blue and phenol red are commonly used; for each of these, an acidic pH turns the indicator a yellow colour. A list of pH indicators is given here.
- *REDUCING AGENTS:* Certain chemicals may stimulate growth by reducing the oxidation-reduction potential in the environment. Cystine and thioglycollate are reducing agents often used for the cultivation of anaerobes and are found in the

Thioglycollate Medium utilized to determine oxygen relationships.

- *SELECTIVE AGENTS:* Antimicrobial agents such as crystal violet, bile salts, brilliant green, potassium tellurite, sodium azide and antibiotics can be employed in selective media to suppress or inhibit the growth of certain groups of microorganisms while allowing growth of desired organisms. These agents are usually bacteriostatic. A list of selective agents often used in media to inhibit gram-positive bacteria is given here.

Chapter 13

# Bacterial Identification

Through the early part of the twentieth century, there appeared to be a general feeling that the same battery of observations and tests could be used to characterize and identify any kind of bacterium. But as different, "exotic" types of bacteria were discovered, it was found that they would tend not to grow in the standard test media nor even in the usual conditions of incubation.

Obligate parasites and strict anaerobes were among the emerging groups of bacteria needing special methods for growth and characterization. By the 1930s, a standard descriptive chart was developed for uniformity in recording the characteristics of the "aerobic saprophytes" (which are equivalent to what we call the "commonly-found chemoheterotrophs" in our general courses today).

As we now know, a huge battery of tests done at once to identify an unknown organism would result in a lot of media and time being wasted dealing with irrelevant tests. (Time and media are money!) Thus we would like to proceed in stages, running those tests which are applicable to what basic knowledge we have about our unknown. That is, a very different set of tests would be run on a gram-negative rod compared to a gram-positive coccus.

There is no medium (differential or otherwise) that can possibly support the growth of all of the different species of bacteria. As an example, many different formulations exist for media to detect glucose fermentation, based on nutrient requirements of various groups of bacteria. Also, when running the standard test for oxygen relationship with Thioglycollate Medium, consider that (1) many organisms (including a lot of chemoheterotrophs!) cannot grow in this medium and (2) the medium does not allow for anaerobic growth which is due to phototrophy (more specifically, metabolism in the presence of light as performed by the non-oxygen-evolving photosynthetic bacteria) or anaerobic respiration (the use of alternate electron acceptors such as nitrate and sulfate). A discussion of this test is on our oxygen relationship page.

As genotypic characterization (determination of the DNA and RNA characteristics of our bacteria) is becoming more widely practiced, we may soon be back to one standard of characterizing and identifying bacteria. This time it will be universally applicable as all bacterial genera and species become uniformly defined according to genotypic uniqueness. We hope that the results of the phenotypic tests we run will correlate with the genotypic characteristics and bring about accurate and useful identification of our organisms.

A few commonly-found and easily-grown chemoheterotrophic genera are sorted out based on various "primary tests" which include the use of Glucose Fermentation Broth and O/F Medium. The benzidine test which has been used effectively in the Bacteriology 324 course tests for the presence of iron-porphyrin compounds such as cytochromes and the true catalase enzyme. Some organisms possess the enzyme cytochrome $a_3$ oxidase as part of the electron transport system in respiration; this enzyme is responsible for a positive reaction in the oxidase test where the dye tetramethyl-p- phenylenediamine is reduced to a purple compound.

## THE API-20E ENTERIC IDENTIFICATION SYSTEM

The API-20E test kit for the identification of enteric bacteria (bioMerieux, Inc., Hazelwoo d, MO) provides an easy way to inoculate and read tests relevant to members of the Family Enterobacteriaceae and associated organisms. A plastic strip holding twenty mini-test tubes is inoculated with a saline suspension of a pure culture (as per manufacturer's directions). This process also rehydrates the dessicated medium in each tube. A few tubes are completely filled (CIT, VP and GEL as seen in the photos below), and some tubes are overlaid with mineral oil such that anaerobic reactions can be carried out (ADH, LDC, ODC, $H_2S$, URE).

After incubation in a humidity chamber for 18-24 hours at 37°C, the colour reactions are read (some with the aid of added reagents), and the reactions (plus the oxidase reaction done separately) are converted to a seven-digit code. The code is fed into the manufacturer's database via touch-tone telephone, and the computer voice gives back the identification, usually as genus and species. The reliability of this system is very high, and one finds systems like these in heavy use in many food and clinical labs.

Discussion and illustration of the API-20E system here does not necessarily constitute any commercial endorsement of this product. It is shown in our laboratory courses as a prime example of a convenient multi-purpose testing method one may encounter out there in the "real world."

*In the following photos:*

- Note especially the colour reactions for amino acid decarboxylations (ADH through ODC) and carbohydrate fermentations (GLU through ARA).
  - The amino acids tested are (in order) arginine, lysine and ornithine. Decarboxylation is shown by an alkaline reaction (red colour of the particular pH indicator used).
  - The carbohydrates tested are glucose, mannitol, inositol, sorbitol, rhamnose, sucrose, melibiose, amygdalin and arabinose. Fermentation is shown by an acid reaction (yellow colour of indicator).
- Hydrogen sulfide production ($H_2S$) and gelatin hydrolysis (GEL) result in a black colour throughout the tube.
- A positive reaction for tryptophan deaminase (TDA) gives a deep brown colour with the addition of ferric chloride; positive results for this test correlate with positive phenylalanine and lysine deaminase reactions which are characteristic of *Proteus, Morganella* and *Providencia.*

*In the first set of reactions:*

- Culture "5B" (isolated from an early stage of sauerkraut fermentation) is identified as *Enterobacter agglomerans* which has been a convenient dumping ground for organisms now being reassigned to better-defined genera and species including the new genus *Pantoea.* This particular isolate produces reddish (lactose +), "pimply" colonies on MacConkey Agar which exude an extremely viscous slime as may be seen here; this appearance is certainly atypical of organisms identified as *E. agglomerans* or *Pantoea* in general.
- Culture "8P44" is identified as *Edwardsiella hoshinae.* The CDC had identified this culture (in 1988) as the ultra-rare Biogroup 1 of *Edwardsiella tarda* which may not be in the API-20E database. This system probably would not be able to differentiate between these two organisms.

## GENOTYPIC IDENTIFICATION

This page summarizes a few basic things regarding 16S ribosomal RNA gene analysis. With this comparative test, differences in the DNA base sequences between different organisms can be determined quantitatively, such that a phylogenetic tree can be constructed to illustrate probable evolutionary relatedness between the organisms.

The nucleotide base sequence of the gene which codes for 16S ribosomal RNA is becoming an important standard for the definition of bacterial species. Comparisons of the sequence between different species suggest the degree to which they are related to each other; a relatively greater or lesser difference between two species suggests a relatively earlier or later time in which they shared a common ancestor.

| culture no. | ONPG | ADH | LDC | ODC | CIT | H2S | URE | TDA | IND | VP | GEL | GLU | MAN | INO | SOR | RHA | SAC | MEL | AMY | ARA | identification |
|---|---|---|---|---|---|---|---|---|---|---|---|---|---|---|---|---|---|---|---|---|---|
| 8101 | + | – | + | + | – | – | – | – | + | – | – | + | + | – | + | + | + | + | – | + | *Escherichia coli* |
| 5B | + | – | – | – | + | – | – | – | – | + | – | + | + | + | + | + | – | – | + | + | *Enterobacter agglomerans* |
| 8P44 | – | – | + | + | – | + | – | – | + | – | – | + | + | – | – | – | + | – | – | + | *Edwardsiella hoshinae* |

A comparison between eleven species of gram-negative bacteria is illustrated on a separate sequence comparison page, where the sequences are aligned such that similarities and differences can be readily seen when one scrolls to the right or left. Gaps and insertions of nucleic acid bases (the result of "frame-shift" mutations occuring over eons of time as the organisms diverge from common ancestors) which affect long stretches of DNA have to be taken into account for a proper alignment.

In an earlier version of the above-mentioned sequence comparison page, when only four species were compared with each other, a relatively short segment stood out as appearing to be "frame-shifted" when comparing Pseudomonas fluorescens with a group of three enterics. This situation is shown as follows with the nucleotide bases of the segment in question shown in red.

| culture no. | ONPG | ADH | LDC | ODC | CIT | H2S | URE | TDA | IND | VP | GEL | GLU | MAN | INO | SOR | RHA | SAC | MEL | AMY | ARA | identification |
|---|---|---|---|---|---|---|---|---|---|---|---|---|---|---|---|---|---|---|---|---|---|
| 8101 | + | − | + | + | − | − | − | − | + | − | − | + | + | − | + | + | + | + | − | + | *Escherichia coli* |
| 5B | + | − | − | − | + | − | − | − | − | + | − | + | + | + | + | + | − | − | + | + | *Enterobacter agglomerans* |
| 8P44 | − | − | + | + | − | + | − | − | + | − | − | + | + | − | − | − | + | − | − | + | *Edwardsiella hoshinae* |

| | |
|---|---|
| *Pseudomonas fluorescens* | ...gctaataccgcatacgtcctacgggagaaagcagggg... |
| Our new organism, shown below as "AH" | ...gctaataccgcataacgtcgcaagaccaaagcggggg... |
| *Budvicia aquatica* | ...gctaataccgcgtaacgtcgaaagaccaaagcggggg... |
| *Edwardsiella tarda* | ...gctaataccgcataacgtcgcaagaccaaagtggggg... |

One can surmise that a frame-shift mutation – if the bases are not misplaced to the extent that the mutation becomes silent or lethal – could be a "cheap" way to effect a major change in the genotype and subsequent phenotype – perhaps resulting in one of those infamous "leaps" in evolution one hears conjectured about from time to time. Even though the specific sequence within a shifted segment of DNA may not be changed, the shift will result in the nucleotide bases being re-grouped into different triplet codes and read accordingly, and the resulting gene may produce a vastly different protein which can change the appearance or function of a cell to a significant extent. So, when sequences between two species are compared, the organisms may appear to be a bit more closely related if these relatively short frame-shifted segments were taken into consideration. (With long stretches of DNA, one would not expect independant genes farther along the chromosome to be affected.)

When a 1308-base stretch of that part of the chromosome which codes for 16S ribosomal RNA was lined up and analyzed ("manually" when I had a little time to kill) to find the extent to which the above four organisms differed from each other, the per cent difference between any two organisms was determined, and the results are summarized as follows:

| | PF | AH | BA | ET |
|---|---|---|---|---|
| PF | | | | |
| AH | 14.8* | | | |
| BA | 14.5 | 3.2 | | |
| ET | 14.9 | 4.3 | 5.0 | |

* An example: The same bases appear in the same sequence, position by position, for each of the two organisms except for 14.8% of the time.

With the per cent differences used to denote probable evolutionary distances between the organisms, a phylogenetic tree was roughed out to illustrate the relationships. The distances between any two organisms, when read along the horizontal lines, corresponds closely to the per cent differences. (The bar at the bottom signifies approximately 1% base difference.)

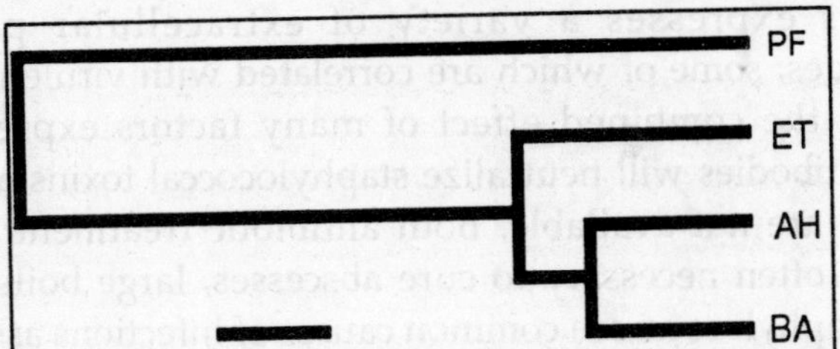


Databases of various gene sequences are found. Genbank's database was used as the source of the above sequences. And rather than having to line up the sequences and determine the differences manually, a set of programs to analyse sequence data and plot trees are available.

# Chapter 14

# Staphylococcus

Bacteria in the genus Staphylococcus are pathogens of man and other mammals. Traditionally they were divided into two groups on the basis of their ability to clot blood plasma (the coagulase reaction). The coagulase-positive staphylococci constitute the most pathogenic species *S aureus*. The coagulase-negative staphylococci (CNS) are now known to comprise over 30 other species.

The CNS are common commensals of skin, although some species can cause infections. It is now obvious that the division of staphylococci into coagulase positive and negative is artificial and indeed, misleading in some cases. Coagulase is a marker for *S aureus* but there is no direct evidence that it is a virulence factor. Also, some natural isolates of *S aureus* are defective in coagulase. Nevertheless, the term is still in widespread use among clinical microbiologists.

*S aureus* expresses a variety of extracellular proteins and polysaccharides, some of which are correlated with virulence. Virulence results from the combined effect of many factors expressed during infection. Antibodies will neutralize staphylococcal toxins and enzymes, but vaccines are not available. Both antibiotic treatment and surgical drainage are often necessary to cure abscesses, large boils and wound infections. Staphylococci are common causes of infections associated with indwelling medical devices. These are difficult to treat with antibiotics alone and often require removal of the device. Some strains that infect hospitalized patients are resistant to most of the antibiotics used to treat infections, vancomycin being the only remaining drug to which resistance has not developed.

## TAXONOMY

DNA-ribosomal RNA (rRNA) hybridization and comparative oligonucleotide analysis of 16S rRNA has demonstrated that staphylococci form a coherent group at the genus level. This group occurs within the broad Bacillus-Lactobacillus-Streptococcus cluster defining Gram-positive bacteria with a low G + C content of DNA.

At least 30 species of staphylococci have been recognized by biochemical analysis and in particular by DNA-DNA hybridization. Eleven of these can be isolated from humans as commensals. *S aureus* (nares) and *S epidermidis* (nares, skin) are common commensals and also have the greatest pathogenic potential. *S saprophyticus* (skin, occasionally) is also a common cause of urinary tract infection. *S haemolyticus, S simulans, S cohnii, S warneri* and *S lugdunensis* can also cause infections in man.

## IDENTIFICATION OF STAPHYLOCOCCI IN THE CLINICAL LABORATORY

*Structure*

Staphylococci are Gram-positive cocci about 0.5 - 1.0 µm in diameter. They grow in clusters, pairs and occasionally in short chains. The clusters arise because staphylococci divide in two planes. The configuration of the cocci helps to distinguish micrococci and staphylococci from streptococci, which usually grow in chains. Observations must be made on cultures grown in broth, because streptococci grown on solid medium may appear as clumps. Several fields should be examined before deciding whether clumps or chains are present.

*Catalase Test*

The catalase test is important in distinguishing streptococci (catalase-negative) staphylococci which are catalase positive. The test is performed by flooding an agar slant or broth culture with several drops of 3% hydrogen peroxide. Catalase-positive cultures bubble at once. The test should not be done on blood agar because blood itself will produce bubbles.

## ISOLATION AND IDENTIFICATION

The presence of staphylococci in a lesion might first be suspected after examination of a direct Gram stain. However, small numbers of bacteria in blood preclude microscopic examination and require culturing first.

The organism is isolated by streaking material from the clinical specimen (or from a blood culture) onto solid media such as blood agar, tryptic soy agar or heart infusion agar. Specimens likely to be contaminated with other microorganisms can be plated on mannitol salt agar containing 7.5% sodium chloride, which allows the halo-tolerant staphylococci to grow. Ideally a Gram stain of the colony should be performed and tests made for catalase and coagulase production, allowing the coagulase-positive *S aureus* to be identified quickly. Another very useful test for *S aureus* is the production of thermostable deoxyribonuclease. *S aureus* can be confirmed by testing colonies for agglutination with latex particles coated with immunoglobulin G and fibrinogen which bind protein A and the clumping factor, respectively, on the bacterial cell surface. These are available from commercial suppliers (e.g., Staphaurex). The most recent latex test (Pastaurex) incorporates monoclonal antibodies to serotype 5 and 8 capsular polysaccharide in order to reduce the number of false negatives. (Some recent clinical isolates of *S aureus* lack production of coagulase and/or clumping factor, which can make identification difficult.)

The association of *S epidermidis* (and to a lesser extent of other coagulase-negative staphylococci) with nosocomial infections associated with indwelling devices means that isolation of these bacteria from blood is likely to be important and not due to chance contamination, particularly if successive blood cultures are positive. Nowadays, identification of *S epidermidis* and other species of *Staphylococcus* is performed using commercial biotype identification kits, such as API Staph Ident, API Staph-Trac, Vitek GPI Card and Microscan Pos Combo. These comprise preformed strips containing test substrates.

## EPIDEMIOLOGY OF *STAPHYLOCOCCUS AUREUS* INFECTIONS

Because *S aureus* is a major cause of nosocomial and community-acquired infections, it is necessary to determine the relatedness of isolates collected during the investigation of an outbreak. Typing systems must be reproducible, discriminatory, and easy to interpret and to use. The traditional method for typing *S aureus* is phage-typing. This method is based on a phenotypic marker with poor reproducibility. Also, it does not type many isolates (20% in a recent survey at the Centre for Disease Control and Prevention), and it requires maintenance of a large number of phage stocks and propagating strains and consequently can be performed only by specialist reference laboratories.

Many molecular typing methods have been applied to the epidemiological analysis of *S aureus*, in particular, of methicillin-resistant strains (MRSA). Plasmid analysis has been used extensively with success, but suffers the disadvantage that plasmids can easily be lost and acquired and are thus inherently unreliable. Methods designed to recognize restriction fragment length polymorphisms (RFLP) using a variety of gene probes, including rRNA genes (ribotyping), have had limited success in the epidemiology of MRSA. In this technique the choice of restriction enzyme used to cleave the genomic DNA, as well as the probes, is crucial. Random primer PCR offers potential for discriminating between strains but a suitable primer has yet to be identified for *S aureus*. The method currently regarded as the most reliable is pulsed field gel electrophoresis, where genomic DNA is cut with a restriction enzyme that generates large fragments of 50-700 kb.

## CLINICAL MANIFESTATIONS OF *S AUREUS*

*S aureus* is notorious for causing boils, furuncles, styes, impetigo and other superficial skin infections in humans. It may also cause more serious infections, particularly in persons debilitated by chronic illness, traumatic injury, burns or immunosuppression. These infections include pneumonia, deep abscesses, osteomyelitis, endocarditis, phlebitis, mastitis and meningitis, and are often associated with hospitalized patients rather than healthy individuals in the community. *S aureus* and *S epidermidis* are common causes of infections associated with indwelling devices such as joint prostheses, cardiovascular devices and artificial heart valves.

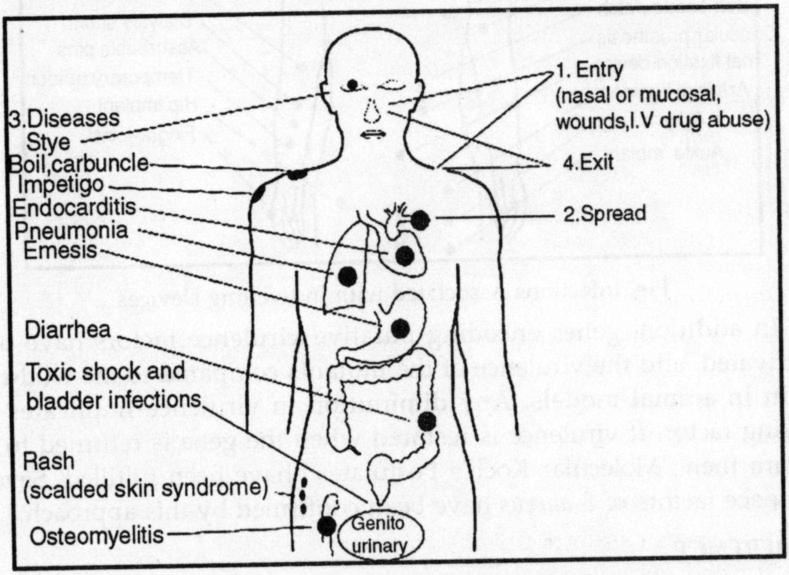


**Fig.** Pathogenesis of Staphylococcal Infections

## PATHOGENESIS OF *S AUREUS* INFECTIONS

*Saureus* expresses many cell surface-associated and extracellular proteins that are potential virulence factors. For the majority of diseases caused by this organism, pathogenesis is multifactorial. Thus it is difficult to determine precisely the role of any given factor. This also reflects the inadequacies of many animal models for staphylococcal diseases.

However, there are correlations between strains isolated from particular diseases and expression of particular factors, which suggests their importance in pathogenesis. With some toxins, symptoms of a human disease can be reproduced in animals with pure proteins. The application of molecular biology has led to recent advances in the understanding of pathogenesis of staphylococcal diseases. Genes encoding potential virulence factors have been cloned and sequenced and proteins purified. This has facilitated studies at the molecular level on their modes of action, both in in vitro and in model systems.

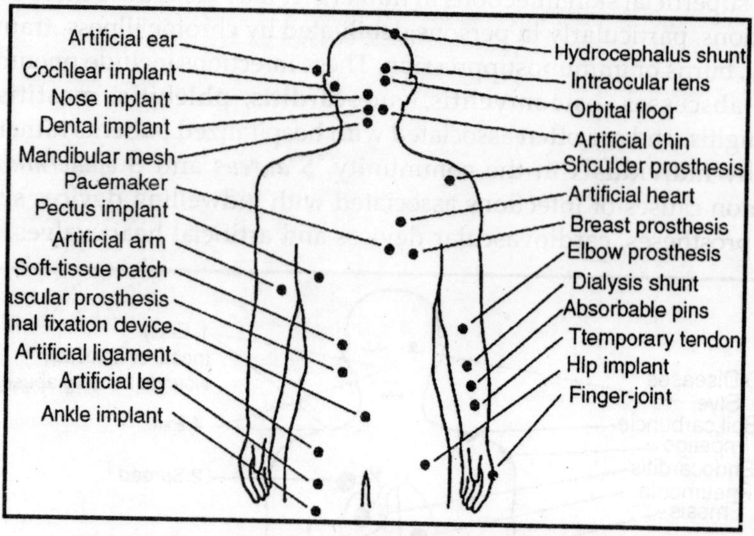


**Fig.** Infections Associated with Indwelling Devices

In addition, genes encoding putative virulence factors have been inactivated, and the virulence of the mutants compared to the wild-type strain in animal models. Any diminution in virulence implicates the missing factor. If virulence is restored when the gene is returned to the mutant then "Molecular Koch's Postulates" have been fulfilled. Several virulence factors of *S aureus* have been confirmed by this approach.

## ADHERENCE

In order to initiate infection the pathogen must gain access to the host and attach to host cells or tissues.

### *Saureus* Adheres to Host Proteins

*S aureus* cells express on their surface proteins that promote attachment to host proteins such as laminin and fibronectin that form part of the extracellular matrix. Fibronectin is present on epithelial and endothelial surfaces as well as being a component of blood clots. In addition, most strains express a fibrinogen/fibrin binding protein (the clumping factor) which promotes attachment to blood clots and traumatized tissue. Most strains of *S aureus* express fibronectin and fibrinogen-binding proteins.

The receptor which promotes attachment to collagen is particularly associated with strains that cause osteomyelitis and septic arthritis.

Interaction with collagen may also be important in promoting bacterial attachment to damaged tissue where the underlying layers have been exposed.

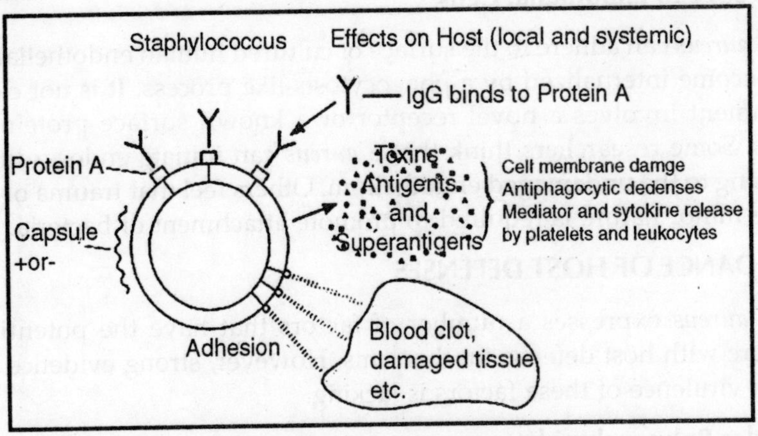


**Fig.** Summary of Virulence Factors of *Staphylococcus Aureus.*

Evidence that these staphylococcal matrix-binding proteins are virulence factors has come from studying defective mutants in vitro adherence assays and in experimental infections. Mutants defective in binding to fibronectin and to fibrinogen have reduced virulence in a rat model for endocarditis, suggesting that bacterial attachment to the sterile vegetations caused by damaging the endothelial surface of the heart valve is promoted by fibronectin and fibrinogen.

Similarly, mutants lacking the collagen-binding protein have reduced virulence in a mouse model for septic arthritis. Furthermore, the soluble ligand-binding domain of the fibrinogen, fibronectin and collagen-binding proteins expressed by recombinant methods strongly blocks interactions of bacterial cells with the corresponding host protein.

### Role of Adherence in Infections Associated with Medical Devices

Infections associated with indwelling medical devices ranging from simple intravenous catheters to prosthetic joints and replacement heart valves can be caused by *S aureus* and *S epidermidis*. Very shortly after biomaterial is implanted in the human body it becomes coated with a complex mixture of host proteins and platelets.

In one model system involving short-term contact between biomaterial and blood, fibrinogen was shown to be the dominant component and was primarily responsible for adherence of *S aureus* in subsequent in vitro assays. In contrast, with material that has been in the body for longer periods (e.g., human intravenous catheters) the fibrinogen is degraded and no longer promotes bacterial attachment. Instead, fibronectin, which remains intact, becomes the predominant ligand promoting attachment.

### Adherence to Endothelial Cells

*S aureus* can adhere to the surface of cultured human endothelial cells and become internalized by a phagocytosis-like process. It is not clear if attachment involves a novel receptor or a known surface protein of *S aureus*. Some researchers think that *S aureus* can initiate endocarditis by attaching to the undamaged endothelium. Others feel that trauma of even a very minor nature is required to promote attachment of bacteria.

## AVOIDANCE OF HOST DEFENSES

*S aureus* expresses a number of factors that have the potential to interfere with host defense mechanisms. However, strong evidence for a role in virulence of these factors is lacking.

### Capsular Polysaccharide

The majority of clinical isolates of *S aureus* express a surface polysaccharide of either serotype 5 or 8. This has been called a microcapsule because it can be visualized only by electron microscopy after antibody labeling, unlike the copious capsules of other bacteria which are visualized by light microscopy. *S aureus* isolated from infections expresses high levels of polysaccharide but rapidly loses it upon laboratory subculture. The function of the capsule is not clear. It may impede phagocytosis, but in in vitro tests this was only demonstrated in the absence of complement. Conversely, comparing wild-type and a capsule defective mutant strain in an endocarditis model suggested that polysaccharide expression actually impeded colonization of damaged heart valves, perhaps by masking adhesins.

### Protein A

Protein A is a surface protein of *S aureus* which binds immunoglobulin G molecules by the Fc region. In serum, bacteria will bind IgG molecules the wrong way round by this non-immune mechanism. In principle this will disrupt opsonization and phagocytosis. Indeed mutants of *S aureus* lacking protein A are more efficiently phagocytozed in vitro, and studies with mutants in infection models suggest that protein A enhances virulence.

### Leukocidin

*S aureus* can express a toxin that specifically acts on polymorphonuclear leukocytes. Phagocytosis is an important defense against staphylococcal infection so leukocidin should be a virulence factor.

### Damage to the Host

*S aureus* can express several different types of protein toxins which are probably responsible for symptoms during infections. Some damage the membranes of erythrocytes, causing hemolysis; but it is unlikely that hemolysis is relevant in vivo. The leukocidin causes membrane damage to leukocytes and is not hemolytic. Systemic release of a-toxin causes septic shock, while enterotoxins and TSST-1 cause toxic shock.

## MEMBRANE DAMAGING TOXINS

### α-toxin

The best characterized and most potent membrane-damaging toxin of *S aureus* is α-toxin. It is expressed as a monomer that binds to the membrane of susceptible cells. Subunits then oligomerize to form hexameric rings with a central pore through which cellular contents leak.

Susceptible cells have a specific receptor for α-toxin which allows low concentrations of toxin to bind, causing small pores through which monovalent cations can pass. At higher concentrations, the toxin reacts non-specifically with membrane lipids, causing larger pores through which divalent cations and small molecules can pass. However, it is doubtful if this is relevant under normal physiological conditions.

In humans, platelets and monocytes are particularly sensitive to α-toxin. They carry high affinity sites which allow toxin to bind at concentrations that are physiologically relevant. A complex series of secondary reactions ensue, causing release of eicosanoids and cytokines which trigger production of inflammatory mediators. These events cause the symptoms of septic shock that occur during severe infections caused by *Saureus*.

The notion that α-toxin is a major virulence factor of *S aureus* is supported by studies with the purified toxin in animals and in organ culture. Also, mutants lacking a-toxin are less virulent in a variety of animal infection models.

### β-toxin

β-toxin is a sphingomyelinase which damages membranes rich in this lipid. The classical test for β-toxin is lysis of sheep erythrocytes. The majority of human isolates of *S aureus* do not express β-toxin. A lysogenic bacteriophage is inserted into the gene that encodes the toxin. This phenomenon is called negative phage conversion. Some of the phages that inactivate the β-toxin gene carry the determinant for an enterotoxin and staphylokinase. In contrast the majority of isolates from bovine mastitis express β-toxin, suggesting that the toxin is important in the pathogenesis of mastitis. This is supported by the fact that β-toxin-deficient mutants have reduced virulence in a mouse model for mastitis.

### δ-toxin

The δ-toxin is a very small peptide toxin produced by most strains of *S aureus*. It is also produced by *S epidermidis* and *S lugdunensis*. The role of δ-toxin in disease is unknown.

### γ-toxin and Leukocidin

The γ-toxin and the leukocidins are two-component protein toxins that damage membranes of susceptible cells. The proteins are expressed separately but act together to damage membranes. There is no evidence that they form multimers prior to insertion into membranes. The γ-toxin locus expresses three proteins. The B and C components form a leukotoxin with poor hemolytic activity, whereas the A and B components are hemolytic and weakly leukotoxic.

The classical Panton and Valentine (PV) leukocidin is distinct from the leukotoxin expressed by the γ-toxin locus. It has potent leukotoxicity and, in contrast to δ-toxin, is non-hemolytic. Only a small fraction of *S aureus* isolates (2% in one survey) express the PV leukocidin, whereas 90% of those isolated from severe dermonecrotic lesions express this toxin. This suggests that PV leukocidin is an important factor in necrotizing skin infections.

PV-leukocidin causes dermonecrosis when injected subcutaneously in rabbits. Furthermore, at a concentration below that causing membrane damage, the toxin releases inflammatory mediators from human neutrophils, leading to degranulation. This could account for the histology of dermonecrotic infections (vasodilation, infiltration and central necrosis).

### Superantigens: Enterotoxins and Toxic shock Syndrome Toxin

*Saureus* can express two different types of toxin with superantigen activity, enterotoxins, of which there are six serotypes (A, B, C, D, E and G) and toxic shock syndrome toxin (TSST-1). Enterotoxins cause diarrhea and vomiting when ingested and are responsible for staphylococcal food poisoning. When expressed systemically, enterotoxins can cause toxic shock syndrome (TSS) - indeed enterotoxins B and C cause 50% of non-menstrual TSS. TSST-1 is very weakly related to enterotoxins and does not have emetic activity. TSST-1 is responsible for 75% of TSS, including all menstrual cases. TSS can occur as a sequel to any staphylococcal infection if an enterotoxin or TSST-1 is released systemically and the host lacks appropriate neutralizing antibodies. Tampon-associated TSS is not a true infection, being caused by growth of *S aureus* in a tampon and absorption of the toxin into the blood stream. TSS came to prominence with the introduction of super-absorbent tampons; and although the number of such cases has decreased dramatically, they still occur despite withdrawal of certain types of tampons from the market.

Superantigens stimulate T cells non-specifically without normal antigenic recognition. Up to one in five T cells may be activated, whereas only 1 in 10,000 are stimulated during antigen presentation. Cytokines are released in large amounts, causing the symptoms of TSS. Superantigens bind directly to class II major histocompatibility complexes of antigen-presenting cells outside the conventional antigen-binding grove. This complex recognizes only the Vb element of the T cell receptor. Thus any T cell with the appropriate Vb element can be stimulated, whereas normally antigen specificity is also required in binding.

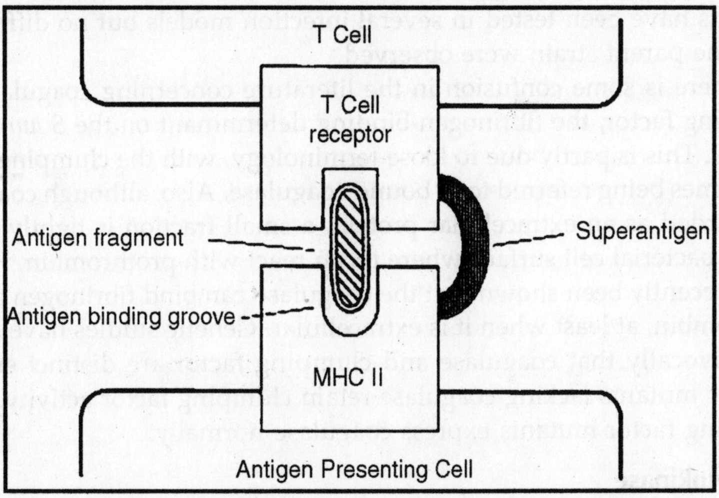


**Fig.** Superantigens and the Non-specific Stimulation of T Cells.

## Epidermolytic (exfoliative) toxin (ET)

This toxin causes the scalded skin syndrome in neonates, with widespread blistering and loss of the epidermis. There are two antigenically distinct forms of the toxin, ETA and ETB. There is evidence that these toxins have protease activity. Both toxins have a sequence similarity with the *S aureus* serine protease, and the three most important amino acids in the active site of the protease are conserved. Furthermore, changing the active site of serine to a glycine completely eliminated toxin activity. However, ETs do not have discernible proteolytic activity but they do have esterase activity. It is not clear how the latter causes epidermal splitting. It is possible that the toxins target a very specific protein which is involved in maintaining the integrity of the epidermis.

## OTHER EXTRACELLULAR PROTEINS

### Coagulase

Coagulase is not an enzyme. It is an extracellular protein which binds to prothrombin in the host to form a complex called staphylothrombin. The protease activity characteristic of thrombin is activated in the complex, resulting in the conversion of fibrinogen to fibrin. This is the basis of the tube coagulase test, in which a clot is formed in plasma after incubation with the *S aureus* broth-culture supernatant. Coagulase is a traditional marker for identifying *S aureus* in the clinical microbiology laboratory. However, there is no evidence that it is a virulence factor, although it is reasonable to speculate that the bacteria could protect themselves from host defenses by causing localized clotting. Notably, coagulase deficient mutants have been tested in several infection models but no differences from the parent strain were observed.

There is some confusion in the literature concerning coagulase and clumping factor, the fibrinogen-binding determinant on the *S aureus* cell surface. This is partly due to loose terminology, with the clumping factor sometimes being referred to as bound coagulase. Also, although coagulase is regarded as an extracellular protein, a small fraction is tightly bound on the bacterial cell surface where it can react with prothrombin. Finally, it has recently been shown that the coagulase can bind fibrinogen as well as thrombin, at least when it is extracellular. Genetic studies have shown unequivocally that coagulase and clumping factor are distinct entities. Specific mutants lacking coagulase retain clumping factor activity, while clumping factor mutants express coagulase normally.

### Staphylokinase

Many strains of *S aureus* express a plasminogen activator called staphylokinase. The genetic determinant is associated with lysogenic bacteriophages. A complex formed between staphylokinase and plasminogen activates plasmin-like proteolytic activity which causes dissolution of fibrin clots. The mechanism is identical to streptokinase, which is used in medicine to treat patients suffering from coronary thrombosis. As with coagulase there is no evidence that staphylokinase is a virulence factor, although it seems reasonable to imagine that localized fibrinolysis might aid in bacterial spreading.

### Enzymes

*S aureus* can express proteases, a lipase, a deoxyribonuclease (DNase) and a fatty acid modifying enzyme (FAME). The first three probably provide nutrients for the bacteria, and it is unlikely that they have anything but a minor role in pathogenesis. However, the FAME enzyme may be important in abscesses, where it could modify anti-bacterial lipids and prolong bacterial survival. The thermostable DNase is an important diagnostic test for identification of *S aureus*.

## Coagulase Negative Staphylococci

Staphylococci other than *S aureus* can cause infections in man. *S epidermidis* is the most important coagulase-negative staphylococcus (CNS) species and is the major cause of infections associated with prosthetic devices and catheters. CNS also cause peritonitis in patients receiving continuous ambulatory peritoneal dialysis and endocarditis in those with prosthetic valves. These infections are not usually nosocomially acquired. Other species such as *S haemolyticus, S warneri, S hominis, S capitis, S intermedius, S schleiferi* and *S simulans* are infrequent pathogens. *S lugdunesis* is a newly recognized species. It is probably more pathogenic than are other CNS species, with cases of endocarditis and other infections being reported. It is likely that the incidence of infections caused by these organisms is underestimated because of difficulties in identification.

Diagnosis of CNS infections is difficult. Infections are often indolent and chronic with few obvious symptoms. This is due to the smaller array of virulence factors and toxins compared to those in the case of *S aureus*. *S epidermidis* is a skin commensal and is one of the most common contaminants of samples sent to the diagnostic laboratory, while *S lugdunensis* is often confused with *S aureus*. Precise identification of CNS species requires the use of expensive test kits, such as the API-Staph.

In contrast to *S aureus,* little is known about mechanisms of pathogenesis of *S epidermidis* infections. Adherence is obviously a crucial step in the initiation of foreign body infections. Much research has been done on the interaction between *S epidermidis* and plastic material used in implants, and a polysaccharide adhesion (PS/A) has been identified. Mutants lacking PS/A are less virulent in an animal model for foreign body infection, and immunization with purified PS/A is protective. Bacteria-plastic interactions are probably important in colonization of catheters through the point of entry. However, host proteins are quickly deposited on implants. *S epidermidis* does not bind to fibrinogen but most isolates bind fibronectin, albeit less avidly than *S aureus*. However, it is not known if a protein analogous to the fibronectin binding protein of *S aureus* is involved.

A characteristic of clinical isolates of *S epidermidis* is the production of "slime." This is a controversial topic. Some feel that slime is an in vitro manifestation of the ability to form a biofilm in vivo, for example on the surface of a prosthetic device, and is thus a virulence marker. In vitro, slime is formed during growth in broth as a biofilm on the surface of the growth vessel. The composition of this slime is probably influenced by the growth medium. One study with defined medium showed that the slime was predominantly secreted teichoic acid, a polymer normally found in the cell wall of staphylococci. Some polysaccharides in slime from bacteria grown on solid medium are derived from the agar.

## RESISTANCE OF STAPHYLOCOCCI TO ANTIMICROBIAL DRUGS

Hospital strains of *S aureus* are often resistant to many different antibiotics. Indeed strains resistant to all clinically useful drugs, apart from the glycopeptides vancomycin and teicoplanin, have been described. The term MRSA refers to methicillin resistance and most methicillin-resistant strains are also multiply resistant. Plasmid-associated vancomycin resistance has been detected in some enterococci and the resistance determinant has been transferred from enterococci to *S aureus* in the laboratory and may occur naturally. *S epidermidis* nosocomial isolates are also often resistant to several antibiotics including methicillin. In addition, *S aureus* expresses resistance to antiseptics and disinfectants, such as quaternary ammonium compounds, which may aid its survival in the hospital environment.

Since the beginning of the antibiotic era *S aureus* has responded to the introduction of new drugs by rapidly acquiring resistance by a variety of genetic mechanisms including (1) acquisition of extrachromosomal plasmids or additional genetic information in the chromosome via transposons or other types of DNA insertion and (2) by mutations in chromosomal genes.

Many plasmid-encoded determinants have recently become inserted into the chromosome at a site associated with the methicillin resistance determinant. There may be an advantage to the organism having resistance determinants in the chromosome because they will be more stable.There are essentially four mechanisms of resistance to antibiotics in bacteria: (1) enzymatic inactivation of the drug, (2) alterations to the drug target to prevent binding, (3) accelerated drug efflux to prevent toxic concentrations accumulating in the cell, and (4) a by-pass mechanism whereby an alternative drug-resistant version of the target is expressed.

**Table. Antimicrobial Resistance**

| *Antimicrobial* | *Resistance Mechanism* | *Genetic Basis* |
|---|---|---|
| Penicillin | β-lactamase. Enzymatic inactivation of penicillin | Plasmid |

| | | |
|---|---|---|
| Methicillin | Expression of new penicillin-rosistant penicillin-binding protein. Bypass | Novel chromosimal locus acquired from unknown source |
| Tetracycline | 1. Efflux from cell | Plasmid |
| | 2. Modification of ribosome | Novel chromosomal locus acquired from unknown source |
| Chloramphenicol | Enzymatic inactivation | Plasmid |
| Erythromycin | Enzymatic modification of ribosomal RNA. Prevents | Plasmid |
| | drug binding to ribosome | Transposon in chromosome |
| Streptomycin | 1. Mutation in ribosomal protein, Prevents drug binding | Mutation in chromosomal gene encoding drug target |
| Kanamycin | Enzymatic inactivation | Plasmid |
| Gentamicin | | Transposon in chromosome |
| Trimethoprim | Alternative dihydofolate reductase. Bypass | Plasmid |
| Mupirocin | Alternative isoleucyl tRNA synthase. Bypass | Plasmid |
| Pluoroquinolones | 1. Altered DNA gyrase | Multation in chromosomal gene encoding drug target |
| | 2. Efflux | Mutation increases expression of natural afflux mechanism |
| Antiseptics | Efflux | Plasmid |

## FUTURE PROSPECTS

### Antimicrobial Drugs

Ever since the first use of penicillin, *S aureus* has shown a remarkable ability to adapt. Resistance has developed to new drugs within a short time of their introduction. Some strains are now resistant to most conventional antibiotics. It is worrisome that there do not seem to be any new antibiotics on the horizon. Any recent developments have been modifications to existing drugs.

The original strategy used by the pharmaceutical industry to find antimicrobial drugs was to screen natural products and synthetic chemicals for antimicrobial activity. The mechanism of action was then investigated.

New approaches are being adopted to find the next generation of antimicrobials. Potential targets such as enzymes involved in an essential function (e.g., in cell division) are identified based on knowledge of bacterial physiology and metabolism. Screening methods are then developed to identify inhibitors of a specific target molecule. In addition, with detailed molecular knowledge of the target molecule, specific inhibitors can be designed.

### Vaccines and New Approaches to Combatting Nosocomial Infections

No vaccine is currently available to combat staphylococcal infections. There may now be a case for considering methods to prevent disease, particularly in hospitalized patients.

Hyperimmune serum from human volunteer donors or humanized monoclonal antibodies directed towards surface components (e.g., capsular polysaccharide or surface protein adhesions) could both prevent bacterial adherence and also promote phagocytosis of bacterial cells. Indeed a prototype vaccine based on capsular polysaccharide from *S aureus* has been administered to volunteers to raise hyperimmune serum, which could be given to patients in hospital before surgery. A vaccine based on fibronectin binding protein induces protective immunity against mastitis in cattle and might also be used as a vaccine in humans.

When the molecular basis of the interactions between the bacterial surface proteins and the host matrix protein ligands are known it might be possible to design compounds that block the interactions and thus prevent bacterial colonization. These could be administered systemically or topically.

# Chapter 15

# Streptococcus

The genus *Streptococcus*, a heterogeneous group of Gram-positive bacteria, has broad significance in medicine and industry. Various streptococci are important ecologically as part of the normal microbial flora of animals and humans; some can also cause diseases that range from subacute to acute or even chronic. Among the significant human diseases attributable to streptococci are scarlet fever, rheumatic heart disease, glomerulonephritis, and pneumococcal pneumonia. Streptococci are essential in industrial and dairy processes and as indicators of pollution.

The nomenclature for streptococci, especially the nomenclature in medical use, has been based largely on serogroup identification of cell wall components rather than on species names. For several decades, interest has focused on two major species that cause severe infections: *S pyogenes* (group A streptococci) and *S pneumoniae* (pneumococci). In 1984, two members were assigned a new genus - the group D enterococcal species (which account for 98% of human enterococcal infections) became *Enterococcus faecalis* (the majority of human clinical isolates) and *E faecium* (associated with a remarkable capacity for antibiotic resistance).

In recent years, increasing attention has been given to other streptococcal species, partly because innovations in serogrouping methods have led to advances in understanding the pathogenetic and epidemiologic significance of these species. A variety of cell-associated and extracellular products are produced by streptococci, but their cause-effect relationship with pathogenesis has not been defined. Some of the other medically important streptococci are *S agalactiae* (group B), an etiologic agent of neonatal disease; *E faecalis* (group D), a major cause of endocarditis, and the viridans streptococci.

Particularly for the viridans streptococci, taxonomy and nomenclature are not yet fully reliable or consistent. Important members of the viridans streptococci, normal commensals, include *S mutans* and *S sanguis* (involved in dental caries), *S mitis* (associated with bacteremia, meningitis, periodontal disease and pneumonia), and *"S milleri"* (associated with suppurative infections in children and adults). There remains persistent taxonomic confusion regarding "*S milleri.*" These and other streptococci of medical importance are listed in Table by serogroup designation, normal ecologic niche, and associated disease.

## CLINICAL MANIFESTATIONS

In humans, diseases associated with the streptococci occur chiefly in the respiratory tract, bloodstream, or as skin infections. Human disease is most commonly associated with Group A streptococci. Acute group A streptococcal disease is most often a respiratory infection (pharyngitis or tonsillitis) or a skin infection (pyoderma). Also medically significant are the late immunologic sequelae, not directly attributable to dissemination of bacteria, of group A infections (rheumatic fever following respiratory infection and glomerulonephritis following respiratory or skin infection) which remain a major worldwide health concern.

Much effort is being directed toward clarifying the risk and mechanisms of these sequelae and identifying rheumatogenic and nephritogenic strains. *S pneumoniae* remains a primary cause of serious focal and systemic infections, the first most common cause of community acquired pneumonia in the United States and of fatal bacterial pneumonia in developing countries. Hemorrhagic shock in association with *S pneumoniae* sepsis in previously healthy children has been reported recently in the United States. Of major biologic importance is a renewed interest in safe and effective streptococcal vaccines.

## STRUCTURE

Both *S pyogenes* and *S pneumoniae* are Gram-positive cocci, nonmotile, and nonsporulating; they usually require complex culture media. *S pyogenes* characteristically is a round-to-ovoid coccus 0.6-1.0 μm in diameter. They divide in one plane and thus occur in pairs, or (especially in liquid media or clinical material) in chains of varying lengths. S *pneumoniae* appears as a 0.5-1.25 μm diplococcus, typically described as lancet-shaped but sometimes difficult to distinguish morphologically from other streptococci. Streptococcal cultures older than the logarithmic phase, which is the most active growth period of a culture, may lose their Gram-positive staining characteristics.

Streptococci divide in a single plane and tend not to separate, causing chain formation. Capsules are antiphagocytic.

Unlike *Staphylococcus*, all streptococci lack the enzyme catalase. Most are facultative anaerobes but some are obligate anaerobes. Streptococci often have a mucoid or smooth colonial morphology, and *S pneumoniae* colonies exhibit a central depression caused by rapid partial autolysis. As *S pneumoniae* colonies age, viability is lost during fermentative growth in the absence of catalase and peroxidase because of the accumulation of peroxide. Some group B and D streptococci produce pigment. Recently, nutritionally deficient streptococci (also known as wall-deficient, L form, thiol-requiring, satelliting, or pyridoxal-dependent) have been recovered from a variety of clinical sources, including blood, abscesses, and oral and urethral ulcers. These variants demonstrate bizarre pleomorphism microscopically and do not grow on routine subculture.

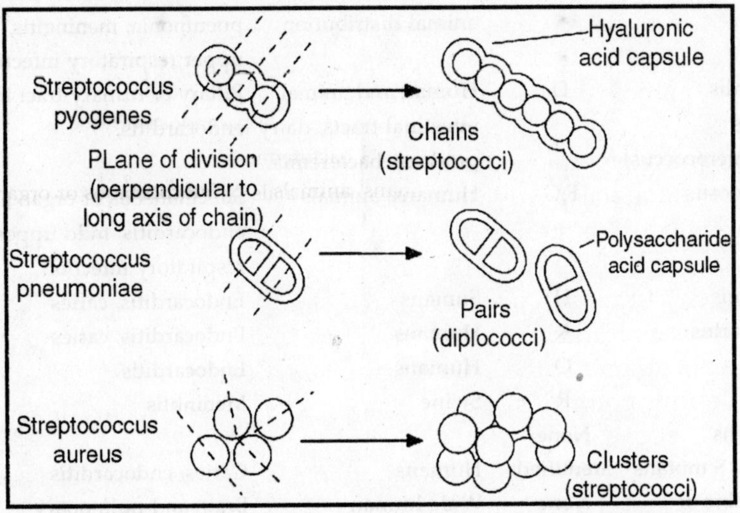


**Fig.** Morphology of the Streptococci in Comparison with Staphylococci.

## CLASSIFICATION, ANTIGENIC TYPES AND EXTRACELLULAR GROWTH PRODUCTS

The type of hemolytic reaction displayed on blood agar has long been used to classify the streptococci. β-Hemolysis is associated with complete lysis of red cells surrounding the colony, whereas α-hemolysis is a partial or "greening" hemolysis associated with reduction of red cell hemoglobin. Nonhemolytic colonies have been termed γ-hemolytic.

Hemolysis is affected by the species and age of red cells as well as by other properties of the base medium. Use of the hemolytic reaction in classification is not completely satisfactory.

Some group A streptococci appear nonhemolytic; group B can manifest α-, β-, or even γ-hemolysis; most *S pneumoniae* are α-hemolytic but can cause β-hemolysis during anaerobic incubation. The viridans group, although linked by the property of α-hemolysis, is actually an extremely diverse group of organisms that does not usually react with Lancefield grouping sera. The taxonomy and biochemical and genetic relationships of these organisms continue to be clarified.

**Table. Medically Important Streptococci.**

| *Type species* | *Lancefield serogroup* | *Normal habital* | *Significant human disease* |
|---|---|---|---|
| S pyogenes | A | Humans, | Acute pharyngitis and others |
| S agalactiae | B | Cattle, humans | Neonatal meningitis and sepsis and infections in adults |
| S equisimills | C | Wide human and animal distribution | Endocarditis, bacteremia, pneumonia, meningitis, mild upper respiratory infection |
| E faecalis S bovis (nonenterococcus) | D | Human and animal intestinal tracts, dairy products bacteremia | Biliary or urinary tract infection, endocarditis, |
| S anginosus | F,G | Humans, animals | Subcutaneous or organ abscesses endocarditis, mild upper respiratory infection |
| S sanguis | H | Sumans | Endocarditis, caries |
| S salivarlus | K | Humans | Endocarditis, caries |
| None | O | Humans | Endocarditis |
| S suis | R | Swine | Menintitis |
| "Viridans" S mitis, S mutans | None identified | Humans | Caries, endocarditis |
| Anaerobic or micro-aerophilic | None identified | Wide human and animal distribution | Brain and pulmonary abscesses, gynecologic infections |
| S pneumoniae | None identified | Humans | Lobar pneumonia and others |

- Strains of the S milleri group (*S constallatus, S intermedius, S anginosus,* minute strains) may possess antigens of groups A, C, F, or G, or no identifiable Lancefield group antigens; a heterogeneous group, genetically related but with a wide variety of phenotypic and biochemical characteristics
- Sisparate grouping undergoing further definition.
- Other viridans strepfococci (*S sanguis, S salivarius "S milleri," S bovis*) have identified group antigens (s); nutritionally variant streptococci may be included in this diverse category.

## Antigenic Types

The cell wall structure of group A streptococci is among the most studied of any bacteria. The cell wall is composed of repeating units of *N-acetylglucosamine* and *N-acetylmuramic* acid, the standard peptidoglycan. For decades, the definitive identification of streptococci has rested on the serologic reactivity of cell wall polysaccharide antigens originally delineated by Rebecca Lancefield. Eighteen group-specific antigens were established. The group A polysaccharide is a polymer of *N-acetylglucosamine* and *rhamnose*. Some group antigens are shared by more than one species; no Lancefield group antigen has been identified for *S pneumoniae* or for some other α- or γ-streptococci. With advances in serologic methods, other streptococci have been shown to possess several established group antigens.

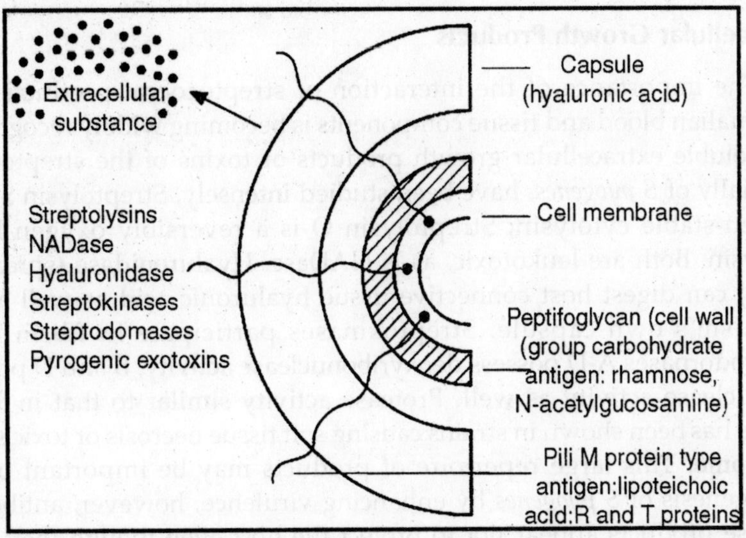


**Fig.** Cell Surface Structure of *S Pyogenes* and Extracellular Substances.

The cell wall also consists of several structural proteins. In group A streptococci, the R and T proteins may serve as epidemiologic markers, but the M proteins are clearly virulence factors associated with resistance to phagocytosis. More than 50 types of *S pyogenes* M proteins have been identified on the basis of antigenic specificity. Both the M proteins and lipoteichoic acid are supported externally to the cell wall on fimbriae, and the lipoteichoic acid, in particular, appears to mediate bacterial attachment to host epithelial cells. M protein, peptidoglycan, N-acetylglucosamine, and group-specific carbohydrate portions of the cell wall have antigenic epitopes similar in size and charge to those of mammalian muscle and connective tissue. Recently emerging strains of increased virulence are distinctly mucoid, rich in M protein and highly encapsulated.

The capsule of *S pyogenes* is composed of hyaluronic acid, which is chemically similar to that of host connective tissue and is therefore nonantigenic. In contrast, the antigenically reactive and chemically distinct capsular polysaccharide of *S pneumoniae* allows the single species to be separated into more than 80 serotypes. The antiphagocytic *S pneumoniae* capsule is the most clearly understood virulence factor of these organisms; type 3 *S pneumoniae,* which produces copious quantities of capsular material, are the most virulent. Unencapsulated *S pneumoniae* are avirulent. The polysaccharide capsule in *S agalactiae* allows differentiation into types Ia, Ib, Ic, II and III.

Finally, the cytoplasmic membrane of *S pyogenes* has antigens similar to those of human cardiac, skeletal, and smooth muscle, heart valve fibroblasts, and neuronal tissues, resulting in a molecular mimicry.

### Extracellular Growth Products

The importance of the interaction of streptococcal products with mammalian blood and tissue components is becoming widely recognized. The soluble extracellular growth products or toxins of the streptococci, especially of *S pyogenes,* have been studied intensely. Streptolysin S is an oxygen-stable cytolysin; Streptolysin O is a reversibly oxygen-labile cytolysin. Both are leukotoxic, as is NADase. Hyaluronidase (spreading factor) can digest host connective tissue hyaluronic acid as well as the organism's own capsule. Streptokinases participate in fibrin lysis. Streptodornases A-D possess deoxyribonuclease activity; B and D possess ribonuclease activity as well. Protease activity similar to that in Staph aureus has been shown in strains causing soft tissue necrosis or toxic shock syndrome. This large repertoire of products may be important in the pathogenesis of *S pyogenes* by enhancing virulence; however, antibodies to these products appear not to protect the host even though they have diagnostic importance.

Three pyrogenic exotoxins of *S pyogenes* (SPEs) are recognized: types A, B, C. These toxins act as superantigens by a mechanism similar to those described for staphylococci, not requiring processing by antigen presenting cells. Rather, they stimulate T cells by binding class II MHC molecules directly and nonspecifically. With superantigens about 20% of T cells may be stimulated (vs 1/10,000 T cells stimulated by conventional antigens) resulting in massive detrimental cytokine release. When *S pyogenes* is lysogenized by certain bacteriophages, the SPEs A or C are produced; nonlysogenized strains are atoxic. SPE B is encoded by the bacterial chromosome. Re-emergence in the late 1980's of these exotoxin-producing strains has been associated with a toxic shock-like syndrome similar in pathogenesis and manifestation to staphylococcal toxic shock syndrome (Ch.12) and other forms of invasive disease associated with severe tissue destruction. SPE's have also been identified from non group A streptococci (groups B, C, F. G) in association with the toxic shock-like syndrome.

Virulence factors in the other streptococcal species, including the enterococci, are less well identified. In group B streptococci, carbohydrate surface antigens associated with antiphagocytosis have been identified, as has neuraminidase, which may play a role in pathogenesis. Among the viridans streptococci, production of the exopolysaccharide (glycocalyx) is associated with the ability to adhere to the cardiac valves and to form vegetations on the valve leaflets.

## Pathogenesis

### Streptococcus Pyogenes *and* Streptococcus Pneumoniae

Streptococci vary widely in pathogenic potential. Despite the remarkable array of cell-associated and extracellular products previously described, no clear scheme of pathogenesis has been worked out. *S pneumoniae* and, to a lesser extent, *S pyogenes* are part of the normal human nasopharyngeal flora.

Their numbers are usually limited by competition from the nasopharyngeal microbial ecosystem and by nonspecific host defense mechanisms, but failure of these mechanisms can result in disease. More often disease results from the acquisition of a new strain following alteration of the normal flora. *S pyogenes* causes inflammatory purulent lesions at the portal of entry, often the upper respiratory tract or the skin. Some strains of streptococci show a predilection for the respiratory tract; others, for the skin. Generally, streptococcal isolates from the pharynx and respiratory tract do not cause skin infections.

Invasion of other portions of the upper or lower respiratory tracts results in infections of the middle ear (otitis media), sinuses (sinusitis), or lungs (pneumonia). In addition, meningitis can occur by direct extension of infection from the middle ear or sinuses to the meninges or by way of bloodstream invasion from the pulmonary focus. Bacteremia can also result in infection of bones (osteomyelitis) or joints (arthritis).

*S pyogenes* (a group A streptococcus) is the leading cause of uncomplicated bacterial pharyngitis and tonsillitis. Indeed, only group A streptococci are sought routinely in cases of pharyngitis, although groups B, C, and G are sometimes identified. *S pyogenes* infections can also result in sinusitis, otitis, mastoiditis, pneumonia with empyema, joint or bone infections, necrotizing fasciitis or myositis, and, more infrequently, in meningitis or endocarditis. *S pyogenes* infections of the skin can be superficial (impetigo) or deep (cellulitis). Although scarlet fever was formerly a severe complication of streptococcal infection, because of antibiotic therapy it is now little more than streptococcal pharyngitis accompanied by rash. Similarly, erysipelas, a form of cellulitis accompanied by fever and systemic toxicity, is less common today. There has, however, been an apparent recent increase in variety, severity and sequelae of *S pyogenes* infections. Because cases of streptococcal disease are not reported to national disease clearinghouses in the US, absolute numbers are not available.

However, the recent resurgence of severe invasive infections has prompted descriptions of "flesh eating bacteria" in the news media. There has been no major change in susceptibility of *S pyogenes* to commonly used antibiotics but rather in the strain variations described above (antigenic types and extracellular growth products). However, a complete explanation for the decline and resurgence is not yet available.

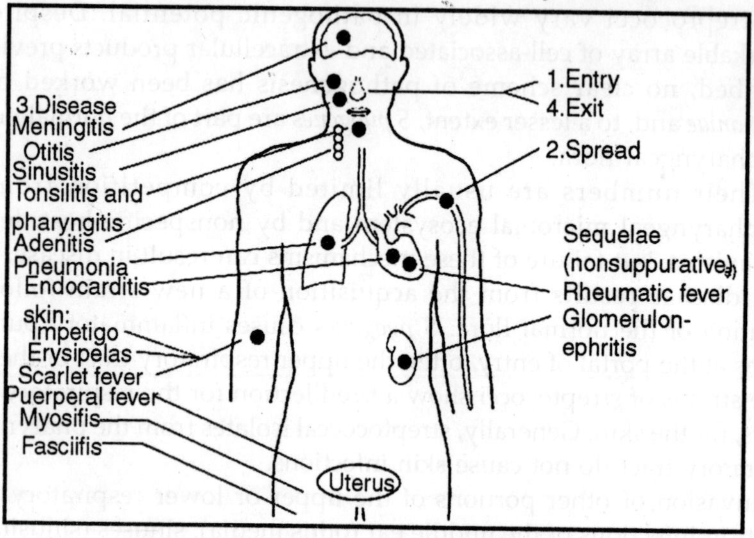


**Fig.** Pathogenesis of *S Pyogenes* Infections.

The capsule of *S pneumoniae* renders it resistant to phagocytosis. The ability to evade this important host defense mechanism allows *S pneumoniae* to survive, multiply, and spread to various organs. The cell wall of S *pneumoniae* contains teichoic acid. The inflammatory response induced by Gram-positive cell walls differs from that induced by the endotoxin of Gram-negative organisms, but does include recruitment of polymorphonuclear neutrophils, changes in permeability and perfusion, cytokine release, and stimulation of platelet-activating factor.

The role of other S pneumoniae moieties in virulence is less clear: protein A, pneumolysin, and peptide permeases. *S pneumoniae* is the leading cause of bacterial pneumonia beyond the neonatal period. Pleural effusion is the most common and empyema (pus in the pleural space) one of the most serious complications of *S pneumoniae.* This organism is also the most common cause of sinusitis, acute bacterial otitis media, and conjunctivitis beyond early childhood. Dissemination from a respiratory focus results in serious disease: outpatient bacteremia in children, meningitis, occasionally acute septic arthritis and bone infections in patients with sickle cell disease and, more rarely, peritonitis (especially in patients with nephrotic syndrome) or endocarditis.

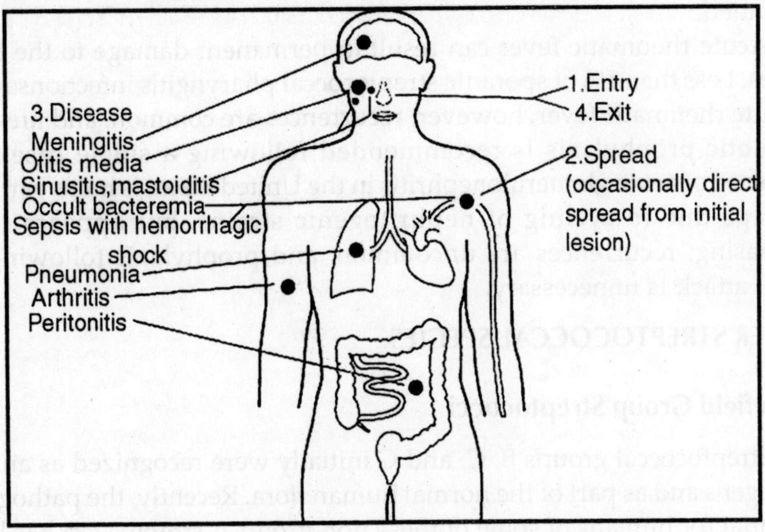


**Fig.** Pathogenesis of *S Pneumoniae* Infections.

### Postinfectious Sequelae

Infection with *S pyogenes* (but not *S pneumoniae*) can give rise to serious nonsuppurative sequelae: acute rheumatic fever and acute glomerulonephritis. These sequelae begin 1-3 weeks after the acute illness, a latent period consistent with an immune-mediated rather than pathogen-disseminated etiology. Whether all *S pyogenes* strains are rheumatogenic is still controversial; however, clearly not all are nephritogenic. These differences in pathogenic potential are not yet understood.

Acute rheumatic fever is a sequela only of pharyngeal infections, but acute glomerulonephritis can follow infections of the pharynx or the skin. Although there is no adequate explanation for the precise pathogenesis of acute rheumatic fever or for its failure to occur after streptococcal pyoderma, an abnormal or enhanced immune response seems essential. Also, persistence of the organism, due perhaps in part to the greater avidity with which the organism adheres to host pharyngeal cells, is associated with an increased likelihood of rheumatic fever.

Acute glomerulonephritis results from deposition of antigen-antibody-complement complexes on the basement membrane of kidney glomeruli. The antigen may be streptococcal in origin or it may be a host tissue species with antigenic determinants similar to those of streptococcal antigen (cross-reactive epitopes for endocardium, sarcolemma, vascular smooth muscle). In the United States, the incidence of acute rheumatic fever had decreased dramatically. Although several areas reported a resurgence in cases in the late 1980's, subsequently, a slow, steady decline continued.

Acute rheumatic fever can result in permanent damage to the heart valves. Less than 1% of sporadic streptococcal pharyngitis infections result in acute rheumatic fever; however, recurrences are common, and life-long antibiotic prophylaxis is recommended following a single case. The incidence of acute glomerulonephritis in the United States is more variable, perhaps due to cycling of nephritogenic strains, but appears to be decreasing; recurrences are uncommon, and prophylaxis following an initial attack is unnecessary.

## OTHER STREPTOCOCCAL SPECIES

### Lancefield Group Streptococci

Streptococcal groups B, C, and G initially were recognized as animal pathogens and as part of the normal human flora. Recently, the pathogenic potential for humans of some of these non-group-A streptococci has been clarified. Group B streptococci, a major cause of bovine mastitis, are a leading cause of neonatal septicemia and meningitis, accounting for a significant changing clinical spectrum of diseases in both pregnant women and their infants. Mortality rates in full-term infants range from 2-8% but in pre-term infants are approximately 30%. Early-onset neonatal disease (associated with sepsis, meningitis and pneumonia at 6d life) is thought to be transmitted vertically from the mother; late-onset (from 7d to 3 mos age) meningitis is acquired horizontally, in some instances as a nosocomial infection.

Group B organisms also have been associated with pneumonia in elderly patients. They are part of the normal oral and vaginal flora and have also been isolated in adult urinary tract infection, chorioamnionitis and endometritis, skin and soft tissue infection, osteomyelitis, meningitis, bacteremia without focus, and endocarditis. Infection in patients with HIV can occur at any age.

Streptococci of groups C and G are associated with mild, as well as severe human disease. None of these groups has been implicated in acute rheumatic fever or acute glomerulonephritis. Group D streptococci are important etiologic agents of urinary tract infections and infections associated with biliary tract procedures, as well as cases of disseminated infection, bacteremia, and endocarditis. Streptococcus bovis bacteremia has been recognized more often in cases of bowel disease.

Group F streptococci are associated with abscess formation and purulent disease. Group R streptococci, well-documented causes of meningitis and septicemia in pigs, also pose a serious health hazard to workers in the pork industry.

### Viridans Streptococci

The biochemically and antigenically diverse group of organisms classified as viridans streptococci, as well as other non-groupable streptococci of the oral and gastrointestinal cavities and urogenital tract, include important etiologic agents of bacterial endocarditis.

Dental manipulation and dental disease with the associated transient bacteremia are the most common predisposing factors in bacterial endocarditis, especially if heart valves have been damaged by previous rheumatic fever or by congenital cyanotic heart disease. *S mutans* and *S sanguis* are odontopathogens responsible for the formation of dental plaque, the dense adhesive microbial mass that colonizes teeth and is linked to caries and other human oral disease. *S mutans* is the more cariogenic of the two species, and its virulence is directly related to its ability to synthesize glucan from fermentable carbohydrates as well as to modify glucan in promoting increased adhesiveness.

### Anaerobes

Like their aerobic counterparts, anaerobic streptococci are part of the normal flora, particularly of the mouth and intestinal tract; they are also part of the normal flora of the upper respiratory and genital tracts and the skin. These anaerobic organisms are linked to a wide variety of serious mixed infections of the female genital tract as well as to brain, pulmonary, and abdominal abscesses.

## HOST DEFENSES

The streptococci are part of the endogenous microbial flora of the nasopharynx. Disease may result from circumvention of the normal Evidence from the Rhesus monkey animal model in dental research shows that IgG may be a more important antibody class than IgA or IgM in protection against caries. Part of the reason may be that IgG is the antibody isotype most efficient at enhancing phagocytosis of *S mutans*. Cell-mediated immunity appears to participate in the protective host response against caries.

## EPIDEMIOLOGY

The streptococci are widely distributed in nature and frequently form part of the normal human flora. Approximately 5-15% of humans carry *S pyogenes* or *S agalactiae* in the nasopharynx. *S pneumoniae* infects humans exclusively, and no reservoir is found in nature. The carrier rate of *S pneumoniae* in the normal human nasopharynx is 20-40%.

All ages, races, and sexes are susceptible to streptococcal disease. Because *S pneumoniae* is a particularly labile organism, sensitive to heat, cold, and drying, horizontal transmission requires close person-to-person contact. Infection is more likely at the extremes of life (<2 yr, > 65 yr), when host resistance is reduced, as described in the preceding section, or after the introduction of a more virulent strain. In the United States, pneumococcal disease is most prevalent during winter, coinciding with increased rates of acquisition but not necessarily of carriage. Alaskan natives have higher rates of invasive pneumococcal disease than do other American populations. The reason for this is unclear.

The incidence of respiratory disease attributed to *S pyogenes* peaks at about 6 years of age, and then again at 13 years of age, and is most common during late winter and early spring in temperate climates. Skin infections are more common among preschool-age children, and are most prevalent in late summer and early fall in temperate climates (when hot, humid weather prevails), and at all times in tropical climates. *S pyogenes* is spread by respiratory droplets or by contact with fomites used by the index individual, either patient or carrier. Skin infections often follow minor skin irritation, such as insect bites. There are occasional reports of streptococcal disease traced to rectal carriers, and of food-borne and vector-born outbreaks. In children, invasive disease with *S pyogenes* may follow varicella, or be associated with burns or malignancy; in adults with surgical or nonsurgical wounds or underlying medical problems, i.e., diabetes, cirrhosis, underlying peripheral vascular disease, or malignancy.

The world prevalence of the serious late sequelae of *S pyogenes* infections (acute rheumatic fever and acute glomerulonephritis) has shifted from temperate to tropical climates. In particular, acute rheumatic fever had ceased to be a major health concern in the US., despite no concomitant decline in group A streptococcal pharyngitis. These diseases previously affected persons with a low standard of living and limited access to medical care. Since 1985, there have been scattered outbreaks of acute rheumatic fever in some regions of the United States. Temporal and geographic clustering provides further evidence for "rheumatogenic" strains. Whether ethnic or racially determined factors affect this shift is not known.

Other streptococcal groups show striking epidemiologic features. An increasing prevalence of non-group-A as compared to group A streptococci in throats has been reported. Studies of the vaginal flora among women of child-bearing age show a *S agalactiae* carrier rate of 15-40%. Vertical transmission of the organisms to neonates of vaginally infected mothers ranges from 40-73%, but the incidence of neonates with disease (in contrast to colonized, healthy neonates) is low, 1-2%. *S suis* has been linked to meningitis among meat handlers. Isolation of *S milleri* or *S bovis* from the bloodstream should raise suspicion of immunosuppression or underlying disease visceral abscess formation or other bowel disease (including colon carcinoma).

In the United States, enterococci are the second most common nosocomial pathogenes associated with both endogenous colonization and patient-to-patient spread. A wide variety of infections results, especially urinary tract and surgical wound infections, with a marked propensity for antibiotic resistance. The widespread usage of newer cephalosporins, which have poor activity against enterococci, allows "break through" of enterococci as clinically significant isolates, the development of resistance in areas of heavy antibiotic use, and a selective advantage to these organisms.

## DIAGNOSIS

### Clinical

It is not usually possible to diagnose streptococcal pharyngitis or tonsillitis on clinical grounds alone. Accurate differentiation from viral pharyngitis is difficult even for the experienced clinician, and therefore the use of bacteriologic methods is essential. However, distinguishing acute streptococcal pharyngitis from the carrier state may be difficult. When documented streptococcal pharyngitis is accompanied by an erythematous punctiform rash, the diagnosis of scarlet fever can be made. With streptococcal toxic shock syndrome, unlike staphylococcal toxic shock syndrome where the organism is elusive, there is often a focal infection or bacteremia.

Criteria for diagnosis of streptococcal toxic shock syndrome include hypotension and shock, isolation of *S pyogenes*, as well as 2 or more of the following: ARDS, renal impairment, liver abnormality, coagulopathy, rash with desquamating soft tissue necrosis. The invasive, potentially fatal *S pyogenes* infections require early recognition, definitive diagnosis, and early aggressive treatment.

Rheumatic fever is a late sequela of pharyngitis and is marked by fever, polyarthritis, and carditis. A combination of clinical and laboratory criteria is used in the diagnosis of acute rheumatic fever. The other late sequela, acute glomerulonephritis, is preceded by pharyngitis or pyoderma; is characterized by fever, blood in the urine (hematuria), and edema; and is sometimes accompanied by hypertension and elevated blood urea nitrogen (azotemia). Pneumococcal pneumonia is a life-threatening disease, often characterized by edema and rapid lobar consolidation.

**Table. Jones Diagnostic Criteria for Acute Rheumatic Fever.**

| *Major* | *Minor* |
|---|---|
| Carditis | Clinical |
| Polyarthritis | Fever |
| Erythema marginatum | Arthralgia |
| Subcutaneous nodules | Chorea |
| Laboratory | Increased erythrocyte sedimentation rate |
| | Increased C-reactive protein level |
| | ECG |
| | Prolonged PR interval |

- In conjunction with culture or serologic evidence of recent streptococcal infection.

## Specimens For Direct Examination and Culture

*S pyogenes* is usually isolated from throat cultures. In cases of cellulitis or erysipelas thought to be caused by *S pyogenes*, aspirates obtained from the advancing edge of the lesion may be diagnostic. *S pneumoniae* is usually isolated from sputum or blood.

Precise streptococcal identification is based on the Gram stain and on biochemical properties, as well as on serologic characteristics when group antigens are present. Table shows biochemical tests that provide sensitive group-specific characteristics permitting presumptive identification of Gram-positive, catalase-negative cocci.

## IDENTIFICATION

Hemolysis should not be used as a stringent identification criterion. Bacitracin susceptibility is a widely used screening method for presumptive identification of *S pyogenes*; however, some *S pyogenes* are resistant to bacitracin (up to 10%) and some group C and G streptococci (about 3-5%) are susceptible to bacitracin. Some of the group *B streptococci* also may be bacitracin sensitive, but are presumptively identified by their properties of hippurate hydrolysis and CAMP positivity. *S pneumoniae* can be separated from other α-hemolytic streptococci on the basis of sensitivity to surfactants, such as bile or optochin (ethylhydrocupreine hydrochloride). These agents activate autolytic enzymes in the organisms that hydrolyze peptidoglycan.

In many instances, presumptive identification is not carried further. Serologic grouping has not been performed as often as it might be because of the lack of available methods and the practical constraints of time and

**Table. Characteristics for the Presumptive Identification of Streptococci of Human Clinical Importance.**

| *Procedure* | *Results of Group:* *A* | *B* | *D (enterococcus)* | *D (nonenterococcus)* | *Non-A,B,D* | *S pneumoniae* | *Viridans* |
|---|---|---|---|---|---|---|---|
| Hemolysis | β | β,α,γ | α,β,γ | α,γ,β | β,α,γ | α | α |
| Bacitracin Sensitivity | + | –(+) | – | – | –(+) | ± | – |
| CAMP test | – | + | – | – | – | – | – |
| Growth at 45°C | – | – | + | + | – | – | + |
| Optochin sensitivity | –(+) | –(+) | – | – | –(+) | + | – |
| Hydrolysis of sodium hippurate | – | + | –(+) | – | – | – | –(+) |
| Tolerance to 6,5 per cent NaCl | – | – | + | – | – | – | – |
| Hydrolysis of esculin presence of 40 per cent bile | – | – | + | + | – | – | –(+) |
| Hydrolysis of pyrrolidonyl naphthylamide | + | – | + | – | – | – | – |

- In general order of frequency.
- Signs in parentheses indicate occasional result.

cost; however, only serologic methods, as listed in Table, provide definitive identification of the streptococci. The Lancefield capillary precipitation test is the classical serologic method. *S pneumoniae,* which lacks a demonstrable group antigen by the Lancefield test, is conventionally identified by the quellung or capsular swelling test that employs type-specific anticapsular antibody. Inspection of Gram-stained sputum remains a reliable predictor for initial antibiotic therapy in community-acquired pneumonia.

**Table. Methods of Serogrouping Streptococci**

| *Nature of Streptococcal Antigen* | *Techniques* |
|---|---|
| Whole cells | Fluorescent antibody |
| | Direct hacterial agglulination |
| | Coagglutination with staphylococcal protein A |
| | Carrier agglutination (antibody-coated latex particles) |
| | Ouellung reaction (for S pneumoniae) |
| Soluble extract | Precipitation (classical capillary of counterimmunoelectrophoresis) |
| | Coagglutination with staphylococcal protein A |
| | Camier agglutination (antibody-coated latex particles) |
| | Enzyme-linked immunosorbent assay (ELISA) |
| | Antigen capture assays |
| | Optical immunoassay |

New methods for serogrouping that show sensitivity and specificity now are being explored. Organisms from throat swabs, incubated for only a few hours in broth, can be examined for the presence of *S pyogenes* using the direct fluorescent antibody or enzyme-linked immunosorbent technique. Additional rapid antigen detection systems for the group carbohydrate have become increasingly popular. However, the sensitivity (70-90%) of these currently available rapid tests for group A streptococcal carbohydrate does not allow exclusion of streptococcal pharyngitis without conventional throat culture (sensitivity of a single throat culture is 90-99%). A third generation assay, the optical immunoassay, is currently being evaluated. *S pneumoniae* can be identified rapidly by counterimmuno-electrophoresis, a modification of the gel precipitin method. The coagglutination test, described in Ch.12, is a more sensitive modification of the conventional direct bacterial agglutination test. The Fc portion of group-specific antibody binds to the protein A of dead staphylococci, leaving the Fab portion free to react with specific streptococcal antigen. The attachment of antibody to other carrier particles in suspension (for example, latex) also is used. The fact that whole streptococcal cells can be used in recently developed methods circumvents the difficulties involved in extracting components that retain appropriate antigenic reactivity. These newer serogrouping methods should make it more practical to identify not only β-hemolytic isolates from the blood or normally sterile sites, but also α-and nonhemolytic strains. It has become increasingly important to identify more of these strains to avoid simply misclassifying them as contaminants. Such information will expand our understanding of the importance of non-group-A streptococci.

### Serologic Titers

Antibodies to some of the extracellular growth products of the streptococci are not protective but can be used in diagnosis. The antistreptolysin O (ASO) titer which peak 2-4 wks after acute infection and anti-NADase titers (which peaks 6-8 weeks after acute infection) are more commonly elevated after pharyngeal infections than after skin infections. In contrast, antihyaluronidase is elevated after skin infections, and anti-DNase B rises after both pharyngeal and skin infections. Titers observed during late sequelae (acute rheumatic fever and acute glomerulonephritis) reflect the site of primary infection. Although it is not as well known as the ASO test, the anti-DNase B test appears superior because high-titer antibody is detected following skin and pharyngeal infections and during the late sequelae. Those titers should be interpreted in terms of the age of the patient and geographic locale.

Although not used in diagnosis, bacteriocin production and phage typing of streptococci are employed in research and epidemiologic studies.

## CONTROL

### Antibiotic Treatment

Penicillin remains the drug of choice for *S pyogenes*. It is safe, inexpensive, and of narrow spectrum, and there is no direct or indirect evidence of loss of efficacy. Prior to the 1990's, *S pneumoniae* was also uniformly sensitive to penicillin but a recent abrupt shift in the usefulness of penicillin has occurred. The group D enterococci are resistant to penicillins, including penicillinase-resistant penicillins such as methicillin, nafcillin, dicloxacillin, and oxacillin, and are becoming increasingly resistant to many other antibiotics. Group B streptococci are often resistant to tetracycline but remain sensitive to the clinically achievable blood levels of penicillin, even though they have penicillin minimal inhibitory concentrations (MIC) considerably higher than those of *S pyogenes*. Although the duration of penicillin therapy varies with the degree of invasiveness, streptococcal pharyngitis is generally adequately treated with 10 days of antibiotic therapy, and pneumococcal pneumonia with 7-14 days. If penicillin allergy occurs, an alternative drug for treating pharyngitis is erythromycin, although sporadic erythromycin and tetracycline resistance has been reported, leaving clindamycin or the newer macrolides as possible treatments. The most important goal of therapy in acute streptococcal pharyngitis is still to prevent rheumatic fever. However, therapy also hastens clinical recovery, avoids suppurative complications and renders the patient non-infectious for others. In addition to antibiotics, the patient with *S pyogenes* myositis or necrotizing fasciitis requires surgical debridement. Lifelong prophylaxis against recurrences of rheumatic fever is achieved with long-acting penicillin or erythromycin. Sulfonamides will not eradicate the streptococcus and thus are not acceptable therapy for streptococcal pharyngitis, but sulfadiazine is effective for preventing recurrent attacks of rheumatic fever. Additional prophylactic coverage before some dental and surgical procedures is necessary in the presence of rheumatic heart disease or prosthetic heart valves. Although streptococcal pharyngitis is usually a benign, self-limited disease, therapy is important to prevent rheumatic fever. There is no convincing evidence that antibiotic therapy prevents glomerulonephritis. Disconcertingly, some patients in recent outbreaks of acute rheumatic fever do not give a history of preceding pharyngitis.

Methods of treating the asymptomatic pharyngeal carrier of *S pyogenes* remain controversial. Recent evidence suggests that up to 20% of children and young adults are carriers, the carrier state involves no risk to the carrier or to others, and it is frequently difficult to eradicate despite the exquisite sensitivity of the organism to penicillin in vitro. A similar failure of antibiotic therapy to eradicate nasopharyngeal carriage or to prevent reinfection with *S pneumoniae* also occurs.

Although antibiotic resistance in *S pneumoniae* is common in many parts of the world, in the United States such strains previously had a geographically limited focus. Recent widespread emergence of *S pneumoniae* resistant to penicillin and other antibiotics has become a microbial threat in the United States as well. Even cefotaxime and ceftriaxone resistance has been documented. Isolates must be carefully screened for susceptibility by oxacillin disc testing, with definitive MIC determination by the E test (A B Biodisk NA, Piscataway, NJ), a convenient and reliable method for detection of resistance to penicillin and extended spectrum cephalosporins.

It is inappropriate to universally treat of pregnant women who are carriers of group B streptococci, or their colonized neonates, for several reasons: the high carrier rate; cost; the associated high risk of penicillin hypersensitivity; the potential increase in infections with penicillin-resistant organisms; the difficulty in altering colonization of women (even when their sexual partners were also treated); and the low risk of neonatal disease. The controversy continues despite recent recommendations for universal screening of preg nant women and selective intrapartum chemoprophylaxis for screen-positive mothers with preterm labour, premature or prolonged rupture of membranes, fever in labour, multiple births or previous infants with group B streptococcal disease.

Clearly, penicillin has reduced the severe morbidity and mortality associated with *S pneumoniae*. The emergence of resistance has now forced re-evaluation of empiric therapy. Clinicians must report clusters of *S pneumoniae* infection and be aware of local patterns of resistance. Penicillin susceptible organisms show MICs 0.06 mg/ml, intermediate strains 0.1-1.0 mg/ml and high level resistant strains 2 mg/ml. For nonmeningeal infection by intermediate strains, parenteral penicillin at high dose can probably be used since the mechanism of resistance involves alteration in penicillin binding proteins (PBP) and saturation. For meningeal infection with intermediate strains or any infection by high level resistant strains only ceftriaxone and cefotaxime retain sufficient activity. Resistance even to these extended spectrum cephalosporins was first reported for the US in 1991. At this writing only vancomycin remains uniformly effective but as discussed below, its use incurs potential for selection of vancomycin resistant enterococci (VRE) or risk of transferring vancomycin resistance from enterococci to *S pneumoniae*.

Currently, no single agent is reliably bactericidal against enterococci. Serious infections with group D enterococci often require a classic synergistic regime combining penicillin or ampicillin with an aminoglycoside, designed to weaken the cell wall with the β-lactam and facilitate entry of the bacteriocidal aminoglycoside. Other β-lactam drugs with good activity against enterococci include piperacillin and imipenem. An alternate drug of choice is vancomycin, but vancomycin-resistant strains of enterococci have been isolated. Nosocomial acquisition of these resistant organisms is of grave concern.

This antibiotic resistance among the streptococci/enterococci is an increasing problem. Studies show that in vitro exchange of resistant DNA can occur in conjugation via plasmids and transposons, or in transduction with bacteriophages. The mechanisms involved in the in vivo genetic exchange are not clearly defined. Evidence is accumulating that other streptococci may be the important donors of resistance markers. Transposon transfer is thought to be the most likely mechanism in *S pneumoniae,* although point mutations also occur. In the setting of heavy β-lactam use, selective pressure is important in emergence of resistant strains. The first penicillin-resistant *S pneumoniae* were reported in 1967 in Australia and in 1974 in North America. In New Guinea, where the first penicillin-resistant strains were reported in 1971, one-third of *S pneumoniae* isolates from patients with severe pneumococcal disease were resistant by 1978. In Hungary in 1992, 69% of *S pneumoniae* isolates were penicillin resistant. This resistance is not β-lactamase mediated but due to alteration in PBP which results in decreased binding of penicillin by the organism, rendering the drug less effective and requiring higher concentrations for saturation. Some strains resistant to erythromycin or tetracycline also have been reported, as well as some multiply resistant strains. In South Africa, outbreaks of infection with strains of *S pneumoniae* resistant to β-lactam antibiotics (penicillins and cephalosporins) as well as to tetracycline, chloramphenicol, erythromycin, streptomycin, clindamycin, sulfonamides, and rifampin were reported in 1977.

Although antibiotic resistance among *S pneumoniae* was infrequent in the United States, a major shift occurred from 1988 to 1990, resulting in the present situation of 15-25% of *S pneumoniae* intermediately or completely resistant to penicillin. Communities with "low prevalence" have 5-10% resistance. Single or multiply resistant strains are transmitted person to person, especially in settings of frequent salivary exchange, antibiotic use and hand-to-hand transmission (as in day care centers) or of crowding (corrections facilities, homeless shelters, nursing homes, military training groups).

Control of the problem of emerging, antibiotic-resistant *S pneumoniae* is multifactorial: 1) surveillance for clusters of invasive disease, resistance and prevalent serotypes; 2) education of physicians and the public about antibiotic use (decrease unnecessary antibiotic use for obviously viral infections and decrease antibiotic prophylaxis for otitis by use of intermittent or expectant dosing or of non β-lactam based prophylaxissulfa. Use topical treatment for impetigo, and short course therapies and narrow spectrum antibiotics); 3) adherence to infection control strategies in day care centers; 4) aggressive promotion of the current 23-valent *S pneumoniae* vaccine and support of efforts to design a new vaccine effective in those <2 years of age, analogous to the eminently successful Haemophilus influenzae type B vaccine where bacterial polysaccharide is conjugated to protein to elicit a T cell-dependent response.

Among the enterococci, resistance to a wide variety of common antibiotics has emerged, with some strains resistant to all currently available antibiotics. There is no clinically proven treatment effective against enterococci multiply resistant to lactams, aminoglycosides, and vancomycin. The emergence of such organisms poses a stunning management dilemma. Resistance among the enterococci can be either intrinsic or acquired (by de novo genetic mutation or acquisition of DNA from resistant organisms). Enterococcal resistance to lactams is also mediated by altered PBP as in pneumococci, allowing cell wall synthesis even in the presence of antibiotic, or much less commonly by β-lactamase. Resistance to aminoglycosides is mediated by decreased uptake or aminoglycoside modifying enzymes, and to vancomycin by decreased cell wall affinity for glycopeptide antibiotics. Further research into the mechanisms of resistance and new class(es) of antibiotics is essential.

A final concern about emerging resistance among the enterococci is the potential for genetic transfer of resistance genes to more virulent pathogens: Staph aureus, *S pneumoniae* and even Gram-negative organisms. So significant is this threat of emerging enterococcal resistance that the Centers for Disease Control and Prevention has issued a document addressing national guidelines. These include recommendations for 1) education of physicians and the public about the impact of vancomycin resistant enterococci (VRE), 2) vigilant surveillance for and detection of VRE, 3) strict enforcement of infection control strategies in hospitals, and 4) prudent vancomycin use or monotherapeutic use of extended spectrum cephalosporins.

In a recent study of vancomycin use in US hospitals, use was about equally divided for treatment of a specific isolate, for prophylaxis, and for empiric coverage. The recommendations discourage vancomycin use for routine surgical prophylaxis, empiric prophylaxis in the patient with febrile neutropenia, the low birth weight infant or patients with vascular or peritoneal catheters, treatment of a single blood culture positive for coagulase-negative staphylococci, primary treatment of antibiotic-associated colitis, attempted eradication of colonization by methicillin-resistant Staph aureus (MRSA), or selective decontamination of the gastrointestinal tract.

## Vaccination

As chemotherapeutic management becomes more difficult because of the threat of resistance, prevention becomes more important. With the introduction of antibiotics, previously successful pneumococcal vaccines fell into disuse. However, although prompt treatment with antibiotics has reduced the serious consequences of *S pneumoniae* infections (pre-antibiotic mortality rate of 30%), the disease incidence remains unchanged, and attention has been redirected to vaccines for *S pneumoniae* as well as for other streptococci. Pneumococcal vaccines (containing the pneumococcal polysaccharides of the most prevalent serotypes) have been licensed in several countries, including the United States. Initial use shows them to be useful and safe, but they remain under-utilized. The spectre of multidrug resistant *S pneumoniae* may provide a new incentive for their use. In 1983, the United States Food and Drug Administration licensed a vaccine containing 23 serotypes, representing coverage against nearly 89% of the pneumococcal isolates submitted to the CDC in the 1987-1988 National Surveillance Study. The population target of pneumococcal vaccines includes those at high risk for serious pneumococcal disease: the elderly (65 and older) and children (2 years of age and older) with sickle cell anemia, with an immunocompromised state (lymphoma, asplenia, myeloma, acquired immunodeficiency syndrome), with nephrotic syndrome, or with chronic cardiopulmonary disease. Vaccines for the other streptococci remain experimental.

Vaccine production for the streptococci presents several formidable problems. For both *S pyogenes* and *S pneumoniae,* a large number of serotypes must be included in effective vaccines since successful selection of a common epitope remains elusive. Continuing surveillance to determine prevalent serotypes is necessary to insure that the vaccine formulations remain appropriate. For *S pyogenes,* it is critical to determine rheumatogenic and nephritogenic strains to limit the required multivalency of the vaccines. Alternatively a newly described conserved portion of M protein is a distant goal. Toxicity has been associated with M protein preparations, but lack of immunogenicity in highly purified preparations of antigens is still a problem. With streptococcal vaccines, the potential risk of antigenic cross-reactivity with cardiac tissue and an associated increased risk of acute rheumatic fever must be appreciated.

In group B neonatal disease chemoprophylaxis does not appear as practical as vaccine control. Passive immunity in group B streptococcal neonatal infection appears protective. Polyvalent hyperimmune gamma globulin and human monoclonal IgM antibody which reacts with multiple serotypes are undergoing efficacy studies. Active immunization of pregnant women with undegraded sialic acid-containing polysaccharide group B antigens is another important aspect of control.

The streptococci are ubiquitous, and their significance in medicine is remarkable. Exciting advances are being made in diagnosis and in understanding the mechanisms of pathogenesis, as well as in control of these well-known organisms. Problems with antibiotic resistance must preclude complacency in dealing with these common pathogens.

Chapter 16

# Neisseria, Moraxella, Kingella and Eikenella

The family Neisseriaceae comprises the genera Neisseria, Moraxella, Kingella, and Acinetobacter. The only significant human pathogens are N gonorrhoeae, the agent of gonorrhea, and N. meningitidis, an agent of acute bacterial meningitidis. N gonorrhoeae infections have a high prevalence and low mortality, whereas N meningitidis infections have a low prevalence and high mortality.

Gonococcal infections are acquired by sexual contact and usually affect mucous membranes of the urethra in men and the endocervix in women, although the infection may disseminate to a variety of tissues. The pathogenic mechanism involves the attachment of the gonococci to nonciliated epithelial cells via pili (fimbriae) and the production of cytotoxic factors (endotoxin). Similarly, the lipopolysaccharide of meningococci is highly toxic, but an additional virulence factor is the antiphagocytic capsule. Both pathogens produce proteases that cleave and inactivate human immunoglobulin A1 (IgA1), a major mucosal immunoglobulin of humans. Many normal individuals harbor meningococci, whereas gonococci are present only if sexual contact with an infected person has occurred. Epidemics of meningococcal meningitis occur sporadically. Gonococcal infections occur frequently and affect large numbers of sexually active people. Other species in this genus are primarily parasites on mucosal surfaces of humans and other animals. Human disease caused by these organisms usually is associated with opportunistic infections in compromised patients.

## *NESSERIA GONORRHOEAE*

### Clinical Manifestations

Gonorrheal infection is generally limited to superficial mucosal surfaces lined with columnar epithelium. The areas most frequently involved are the cervix, urethra, rectum, pharynx, and conjunctiva. Squamous epithelium, which lines the adult vagina, is not susceptible to infection by the gonococcus. However, the prepubertal vaginal epithelium, which has not been keratinized under the influence of estrogen, may be infected. Hence, gonorrhea in young girls may present as vulvovaginitis. Mucosal infections are usually characterized by a marked local neutrophilic response (purulent discharge).

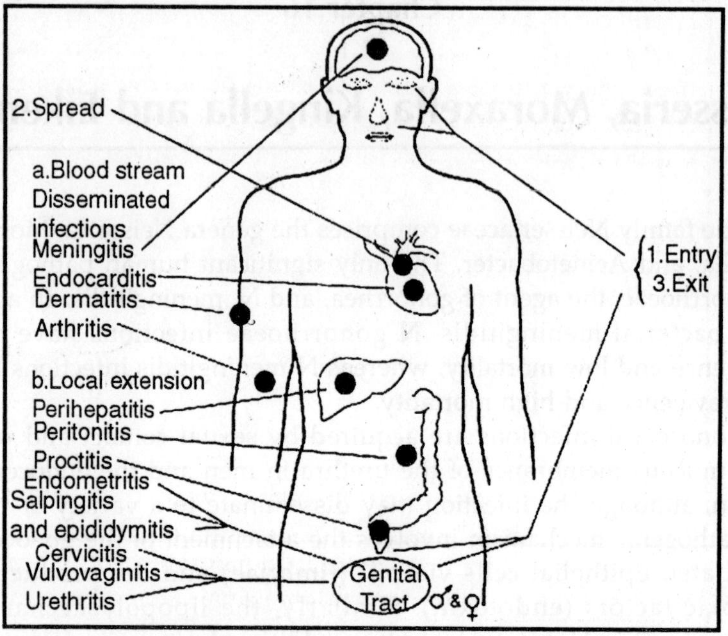


**Fig.** Clinical Manifestations of N Gonorrhoeae Infection.

The most common symptom of uncomplicated gonorrhea is a discharge that may range from a scanty, clear, or cloudy fluid to one that is copious and purulent. Dysuria is often present. Men with asymptomatic urethritis are an important reservoir for transmission. In addition, such men and those who ignore their symptoms are at increased risk for developing complications.

Endocervical infection is the most common form of uncomplicated gonorrhea in women. Such infections are usually characterized by vaginal discharge and sometimes by dysuria (because of coexistent urethritis). The cervical os may be erythematous and friable, with a purulent exudate. About 50 per cent of women with cervical infections are asymptomatic. Local complications include abscesses in Bartholin's and Skene's glands.

Rectal infections with N gonorrhoeae occur in about one-third of women with cervical infection. They most often result from autoinoculation with cervical discharge and are rarely symptomatic. Rectal infections in homosexual men usually result from anal intercourse and are more often symptomatic. The symptoms and signs of gonococcal proctitis range from mild burning on defecation to itching to severe tenesmus and from mucopurulent discharge to frank blood in the stools.

Pharyngeal infections are diagnosed most often in women and homosexual men with a history of fellatio. Such infections may be a focal source of gonococcemia.

Ocular infections can have serious consequences (corneal scarring or perforation); prompt diagnosis and treatment are therefore important. Ocular infections (ophthalmia neonatorum) occur most commonly in newborns who are exposed to infected secretions in the birth canal. Keratoconjunctivitis is occasionally seen in adults as a result of autoinoculation.

Disseminated gonococcal infections result from gonococcal bacteremia. Asymptomatic infections of the pharynx, urethra, or cervix often serve as focal sources for bacteremia. The most common form of disseminated gonococcal infection is the dermatitis-arthritis syndrome. It is characterized by fever, chills, skin lesions, and arthralgias (usually involving the hands, feet, and elbows), which are due to periarticular inflammation of the tendon sheaths. Occasionally, a patient develops a septic joint with effusion. Skin lesions may be macular, pustular, centrally necrotic, or hemorrhagic. Rarely, disseminated gonococcal infection causes endocarditis or meningitis. Gonococci may ascend from the endocervical canal through the endometrium to the fallopian tubes and ultimately to the pelvic peritoneum, resulting in endometritis, salpingitis, and finally, peritonitis. Women usually present with pelvic and abdominal pain, fever, chills, and cervical motion tenderness. This complex of signs and symptoms is referred to as pelvic inflammatory disease (PID). This disease may also be caused by other sexually transmitted organisms (e.g., Chlamydia trachomatis) as well as by non-sexually transmitted bacteria that are part of the normal vaginal flora. Complications of pelvic inflammatory disease include tubo-ovarian abscesses, pelvic peritonitis, or Fitz-Hugh and Curtis syndrome, which is an inflammation of Glisson's capsule of the liver. As many as 15 per cent of women with uncomplicated cervical infections may develop pelvic inflammatory disease. The disease may have serious consequences, including an increased probability of infertility and ectopic pregnancy.

**Structure**

Neisseria species are Gram-negative cocci, 0.6 to 1.0 µm in diameter. The organisms are usually seen in pairs with the adjacent sides flattened. Pili, hairlike filamentous appendages extend several micrometers from the cell surface and have a role in adherence. The outer membrane is composed of proteins, phospholipids, and lipopolysaccharide (LPS). Features that distinguish gonococcal LPS from enteric LPS are the highly branched basal oligosaccharide structure and the absence of repeating O-antigen subunits. For these reasons gonococcal LPS, as well as that of other mucosal pathogens, is referred to as lipooligosaccharide (LOS). Gonococci characteristically release outer membrane fragments (blebs) during growth. These blebs contain LOS and may have a role in pathogenesis.

### Classification and Antigenic Types

The gonococcus is an obligate human pathogen. It is one of two Neisseria species that cause significant human infections. The genus also includes several nonpathogenic species, which may be part of the normal flora and therefore can be confused with N gonorrhoeae. Gonococcal strains can be characterized according to their nutritional requirements (auxotyping). A panel of monoclonal antibodies specific for epitopes on protein I have also been used to type strains. Strains exhibiting specific reaction patterns are termed serovars. A combined auxotype-serovar classification provides greater resolution among gonococcal isolates and is useful in epidemiologic investigations.

### Pathogenesis

Our knowledge of the molecular basis of gonococcal pathogenesis is incomplete. Attachment of gonococci to mucosal cells is mediated in part by pili, although nonspecific factors such as surface charge and hydrophobicity may be important. Pili undergo both phase and antigenic variation. Opa proteins (protein II), which are located in the outer membrane, are also involved in attachment to host cells. Gonococci attach only to microvilli of nonciliated columnar epithelial cells; attachment to ciliated cells is not observed.

Gonococci can invade columnar epithelial cells, although they do not invade ciliated columnar epithelium of the genitourinary tract.

Much of our knowledge of gonococcal invasion comes from studies with tissue culture cells and human fallopian tube organ culture. After gonococci attach to the nonciliated epithelial cells of the fallopian tube, they are surrounded by the microvilli, which draw them to the surface of the mucosal cell. The gonococci appear to enter the epithelial cells by a process called parasite-directed endocytosis.

This process seems to be initiated by microbial factors because it does not occur unless the gonococci are viable and because it involves host cells that are not normally phagocytic. An unidentified factor in serum enhances engulfment of gonococci. The process is inhibited by drugs that block the actions of the microtubule (demecolcine) and microfilament (cytochalasin B) systems. During endocytosis the membrane of the mucosal cell retracts, pinching off a membrane-bound vacuole that contains gonococci; this vacuole is rapidly transported to the base of the cell, where gonococci are released by exocytosis into the subepithelial tissue. Gonococci are not destroyed within the phagocytic vacuole; it is not clear whether they replicate in the vacuoles.

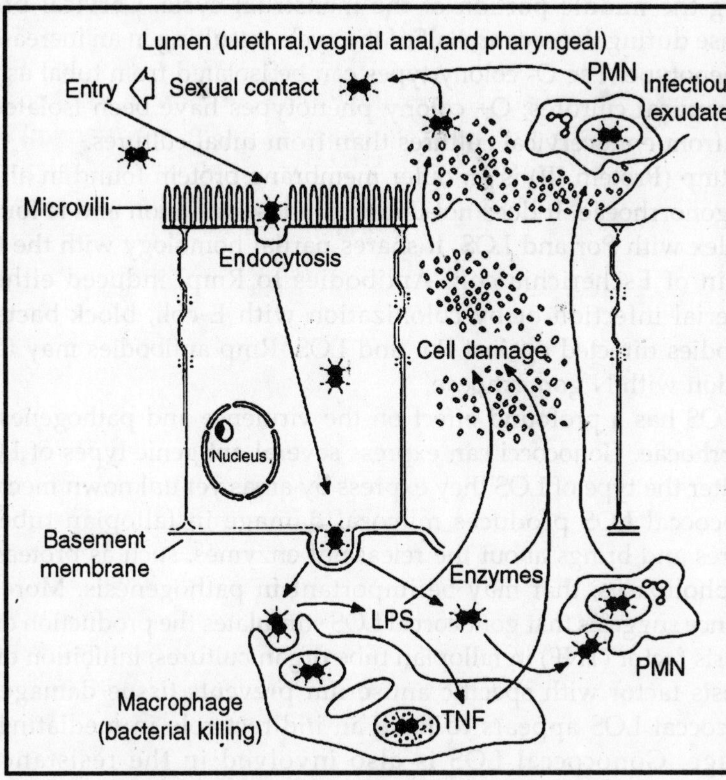


**Fig.** Pathogenesis of Uncomplicated Gonorrhea.

The major porin protein of the gonococcal outer membrane, Por (protein I), has been proposed as a candidate invasin (a substance that helps mediate invasion into a host cell). The insertion of Por into neutrophils treated with the chemotactic peptide, fMLP and leukotriene B4, inhibits degranulation but not the generation of the superoxide anion. The significance of these observations with respect to the pathogenesis of gonorrhea remains to be determined. Each gonococcal strain expresses only one type of Por; however, the Por of different strains may exhibit antigenic differences.

Gonococci can produce one or several outer membrane proteins called Opa proteins (proteins II). These proteins are subject to phase variation and are usually found on cells from colonies possessing an opaque phenotype (O+). At any one time, a gonococcus may express zero, one, or several different Opa proteins, though each strain has 10 or more genes for different Opas. Trypsin-like proteases present in cervical mucus may help select for protease-resistant transparent (O-) colony phenotypes. O+ colony phenotypes (protease sensitive) predominate in cultures taken during the middle portion of the menstrual cycle. Cervical proteases increase during the second half of the cycle, resulting in an increase in the O- phenotype. The O- colony types can be isolated from tubal as well as endocervical cultures; O+ colony phenotypes have been isolated more often from endocervical cultures than from tubal cultures.

Rmp (Protein III) is an outer membrane protein found in all strains of N gonorrhoeae. It does not undergo phase variation and is found in a complex with Por and LOS. It shares partial homology with the Omp A protein of Escherichia coli. Antibodies to Rmp, induced either by a neisserial infection or by colonization with E coli, block bactericidal antibodies directed against Por and LOS. Rmp antibodies may facilitate infection with N gonorrhoeae.

LOS has a profound effect on the virulence and pathogenesis of N gonorrhoeae. Gonococci can express several antigenic types of LOS and can alter the type of LOS they express by an as yet unknown mechanism. Gonococcal LOS produces mucosal damage in fallopian tube organ cultures and brings about the release of enzymes, such as proteases and phospholipases, that may be important in pathogenesis. More recent evidence suggests that gonococcal LOS stimulates the production of tumor necrosis factor (TNF) in fallopian tube organ cultures; inhibition of tumor necrosis factor with specific antiserum prevents tissue damage. Thus, gonococcal LOS appears to have an indirect role in mediating tissue damage. Gonococcal LOS is also involved in the resistance of N gonorrhoeae to the bactericidal activity of normal human serum. Oligosaccharides containing epitopes defined by specific monoclonal antibodies are associated with a serum-resistant phenotype.

Gonococci can utilize host-derived cytidine monophospho-N-acetylneuraminic acid (CMP-NANA) in vivo to sialylate the oligosaccharide component of its LOS, converting a serum-sensitive organism to a serum-resistant one. When such organisms are grown in vitro without CMP-NANA, their resistance to killing by normal human serum is rapidly lost. Organisms with non-sialylated LOS are more invasive than those with sialylated LOS. There is antigenic similarity between neisserial LOS and antigens present on human erthyrocytes. This similarity to self may preclude an effective immune response to these LPS antigens.

Gonococci are highly autolytic and release peptidoglycan fragments during growth. These fragments, released by bacterial and/or host peptidoglycan hydrolases, are toxic for fallopian tube mucosa and may contribute to the intense inflammatory reactions characteristic of gonococcal disease.

N gonorrhoeae is highly efficient at utilizing transferrin-bound iron for in vitro growth; many strains can also utilize lactoferrin-bound iron. Gonococci (and meningococci) bind only human transferrin and lactoferrin. This specificity is thought to be the reason these organisms are exclusively human pathogens. Nevertheless, the role of transferrin- and lactoferrin-bound iron in in vivo growth is unknown. Gonococci express several new proteins when grown under iron-restricted conditions similar to the conditions occurring in the host.

Some of these proteins function as receptors for transferrin, lactoferrin, heme, and hemoglobin; others function in the transport of iron into the cell. Gonococci cannot grow anaerobically unless low concentrations of the alternative electron acceptor nitrite are present. Under these conditions they produce novel proteins. These proteins are apparently produced during an infection because antibodies against them are present in the serum specimens of patients with uncomplicated gonorrhea, disseminated gonococcal infection, or pelvic inflammatory disease. These data suggest that some gonococci in the host are growing under anaerobic conditions. Further studies will determine the relevance of these proteins to pathogenesis.

Strains of N gonorrhoeae (and N meningitidis) produce two distinct extracellular IgA1 proteases, which cleave the heavy chain of human immunoglobulin A1 (IgA1) at different points within the hinge region. Type 1 protease cleaves a prolyl-seryl peptide bond and type 2 protease cleaves a prolyl-threonyl bond in the hinge region of the heavy chain. This region is missing in human IgA2, and so this isotype is not susceptible to cleavage. Each gonococcal or meningococcal isolate elaborates only one of these two enzymes. Split products of IgA1 have been found in the genital secretions of women with gonorrhea, suggesting that the gonococcal IgA1 protease is present and active during genital infection. Fab fragments of IgA1 may bind to the gonococcal cell surface and block the Fc-mediated functions of intact immunoglobulins.

### Host Defenses

Not everyone exposed to N gonorrhoeae acquires the disease. This may be due to variations in the size or virulence of the inoculum, to nonspecific resistance, or to specific immunity. A 50 per cent infective dose (ID50) of about 1,000 organisms has been established, based on the experimental urethral inoculation of male volunteers. There is no reliable ID50 for women, although it is assumed to be similar.

Nonspecific factors have been implicated in natural resistance to gonococcal infection. In women, changes in the genital pH and hormones may increase resistance to infection at certain times of the menstrual cycle. Urinary solutes exhibit bactericidal and bacteriostatic activity of N gonorrhoeae. Factors in urine that seem to be important are pH, osmolarity, and the concentration of urea. The variability in the susceptibility of gonococcal strains to the bactericidal and bacteriostatic properties of urine is thought to be one of the reasons some men do not develop a gonococcal infection when exposed.

Most uninfected individuals have serum antibodies that react with gonococcal antigens. These antibodies probably result from colonization or infection with various Gram-negative bacteria that possess cross-reactive antigens. These "natural" antibodies differ, both qualitatively and quantitatively, from person to person, but may be important in an individual's natural resistance or susceptibility to infection.

Infection with N gonorrhoeae stimulates both mucosal and systemic antibodies to a variety of gonococcal antigens. Mucosal antibodies are primarily IgA and IgG. In genital secretions, antibodies have been identified that react with Por, Opa and LOS, and some of the iron-regulated proteins. Vaccine trials have suggested that antipilus antibodies inhibit the pilus-mediated attachment of the homologous gonococcal strain. Complement is present in endocervical secretions, but in a much lower concentration than in blood. However, there is little evidence to support a role for a complement-mediated bactericidal defense mechanism on the genital mucosa. In general, the IgA response is brief and declines rapidly after treatment; IgG levels decline more slowly.

More information is available about the function of systemic humoral immune mechanisms in gonococcal infection. Gonococcal antigens such as pili, Por, Opa, Rmp, and LOS elicit a serum antibody response during an infection. Antipilus antibody levels tend to be higher in women than in men and are related to the number of previous infections. The predominant IgG subclass that reacts with a variety of gonococcal antigens is IgG3, followed by IgG1 and IgG4. IgG2 is minimal, suggesting that polysaccharides are not important in the immune response to gonococcal infection. Anti-Por antibodies may be bactericidal for the gonococcus. IgG that reacts with Rmp blocks the bactericidal activity of antibodies directed against Por and LOS. Genital infection with N gonorrhoeae stimulates a serum antibody response against the LOS of the infecting strain. Disseminated gonococcal infection results in higher levels of anti-LOS antibody than do genital infections.

Strains that cause uncomplicated genital infections usually are killed by normal human serum and are termed serum sensitive. This bactericidal activity is mediated by IgM and IgG that recognize sites on the LOS. Strains that cause disseminated infections are not killed by most normal human serum and are referred to as serum resistant. Resistance is mediated, in part, by IgA that blocks the IgG-mediated bactericidal activity of the serum. Serum specimens from convalescent patients with disseminating infections contain bactericidal IgG to the LOS of the infecting strain.

Individuals with inherited complement deficiencies have a markedly increased risk of acquiring systemic neisserial infections and are subject to recurring episodes of systemic gonococcal and meningococcal infections, indicating that the complement system is important in host defense.Gonococci activate complement by both the classic and alternative pathways. Complement activation by gonococci leads to the formation of the C5b-9 complex (membrane attack complex) on the outer membrane. In normal human serum, similar numbers of C5b-9 complexes are deposited on serum-sensitive and serum-resistant organisms, but the membrane attack complex is not functional on serum-resistant organisms. Other complement-mediated functions, such as opsonophagocytosis and chemotaxis, are more efficient with serum-sensitive than with serum-resistant gonococci. This may be a significant factor in the pathogenesis of disseminated gonococcal infection and probably contributes to the relative lack of genital symptoms observed with this disease.

Normal human serum contains opsonic anti-Por IgG. Antibodies to various surface-exposed antigens are also present in cervical and urethral secretions of patients with gonorrhea and probably contribute to the opsonophagocytosis of the organism. Opa is important in gonococcus-neutrophil interactions. Gonococci expressing certain Opas interact with neutrophils in the absence of antibodies. Once phagocytosed, gonococci are killed by both oxygen-dependent and oxygen-independent mechanisms. The survival of gonococci within neutrophils has been the subject of considerable controversy, with no clear-cut answer yet available. The opsonization and phagocytosis of gonococci are comparatively more important in mucosal infections than in protection from systemic gonococcal (and meningococcal) infections.

## Epidemiology

The only natural host for N gonorrhoeae is the human. Gonorrhea has all but disappeared in Scandinavia and several other European countries. In the United States, gonorrhea remains the most frequently reported infectious disease. Between 1977 and 1993, the number of reported cases decreased 56 per cent, from 1 million to 439,673 cases per year. The Centers for Disease Control (CDC) estimates that there are two unreported cases for every reported case of gonorrhea. Gonorrhea is transmitted almost exclusively by sexual contact. The highest rates occur in women between the ages of 15 and 19 years and in men 20 and 24 years of age. Persons who have multiple sex partners are at highest risk. Rates of gonorrhea are higher in males and in minority and inner-city populations.

Gonorrhea is usually contracted from a sex partner who is either asymptomatic or has only minimal symptoms. It is estimated that the efficiency of transmission after one exposure is about 35 per cent from an infected woman to an uninfected man and 50 to 60 per cent from an infected man to an uninfected woman. More than 90 per cent of men with urethral gonorrhea will develop symptoms within 5 days; fewer than 50 per cent of women with anogenital gonorrhea will do so. Women with asymptomatic infections are at higher risk of developing pelvic inflammatory disease and disseminated gonococcal infection.

## Diagnosis

Gonococcal infection produces several common clinical syndromes that have multiple causes or that mimic other conditions. Laboratory tests are often required to differentiate among the etiologic agents causing urethritis or cervicitis. The etiologic diagnosis of salpingitis and pelvic peritonitis is quite difficult because mixed infections are common and laparoscopy is required to obtain appropriate cultures. Gonococcal perihepatitis may mimic acute cholecystitis. All of the above syndromes are also caused by C trachomatis, a sexually transmitted bacterium that causes more infections in the United States than N gonorrhoeae. The gonococcal arthritis-dermatitis syndrome, must be, differentiated from meningococcemia and Reiter syndrome, in particular, and from other causes of septic arthritis.

Customarily, the laboratory diagnosis of gonorrhea is made presumptively and then confirmed; the latter process involves identifying characteristics that distinguish N gonorrhoeae from other Neisseria spp. that may be present in the specimen. Nonpathogenic Neisseria are normal inhabitants of the oropharynx and nasopharynx and occasionally are isolated from other sites infected by N gonorrhoeae. A presumptive diagnosis of gonorrhea may be made from Gram-stained smears of urethral, cervical, and rectal specimens if Gram-negative diplococci are observed within leukocytes; it is equivocal if only extracellular Gram-negative diplococci are seen and negative if no Gram-negative diplococci are seen. Gram stain diagnosis has a sensitivity and specificity of >95 per cent in men with symptomatic urethritis. The specificity of Gram stain diagnosis in women is also high if the cervix is wiped clean to remove cervical secretions before collecting the specimen; however, the sensitivity is only about 50 per cent. The sensitivity and specificity of the Gram stain for rectal specimens are lower than with cervical specimens.

Specimens for the laboratory diagnosis of gonorrhea should be collected before treating the patient. Ideally, specimens should be inoculated onto appropriate media and incubated immediately after collection at 35 to 36.5°C in a $CO_2$-enriched atmosphere, which can be obtained by using a candle extinction jar or a $CO_2$ incubator. Urethral specimens are normally obtained from heterosexual men; urethral, rectal, and pharyngeal specimens are normally obtained from homosexual men; and cervical and rectal specimens are normally obtained from women. Specimens are collected with cotton, polyester, or calcium alginate swabs. When appropriate, specimens may also be obtained from the urethra and from Bartholin's and Skene's glands of infected women. Blood cultures should be performed for patients with suspected disseminated infection. Synovial fluid cultures should be performed for patients with septic arthritis.

Urethral, cervical, and pharyngeal specimens are inoculated onto selective medium such as modified Thayer-Martin, Martin-Lewis, or NYC medium. These are complex media that contain antimicrobial and antifungal agents to inhibit the growth of unwanted organisms. Rectal specimens should be inoculated onto modified Thayer-Martin medium which contains trimethoprim lactate to inhibit the growth and swarming of Proteus species. Specimens collected from normally sterile sites such as blood, synovial fluid, and conjunctivae may be inoculated onto a nonselective medium such as chocolate agar.

The combination of oxidase-positive colonies and Gram-negative diplococci provides a presumptive identification of N gonorrhoeae. Fluorescent-antibody staining, coagglutination, specific biochemical tests, and DNA probes may be used for confirmation. DNA probes have also been used to detect gonococci in urethral and cervical specimens. A commercial test based on this approach is available. Serologic tests for uncomplicated gonorrhea have not proved satisfactory.

## Control

There is no effective vaccine to prevent gonorrhea. Candidate vaccines consisting of pilus protein or Por are of little benefit. The development of an effective vaccine has been hampered by the lack of a suitable animal model and the fact that an effective immune response has never been demonstrated. Condoms are effective in preventing the transmission of gonorrhea.

Contact tracing to identify source contacts (i.e., those who infected the index patient) has been useful in identifying asymptomatic individuals or those with ignored symptoms. Contact tracing has also been used to identify contacts who were exposed to the index patient and who may have become infected.

The evolution of antimicrobial resistance in N gonorrhoeae may ultimately affect the control of gonorrhea. Strains with multiple chromosomal resistance to penicillin, tetracycline, erythromycin, and cefoxitin have been identified in the United States and most other parts of the world. Sporadic high-level resistance to spectinomycin and fluoroquinolones have been reported.

Penicillinase-producing strains of N gonorrhoeae were first described in 1976. Five related β-lactamase plasmids of different sizes have been identified in these strains. The strains cause more than one-half of all gonococcal infections in parts of Africa and Asia. Their prevalence has increased dramatically in the United States since 1984 and has affected nearly every major metropolitan area.

Plasmid-mediated high-level resistance of N gonorrhoeae to tetracycline was first described in 1986 and has now been reported in most parts of the world. This resistance is due to the presence of the streptococcal tetM determinant on a gonococcal conjugative plasmid.

The current CDC Treatment Guidelines recommend treatment of all gonococcal infections with antibiotic regimens effective against resistant strains. The recommended antimicrobial agents are ceftriaxone, cefixime, ciprofloxacin, or oflaxacin. Since a significant proportion of patients with gonorrhea are also infected with C trachomatis, doxycycline or erythromycin has been added to treat this concomitant infection.

## *NEISSERIA MENINGITIDES*

### Clinical Presentation

N meningitidis infection results from the bloodborne dissemination (meningococcemia) of the meningococcus, usually following an asymptomatic or mildly symptomatic nasopharyngeal carrier state or a mild rhinopharyngitis. The mildest form is a transient bacteremic illness characterized by a fever and malaise; symptoms resolve spontaneously in 1 to 2 days. Acute meningococcemia is more serious and is often complicated by meningitis. The manifestations of meningococcal meningitis are similar to acute bacterial meningitis caused by organisms such as Streptococcus pneumoniae, Haemophilus influenzae, and E coli. The manifestations result from both infection and increased intracranial pressure. Chills, fever, malaise, and headache are the usual manifestations of infection; headache, vomiting, and rarely, papilledema may result from increased intracranial pressure. Signs of meningeal inflammation are also present. The onset of meningococcal meningitis may be abrupt or insidious.

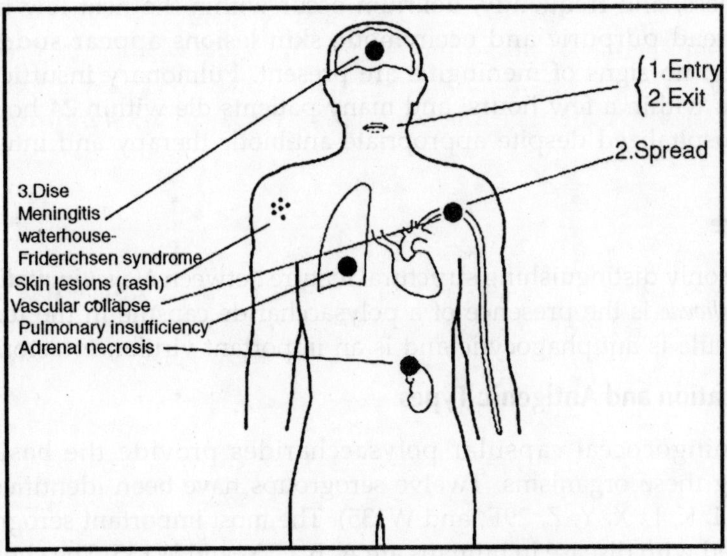


**Fig.** Clinical Manifestations of N Meningitidis Infection.

Infants with meningococcal meningitis rarely display signs of meningeal irritation. Irritability and refusal to take food are typical; vomiting occurs early in the disease and may lead to dehydration. Fever is typically absent in children younger than 2 months of age. Hypothermia is more common in neonates. As the disease progresses, apneic episodes, seizures, disturbances in motor tone, and coma may develop.

In older children and adults, speciric symptoms and signs are usually present, with fever and altered mental status the most consistent findings. Headache is an early, prominent complaint and is usually very severe. Nausea, vomiting, and photophobia are also common symptoms.

Neurologic signs are common; approximately one-third of patients have convulsions or coma when first seen by a physician. Signs of meningeal irritation such as cervical rigidity (Brudzinski sign), thoracolumbar rigidity, hamstring spasm (Kernig sign), and exaggerated reflexes are common.

Petechiae (minute hemorrhagic spots in the skin) or purpura (hemorrhages into the skin) occurs from the first to the third day of illness in 30 to 60 per cent of patients with meningococcal disease, with or without meningitis. The lesions may be more prominent in areas of the skin subjected to pressure, such as the axillary folds, the belt line, or the back.

Fulminant meningococcemia (Waterhouse-Friderichsen syndrome) occurs in 5 to 15 per cent of patients with meningococcal disease and has a high mortality rate. It begins abruptly with sudden high fever, chills, myalgias, weakness, nausea, vomiting, and headache. Apprehension, restlessness, and frequently, delirium occur within the next few hours. Widespread purpuric and ecchymotic skin lesions appear suddenly. Typically, no signs of meningitis are present. Pulmonary insufficiency develops within a few hours, and many patients die within 24 hours of being hospitalized despite appropriate antibiotic therapy and intensive care.

## Structure

The only distinguishing structural feature between *N meningitidis* and *N gonorrhoeae* is the presence of a polysaccharide capsule in the former. The capsule is antiphagocytic and is an important virulence factor.

## Classification and Antigenic Types

Meningococcal capsular polysaccharides provide the basis for grouping these organisms. Twelve serogroups have been identified (A, B, C, H, I, K, L, X, Y, Z, 29E, and W135). The most important serogroups associated with disease in humans are A, B, C, Y, and W135. The chemical composition of these capsular polysaccharides, where known, is listed in Table. The prominent outer membrane proteins of N meningitidis have been designated class 1 through class 5.

The class 2 and 3 proteins function as porins and are analogous to gonococcal Por. The class 4 and 5 proteins are analogous to gonococcal Rmp and Opa, respectively. Serogroup B and C meningococci have been further subdivided on the basis of serotype determinants located on the class 2 and 3 proteins. A few serotypes are associated with most cases of meningococcal disease, whereas other serotypes within the same serogroup rarely caused disease. All known group A strains have the same protein serotype antigens in the outer membrane. Another serotyping system is based on the antigenic diversity of meningococcal LOS. The LOS types are independent of the protein serotypes, although certain combinations frequently occur together.

**Table. Chemical Composition of *N. meningitidis* Capsular Polysaccharides**

| *Serogroup* | *Structural Repeating Unit* |
|---|---|
| Serogroup A°(homopolymer) | ManNA – (1 – P $\xrightarrow{\alpha}$ 6) –<br>3<br>\|<br>OAc |
| Serogroup B(homopolymer | NeuNAc – (2 $\xrightarrow{\alpha}$ B) – |

| Serogroup | Structure |
|---|---|
| Serogroup C°(homopolymer) | NeuNAc – (2 $\xrightarrow{\alpha}$ 9) – (7-OAc, 8-OAc) |
| Serogroup H (monosaccharide-glycerol repeating unit) | → 4)α – D – Gal – (1 → 2) – Gro – (3 – P → |
| Serogroup I (disaccharide repeating unit) | → 4)α – L – GulNAcA(1 → 3)3 – D – ManNAcA(→ (4 – OAc) |
| Serogroup L(trisaccharide repeating unit | → 3) – Β – D – GlcNAc – (1 → 3)Β – D – GlcNAc – (1 → 3)α – D – GlucNAc – (1 – P– |
| Serogroup W135 (disaccharide repeating unit) | 6 – D – Gal(1 $\xrightarrow{\alpha}$ 4) – NeuNAc(2 $\xrightarrow{\alpha}$ 6) – |
| Serogroup X(homopolymer | GlcNAc(1 – P → 4) – |
| Serogroup Y (BO) (disaccharide repeating unit) | 6 – D – Glc(1 $\xrightarrow{\alpha}$ 4) – NeuNAc(2 $\xrightarrow{\alpha}$ 6) (OAc) |
| Serogroup Z (onosaccharide giycerol repeating unit) | D – GalNAc(1 $\xrightarrow{\alpha}$ 1) – Gro – (3'.P $\xrightarrow{\alpha}$ 4) |
| Serogroup 29E (disaccharide repeating | D – GalNAc(1 $\xrightarrow{\beta}$ 7) – KDO(2 $\xrightarrow{\alpha}$ 3) – (4.5-OAc) |

## Pathogenesis

The human nasopharynx is the only known reservoir of *N meningitidis*. Meningococci are spread via respiratory droplets, and transmission requires aspiration of infective particles. Meningococci attach to the nonciliated columnar epithelial cells of the nasopharynx. Attachment is mediated by pili and possibly by other outer membrane components. Invasion of the mucosal cells occurs by a mechanism similar to that observed with gonococci. However, once internalized, meningococci remain in an apical location within the epithelial cell; the route by which they gain access to the subepithelial space remains unclear. Trimers of class 2 and 3 proteins have the ability to translocate from intact cells and insert into eukaryotic cell membranes to form voltage-dependent channels. This process may be important in invasion.

Purified meningococcal LOS is highly toxic and is as lethal for mice as the LOS from E coli or Salmonella typhimurium; however, meningococcal LOS is 5 to 10 times more effective than enteric LPS in eliciting a dermal Shwartzman reaction in rabbits. Meningococcal LPS suppresses leukotriene B4 synthesis in human polymorphonuclear leukocytes. The loss of leukotriene B4 deprives the leukocytes of a strong chemokinetic and chemotactic factor. The events after bloodstream invasion are unclear. Relatively little information is known about how the meningococcus enters the central nervous system.

### Host Defenses

The integrity of the pharyngeal and respiratory epithelium may be important in protection from invasive disease. Chronic irritation of the mucosa due to dust or low humidity, or damage to the mucosa resulting from a concurrent viral or mycoplasmal upper respiratory infection, may be predisposing factors for invasive disease.

The presence of serum bactericidal IgG and IgM is probably the most important host factor in preventing invasive disease. These antibodies are directed against both capsular and noncapsular surface antigens. The antibodies are produced in response to colonization with carrier strains of *N meningitidis, N lactamica,* or other nonpathogenic *Neisseria* species. Protective antibodies are also stimulated by cross-reacting antigens on other bacterial species. The role of bactericidal antibodies in prevention of invasive disease explains why high attack rates are seen in infants from 6 to 9 months old, the age at which maternally acquired antibodies are being lost.

The immunity conferred by specific antibody may not be absolute. Illness has been documented in individuals with levels of antibodies considered to be protective. It has been postulated that the activity of the bactericidal antibodies might be blocked by IgA, induced by other meningococcal strains, or by cross-reacting antigens on enteric or other respiratory bacteria. Since IgA does not bind complement, it may block binding sites for the bactericidal IgG and IgM. Persons with complement deficiencies (C5, C6, C7, or C8) may develop meningococcemia despite protective antibody. This underscores the importance of the complement system in protection from meningococcal disease.

### Epidemiology

The meningococcus usually inhabits the human nasopharynx without causing detectable disease. This carrier state may last for a few days to months and is important because it not only provides a reservoir for meningococcal infection but also enhances host immunity. Between 5 and 30 per cent of normal individuals are carriers at any given time, yet few develop meningococcal disease. Even during epidemics of meningococcal meningitis in military recruits, when the carrier rate may reach 95 per cent, the incidence of systemic disease is less than 1 per cent. Meningococcal carriage rates are highest in older children and young adults, but the attack rates are higher in children, peaking at 5 years of age (group B) and 4 to 14 years of age (group C). The low incidence of disseminated disease following colonization suggests that host rather than bacterial factors play an important determining role.

Meningococcal meningitis occurs sporadically and in epidemics, with the highest incidence during late winter and early spring. Most epidemics are caused by group A strains, but small outbreaks have occurred with group B and C strains. Sporadic cases generally are caused by group B, C, and Y strains. Whenever group A strains become prevalent in the population, the incidence of meningitis increases markedly.

**Diagnosis**

The most characteristic manifestation of meningococcemia is the skin rash, which is essential for its recognition. Petechiae are the most common type of skin lesion. Ill-defined pink macules and maculopapular lesions also occur. Lesions are sparsely distributed over the body. They tend to occur in crops and on any part of the body; however, the face is usually spared and involvement of the palms and soles is less common. The skin rash may progress from a few ill-defined lesions to a widespread eruption within a few hours.

Acute bacterial meningitis has characteristic signs and symptoms. Except in epidemic situations, it is difficult to identify the causative agent without laboratory tests.

In cases of suspected meningococcal disease, specimens of blood, cerebrospinal fluid, and nasopharyngeal secretions should be collected before administration of any antimicrobial agents and examined for the presence of *N meningitidis*. Success in isolation is reduced by prior therapy; however, the microscopic diagnosis is not significantly affected. The cerebrospinal fluid should be concentrated by centrifugation and a portion of the sediment cultured on chocolate or blood agar. The plates should be incubated in a candle jar or $CO_2$ incubator. The presence of oxidase-positive colonies and Gram-negative diplococci provides a presumptive identification of *N meningitidis*. Production of acid from glucose and maltose but not sucrose, lactose, or fructose may be used for confirmation. The serologic group may be determined by a slide agglutination test, using first polyvalent and then monovalent antisera.

Nasopharyngeal specimens must be obtained from the posterior nasopharyngeal wall behind the soft palate and then should be inoculated onto a selective medium such as Thayer-Martin medium and processed as above.

Blood specimens are inoculated in 10- to 15-ml aliquots onto each of three blood bottles to give a final concentration of 10% (vol/vol). Evacuated bottles should be vented. Some strains of N meningitidis are inhibited by the sodium polyanetholsulfonate contained in blood medium. Toxicity may be overcome by the addition of gelatin. Sodium amylosulfate is not toxic for the meningococcus. Blood cultures are subcultured blindly onto chocolate or blood agar for confirmation.

Gram-stained smears of cerebrospinal fluid may be diagnostic; however, finding neisseriae in these smears is often more difficult than finding the strains that cause pneumococcal meningitis. Quellung tests may be of value.

## Control

Group A, C, Y, and W135 capsular polysaccharide vaccines are available and can be used to control outbreaks due to the meningococcal serogroups covered by the vaccine. The A, C, AC, and ACYW135 polysaccharide formulations are currently licensed in the United States. The polysaccharide vaccines are ineffective in young children, and the duration of protection is limited in children vaccinated at 1 to 4 years of age. Routine vaccination of the civilian population in industrialized countries is not currently recommended because the risk of infection is low and most endemic disease occurs in young children. The group B capsular polysaccharide is a homopolymer of sialic acid and is not immunogenic in humans. A group B meningococcal vaccine consisting of outer membrane protein antigens has recently been developed but is not licensed in the United States.

Meningococcal disease arises from association with infected individuals, as evidenced by the 500- to 800-fold greater attack rate among household contacts than among the general population. Because such household members are at high risk, they require chemoprophylaxis. Sulfonamides were the chemoprophylactic agent of choice until the emergence of sulfonamide-resistant meningococci. At present, approximately 25 per cent of clinical isolates of N meningitidis in the United States are resistant to sulfonamides; rifampin is therefore the chemoprophylactic agent of choice.

Penicillin is the drug of choice to treat meningococcemia and meningococcal meningitis. Although penicillin does not penetrate the normal blood-brain barrier, it readily penetrates the blood-brain barrier when the meninges are acutely inflamed. Either chloramphenicol or a third-generation cephalosporin such as cefotaxime or ceftriaxone is used in persons allergic to penicillins.

## MORAXELLA

Moraxella species are parasites of the mucous membranes of humans and other warm-blooded animals. Many species are nonpathogenic. M lacunata can be isolated from the eyes and may cause conjunctivitis in humans living under conditions of poor hygiene. M nonliquefaciens is found in the upper respiratory tract, especially the nose, and may be a secondary invader in respiratory infections. M urethralis can be isolated from urine and the female genital tract. Some strains formerly designated as Mima polymorpha subsp oxidans belong in this species. These organisms can be mistaken for N gonorrhoeae unless appropriate biochemical characteristics are determined.

M catarrhalis organisms are cocci that morphologically resemble Neisseria cells. Other relevant characteristics are presented in Table. M catarrhalis was formerly placed in the genus Neisseria; however, studies of DNA base content, fatty acid composition, and genetic transformation showed that this organism did not belong in that genus. M catarrhalis is a member of the normal flora in 40-50% of normal school children; however, it should be considered more than a harmless commensal of the mucous membranes of humans.

It is an infrequent, yet significant, cause of severe systemic infections such as pneumonia, meningitis, and endocarditis. It is an important cause of lower respiratory tract infections in adults with chronic lung disease and a common cause of otitis media, sinusitis, and conjunctivitis in otherwise healthy children and adults. M catarrhalis may cause clinical syndromes indistinguishable from those caused by gonococci, and so it is important to distinguish these organisms from one another. Many strains produce β-lactamase.

## KINGELLA

Kingella kingae and K denitrificans are oxidase-positive non-motile organisms that are hemolytic when grown on blood agar. They are gram-negative rods, but may resemble coccobacilli or diplococci. They are part of the normal oral flora and occasionally cause infections of bone, joints, and tendons. The organism may enter the circulation with minor oral trauma such as tooth brushing. It is susceptible to penicillin, ampicillin, and erythromycin.

## EIKENELLA

Eikenella corrodens is a small oxidase-positive, fastidious gram-negative rod, which requires carbon dioxide for growth. Many isolates form pits in agar during growth on solid medium. E. corrodens is part of the gingival and bowel flora in 40-70% of humans and may be found in mixed flora infections associated with contamination from these sites. It occurs frequently in infections from human bites. E corrodens is resistant to clindamycin, but susceptible to ampicillin and third generation cephalosporins.

# Chapter 17

# Bacillus

Bacillus species are aerobic, sporulating, rod-shaped bacteria that are ubiquitous in nature. Bacillus anthracis, the agent of anthrax, is the only obligate Bacillus pathogen in vertebrates. Bacillus larvae, B lentimorbus, B popilliae, B sphaericus, and B thuringiensis are pathogens of specific groups of insects. A number of other species, in particular B cereus, are occasional pathogens of humans and livestock, but the large majority of Bacillus species are harmless saprophytes.

Anthrax has afflicted humans throughout recorded history. The fifth and sixth plagues of Egypt described in Exodus are widely believed to have been anthrax. The disease was featured in the writings of Virgil in 25 BC and was familiar in medieval times as the Black Bane. It was from studies on anthrax that Koch established his famous postulates in 1876, and vaccines against anthraxthe best known being that of Pasteur (1881)were among the first bacterial vaccines developed.

Bacillus species are used in many medical, pharmaceutical, agricultural, and industrial processes that take advantage of their wide range of physiologic characteristics and their ability to produce a host of enzymes, antibiotics, and other metabolites. Bacitracin and polymyxin are two well-known antibiotics obtained from Bacillus species. Several species are used as standards in medical and pharmaceutical assays.

The spores of the obligate thermophile B stearothermophilus are used to test heat sterilization procedures, and B subtilis subsp globigii, which is resistant to heat, chemicals, and radiation, is widely used to validate alternative sterilization and fumigation procedures. Certain Bacillus species are important in the natural or artificial degradation of waste products. Some Bacillus insect pathogens are used as the active ingredients of insecticides.

Because the spores of many Bacillus species are resistant to heat, radiation, disinfectants, and desiccation, they are difficult to eliminate from medical and pharmaceutical materials and are a frequent cause of contamination. Bacillus species are well known in the food industries as troublesome spoilage organisms.

## CLINICAL MANIFESTIONS

Although anthrax remains the best-known Bacillus disease, in recent years other Bacillus species have been increasingly implicated in a wide range of infections including abscesses, bacteremia/septicemia, wound and burn infections, ear infections, endocarditis, meningitis, ophthalmitis, osteomyelitis, peritonitis, and respiratory and urinary tract infections. Most of these occur as secondary or mixed infections or inimmunodeficient or otherwise immunocompromised hosts (such as alcoholics and diabetics), but a significant proportion are primary infections in otherwise healthy individuals.

Some of these infections are severe or lethal. Of the species listed in Table, most frequently implicated in these types of infection is B cereus, followed by B licheniformis and B subtilis. Bacillus alvei, B brevis, B circulans, B coagulans, B macerans, B pumilus, B sphaericus, and B thuringiensis cause occasional infections. As secondary invaders, Bacillus species may exacerbate preexisting infections by producing either tissue-damaging toxins or metabolites such as penicillinase that interfere with treatment.

Bacillus cereus is well known as an agent of food poisoning, and a number of other Bacillus species, particularly B subtilis and B licheniformis, are also incriminated periodically in this capacity.

## ANTHRAX

Anthrax is primarily a disease of herbivores. Humans acquire it as a result of contact with infected animals or animal products. In humans the disease takes one of three forms, depending on the route of infection. Cutaneous anthrax, which accounts for more than 95 per cent of cases worldwide, results from infection through skin lesions; intestinal anthrax results from ingestion of spores, usually in infected meat; and pulmonary anthrax results from inhalation of spores.

Cutaneous anthrax usually occurs through contamination of a cut or abrasion, although in some countries biting flies may also transmit the disease. After a 2- to 3-day incubation period, a small pimple or papule appears at the inoculation site. A surrounding ring of vesicles develops. Over the next few days, the central papule ulcerates, dries, and blackens to form the characteristic eschar. The lesion is painless and is surrounded by marked edema that may extend for some distance. Pus and pain appear only if the lesion becomes infected by a pyogenic organism. Similarly, marked lymphangitis and fever usually point to a secondary infection. In most cases the disease remains limited to the initial lesion and resolves spontaneously. The main dangers are that a lesion on the face or neck may swell to occlude the airway or may give rise to secondary meningitis.

If host defenses fail to contain the infection, however, fulminating septicemia develops. Approximately 20 per cent of untreated cases of cutaneous anthrax progress to fatal septicemia. However, B anthracis is susceptible to penicillin and other common antibiotics, so effective treatment is almost always available.

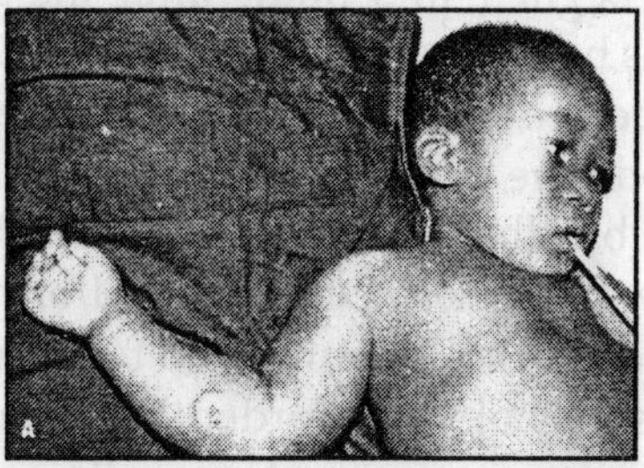

Intestinal anthrax is analogous to cutaneous anthrax but occurs on the intestinal mucosa. As in cutaneous anthrax, the organisms probably invade the mucosa through a preexisting lesion. organisms spread from the mucosal lesion to the lymphatic system. In pulmonary anthrax, inhaled spores are transported by alveolar macrophages to the mediastinal lymph nodes, where they germinate and multiply to initiate systemic disease. Gastrointestinal and pulmonary anthrax are both more dangerous than the cutaneous form because they are usually identified too late for treatment to be effective.

Herbivorous animals, the primary hosts of B anthracis, contract the infection by ingesting spores on forage plants; the spores are derived from soil or dust or are deposited on leaves by flies after feeding on an anthrax-infected carcass. If the spores enter a lesion in the gastrointestinal mucosa, they germinate and are taken into the bloodstream and lymphatics, finally producing systemic anthrax, which is usually fatal.

Symptoms prior to fulminant systemic anthrax may be absent or mild, consisting, for example, of malaise, low fever, and mild gastrointestinal symptoms in the case of gastrointestinal disease. During this phase the organism is multiplying and producing toxin in the regional lymph nodes and spleen. Released toxin causes breakdown of these organs probably of the spleen in particular. This causes the sudden onset of hyperacute illness with dyspnea, cyanosis, high fever, and disorientation, which progress in a few hours to shock, coma, and death. Although symptoms vary somewhat with the host species, this final acute phase is marked by a high-grade bacteremia. In humans, blood cultures are not always positive.

## BACILLUS FOOD POISONING

Bacillus cereus can cause two distinct types of food poisoning. The diarrheal type is characterized by diarrhea and abdominal pain occurring 8 to 16 hours after consumption of the contaminated food. It is associated with a variety of foods, including meat and vegetable dishes, sauces, pastas, desserts, and dairy products. In emetic disease, on the other hand, nausea and vomiting begin 1 to 5 hours after the contaminated food is eaten. Boiled rice that is held for prolonged periods at ambient temperature and then quick-fried before serving is the usual offender, although dairy products or other foods are occasionally responsible. The symptoms of food poisoning caused by other Bacillus species (B subtilis, B licheniformis, and others) are less well defined. Diarrhea and/or nausea occurs 1 to 14 hours after consumption of the contaminated food. A wide variety of food types have proved responsible in recorded instances.

A Bacillus food poisoning episode usually occurs because spores survive cooking or pasteurization and then germinate and multiply when the food is inadequately refrigerated. The symptoms of B cereus food poisoning are caused by a toxin or toxins produced in the food during this multiplication. Toxins have not yet been identified for other Bacillus species that cause food poisoning.

## STRUCTURE AND CLASSIFICATION

The family Bacillaceae, consisting of rod-shaped bacteria that form endospores, has two principal subdivisions: the anaerobic spore-forming bacteria of the genus Clostridium, and the aerobic or facultatively anaerobic spore-forming bacteria of the genus Bacillus frequently known as ASB (aerobic spore-bearers). Bacterial cells of Bacillus cultures are Gram positive when young, but in some species become Gram negative as they age.

Most Bacillus species are saprophytes. Table lists the identifying characteristics of some of the species most likely to be encountered by the physician. Not only are Bacillus endospores resistant to hostile physical and chemical conditions, but also various species have unusual physiologic properties that enable them to survive or thrive in harsh environments, ranging from desert sands and hot springs to Arctic soils and from fresh waters to marine sediments. The genus includes thermophilic, psychrophilic, acidophilic, alkaliphilic, halotolerant, and halophilic representatives, which are capable of growing at temperatures, pH values, and salt concentrations at which few other organisms could survive.

Figure shows the structure of a generalized Bacillus endospore (details of the structure differ from species to species). One spore is produced per vegetative cell. The central protoplast, or germ cell, carries the constituents of the future vegetative cell, accompanied by dipicolinic acid, which is essential to the heat resistance of the spore. Surrounding the protoplast is a cortex consisting largely of peptidoglycan (murein), which is also important in the heat and radiation resistance of the spore. The inner layer, the cortical membrane or protoplast wall, becomes the cell wall of the new vegetative cell when the spore germinates. The spore coats, which constitute up to 50 per cent of the volume of the spore, protect it from chemicals, enzymes, etc.

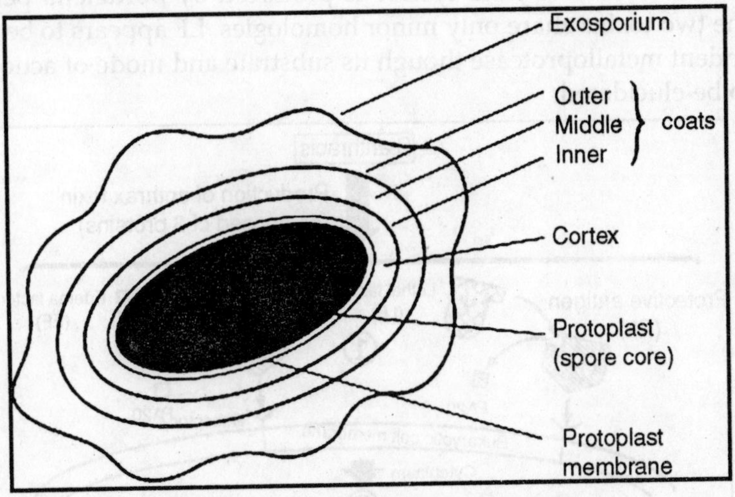


**Fig.** Cross Section of a Bacillus Spore.

The events involved in sporulation of vegetative cells and in germination of spores are complex and are influenced by factors such as temperature, pH, and the availability of certain divalent cations and carbon- and nitrogen-containing compounds. Spores formed under different conditions have different stabilities and degrees of resistance to heat, radiation, chemicals, desiccation, and other hostile conditions.

## PATHOGENISIS

The pathogenicity of B anthracis depends on two virulence factors: a poly-y-D-glutamic acid polypeptide capsule, which protects it from phagocytosis by the defensive phagocytes of the host, and a toxin produced in the log phase of growth. This toxin consists of three proteins: protective antigen (PA) (82. 7 kDa), lethal factor (LF) (90.2 kDa), and edema factor (EF) (88.9 kDa). Host proteases in the blood and on the eukaryotic cell surface activate protective antigen by cutting off a 20-kDa segment, exposing a binding site for LF and EF. The activated 63 kDa PA polypeptide binds to specific receptors on the host cell surface, thereby creating a secondary binding site for which LF and EF compete. The complex (PA+LF or PA+EF) is internalized by endocytosis and, following acidification of the endosome, the LF or EF cross the membrane into the cytosol via PA-mediated ion-conductive channels. This is analogous to the A-B structure-function model of cholera toxin with PA behaving as the B (binding) moiety. EF, responsible for the characteristic edema of anthrax, is a calmodulin-dependent adenylate cyclase. (Calmodulin is the major intracellular calcium receptor in eukaryotic cells.) The only other known bacterial adenylate cyclase is produced by Bordetella pertussis, but the two toxins share only minor homologies. LF appears to be a zinc-dependent metalloprotease though its substrate and mode of action have yet to be elucidated.

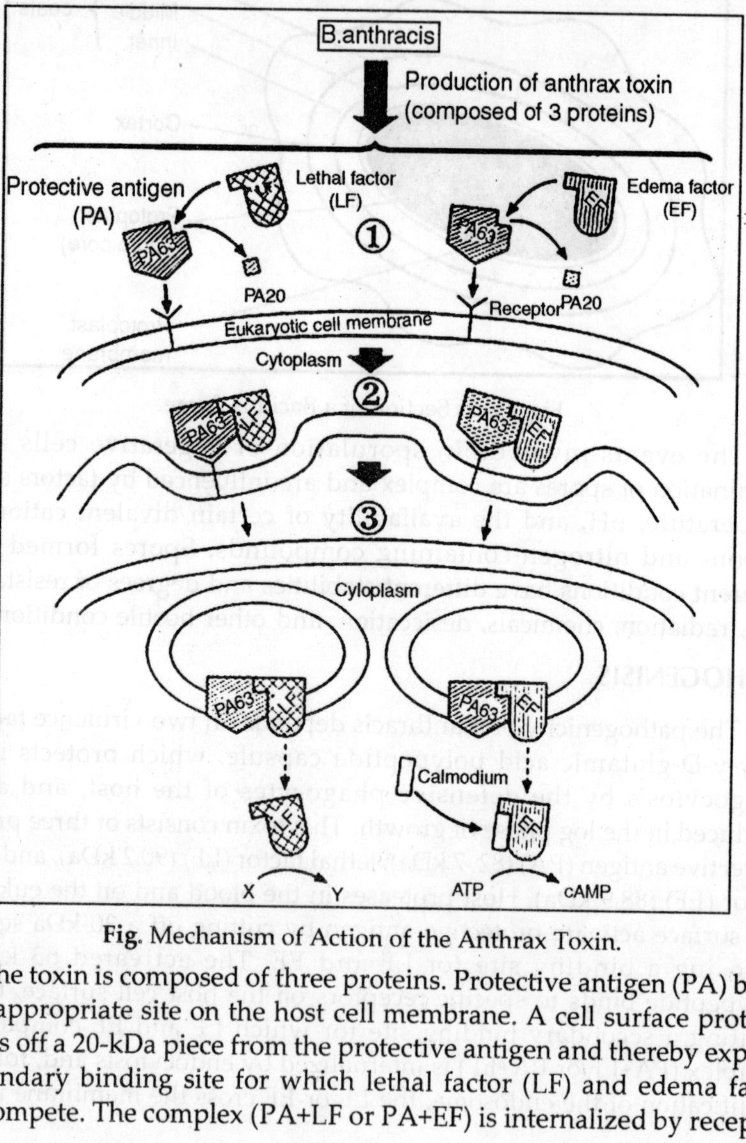


**Fig.** Mechanism of Action of the Anthrax Toxin.

The toxin is composed of three proteins. Protective antigen (PA) binds to an appropriate site on the host cell membrane. A cell surface protease cleaves off a 20-kDa piece from the protective antigen and thereby exposes a secondary binding site for which lethal factor (LF) and edema factor (EF) compete. The complex (PA+LF or PA+EF) is internalized by receptor-mediated endocytosis, and acidification of the endosome results in the transfer of the LF or EF across the endosome membrane into the cytosol where they carry out their catalytic actions. (Model by S.H. Leppla, Ph.D., Laboratory of Microbial Ecology, National Institutes of Health, Bethesda, MD.)

The toxin and capsule of B anthracis are encoded on two large plasmids called pXO 1 (110 MDa) and pX02 (60 MDa), respectively. Strains lacking either of these plasmids have greatly reduced virulence. The attenuated live vaccine strain developed by Sterne in 1937, which is still the basis of most anthrax vaccines for livestock, lacks pX02 and is therefore Cap- Tox+. The protection afforded by such vaccines apparently is related primarily to antibodies specific for the protective antigen component of the toxin. In contrast, the attenuated vaccine strains developed by Pasteur 110 years ago were inadvertently cured of pXO1 (by subculturing at 42° to 43°C); these Pasteur strains are therefore Cap+ Tox-. Strains of this type do not induce protective immunity; the partial effectiveness of Pasteur's vaccines is now believed to have been due to the residual uncured (Cap+ Tox+) cells they contained, and this would also explain the partial virulence of these strains.

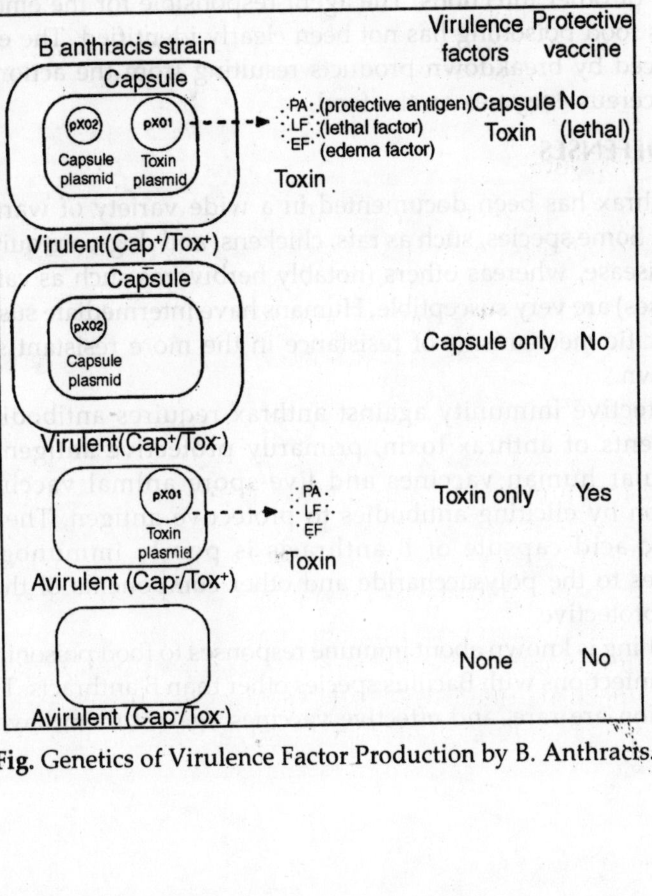


**Fig.** Genetics of Virulence Factor Production by B. Anthracis.

Plasmids pX01 and pX02 encode, respectively, the anthrax toxin and capsule. Curing the bacteria of pX01 produces an encapsulated, nontoxigenic strain that is nonprotective.

Curing of pX02 produces a toxigenic nonencapsulating strain that can be used as a protective vaccine. Production of protective antigen is essential for a strain to be protective.

The only other Bacillus species for which virulence factors have been identified is B cereus. A 38 to 46-kDa protein complex has been shown in animal models to cause necrosis of the skin or intestinal mucosa, to induce fluid accumulation in the intestine, and to be a lethal toxin. This protein is believed to be responsible for the necrotic and toxemic nature of severe B cereus infections and for the diarrheal form of food poisoning.

Bacillus cereus also produces two hemolysins; one of these, cereolysin (58 kDa), is a potent necrotic and lethal toxin. Although this toxin is neutralized by serum cholesterol, it probably contributes to the pathogenesis of B cereus infections.

Little is known about the other hemolysin at present. Phospholipases produced by B cereus may act as exacerbating factors by degrading host cell membranes following exposure of their phospholipid substrates in wounds or other infections. The agent responsible for the emetic type of B cereus food poisoning has not been clearly identified. The emesis may be induced by breakdown products resulting from the action of one or more B cereus enzymes on the food.

## HOST DEFENSES

Anthrax has been documented in a wide variety of warm-blooded animals. Some species, such as rats, chickens, and dogs, are quite resistant to the disease, whereas others (notably herbivores such as cattle, sheep, and horses) are very susceptible. Humans have intermediate susceptibility. The specific mechanisms of resistance in the more resistant species are not known.

Protective immunity against anthrax requires antibodies against components of anthrax toxin, primarily protective antigen. Both the noncellular human vaccines and live-spore animal vaccines confer protection by eliciting antibodies to protective antigen. The poly-g-D-glutamic acid capsule of B anthracis is poorly immunogenic, and antibodies to the polysaccharide and other components of the cell wall are not protective.

Nothing is known about immune responses to food poisoning or other types of infections with Bacillus species other than B anthracis. These types of infection are rare, and effective vaccines against them have not been developed.

## EPIDEMIOLOGY

The ultimate reservoir of B anthracis is contaminated soil, in which spores remain viable for long periods. Herbivores, the primary hosts, become infected when foraging in a contaminated region. Because the organism does not depend on an animal reservoir, it cannot readily be eradicated from a region, and anthrax remains endemic in many countries. Humans become infected almost exclusively through contact with infected animals or animal products. Human anthrax is traditionally classified as either nonindustrial or industrial anthrax, depending on whether the disease is acquired directly from animals or indirectly during handling of contaminated animal products. Nonindustrial anthrax usually affects people who work with animals or animal carcasses, such as farmers, veterinarians, knackers, and butchers, and is almost always cutaneous. Industrial anthrax, acquired from handling contaminated hair, hides, wool, bone meal, or other animal products, has a higher chance of being pulmonary as a result of the inhalation of spore-laden dust.

The development of an effective animal vaccine in the 1930s, together with improved factory hygiene, introduction of procedures for sterilizing imported animal products, replacement of animal products with man-made alternatives, and the availability since the mid-1960s of a human vaccine, has resulted in a greatly reduced incidence of the disease in North America. Human anthrax is now very rare in the United States. However, major epidemics still break out in endemic countries, normally following an outbreak in livestock. Nonendemic countries must remain alert for episodes of anthrax arising from imported animal products.

## DIAGNOSIS

The clinical diagnosis of anthrax is confirmed by directly visualizing or culturing the anthrax bacilli. Fresh smears of vesicular fluid, fluid from under the eschar, blood, lymph node or spleen aspirates, or (in meningitic cases) cerebrospinal fluid are stained with polychrome methylene blue (M'Fadyean's stain) and examined for the characteristic square-ended, blue-black bacilli surrounded by a pink capsule. (It should be remembered that B anthracis organisms are not invariably detected in stained blood smears of humans dying of anthrax.) Alternatively, the bacilli may be cultured from these specimens and checked for sensitivity to the anthrax gamma phage, for penicillin sensitivity, and for capsule formation. Colonies grown overnight at 37°C on blood agar are gray or white, nonhemolytic, with a dry, ground-glass appearance; they are at least 3 mm in diameter and sometimes have tails. Capsules can be seen in polychrome methylene blue-stained smears of cultures grown on nutrient agar containing 0.7 per cent sodium bicarbonate and incubated overnight under $CO_2$ (e.g., in a candle jar); encapsulated colonies are mucoid. Alternatively, 2 ml of blood (such as commercial defibrinated horse blood) inoculated with a pinhead quantity of material from a suspected colony and incubated at 37°C yields readily demonstrable encapsulated bacilli in 6 hours. Culturing may be unsuccessful if the patient has been treated with antibiotics.

Isolation of *B anthracis* from old specimens or from animal or environmental material being examined for public health purposes is more difficult, particularly if, as is often the case, *B cereus* or other Bacillus species are present in substantial numbers. The specimen should be examined both unheated and heated to 60°C to 65°C for 15 min with subculture to both blood or nutrient agar and specialized selective agars. Very rarely it may be necessary to use mouse or guinea pig inoculation to isolate B anthracis. Up to about 0.2 ml of the specimen (or an aqueous extract of the specimen) is injected subcutaneously into a mouse, or intramuscularly or subcutaneously in a guinea pig (more sensitive than a mouse); the encapsulated bacilli can be seen in a smear of blood aspirated from the heart of the animal at death, and the bacteria are readily observed in and isolated from this blood. If soil samples are being used, the animals should be injected 24 hours earlier with tetanus and gas gangrene antitoxin.

When a specimen from an individual not suspected clinically of having anthrax yields substantial numbers of Gram-positive bacilli, the specimen should be cultured and tested as shown in Figure to determine the Bacillus species present. The most common Bacillus species may be identified by the characteristics in Table. Incrimination of a Bacillus species as the cause of an infection is usually based on its presence in large numbers at the infection site, especially in the absence of other known pathogens. Since Bacillus species are common environmental organisms, their presence in small numbers is not generally considered significant. For this reason, the use of selective or enrichment systems for isolating clinically relevant, nonanthrax Bacillus species is confined to just a few situations, such as the retrospective examination of feces several days after a food poisoning incident (by which time the offending Bacillus organism may be present in only small numbers).

## CONTROL

To comprehend the strategies used to control anthrax, it is important to understand the cycle of infection in susceptible animals. As a susceptible animal with anthrax approaches death, its blood contains as many as $10^9$ bacilli/ml (depending on the species). Necrosis of the walls of small blood vessels during the acute phase of the illness leads to hemorrhages and to characteristic bloody exudations from the mouth, nose, and anusa highly diagnostic sign. These exudates carry vast numbers of the bacilli, which sporulate on exposure to air and produce a heavily contaminated environmental site that is potentially capable of infecting other animals for many years.

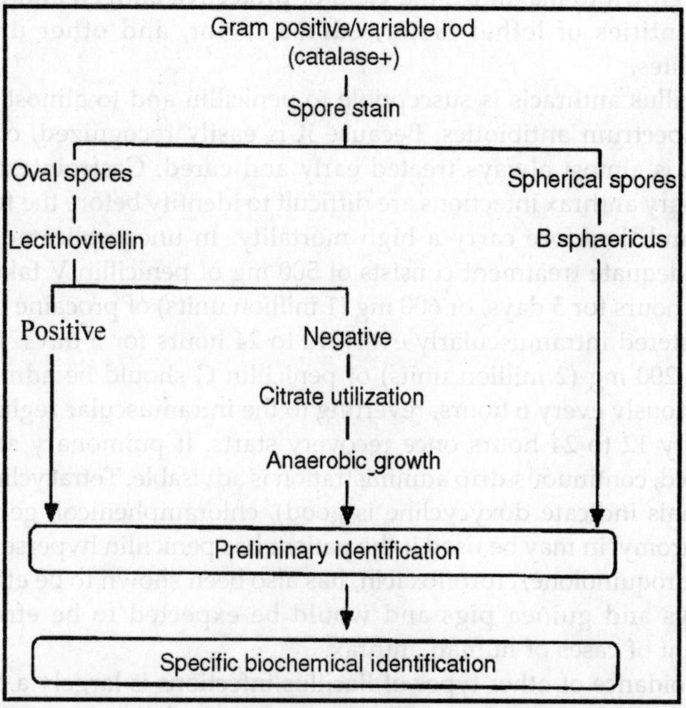


**Fig.** Flow Chart for Identification of Principal Bacillus Species.

Because sporulation of B anthracis requires oxygen and therefore does not occur inside a closed carcass, regulations in most countries forbid postmortem examination of animals when anthrax is suspected. The vegetative cells in the carcass are killed in a few days by the process of putrefaction. Nevertheless, in the case of livestock, legislation invariably requires that the carcass be burned or buried in quicklime (calcium oxide). However, it is becoming increasingly apparent in the new era of sensitivity about environmental contamination that implementation in the past (sometimes many decades) of the order to bury in quicklime has left us a legacy of burial sites which are contaminated with viable anthrax spores and it is to be hoped that this instruction will be removed from veterinary public health orders.

Livestock in endemic areas are effectively protected by yearly inoculations with a vaccine made from spores of a live attenuated strain. Noncellular vaccines for human use are available for individuals in high-risk occupations. They appear to have contributed to the decline in incidence of industrial anthrax since they became available in the 1960s, but animal studies suggest that there are limitations to their ability to protect against anthrax. The human vaccine available in the United States is an aluminum hydroxide-adsorbed cell-free filtrate of a B anthracis culture grown to maximize the yield of protective antigen and minimize the quantities of lethal factor, edema factor, and other unwanted metabolites.

Bacillus anthracis is susceptible to penicillin and to almost all other broad-spectrum antibiotics. Because it is easily recognized, cutaneous anthrax is almost always treated early and cured. Gastrointestinal and pulmonary anthrax infections are difficult to identify before the fulminant phase and therefore carry a high mortality. In uncomplicated anthrax cases, adequate treatment consists of 500 mg of penicillin V taken orally every 6 hours for 5 days, or 600 mg (1 million units) of procaine penicillin administered intramuscularly every 12 to 24 hours for 5 days. In severe cases, 1,200 mg (2 million units) of penicillin G should be administered intravenously every 6 hours, reverting to the intramuscular regime of 600 mg every 12 to 24 hours once recovery starts. If pulmonary anthrax is suspected, continuous-drip administration is advisable. Tetracyclines (tests in animals indicate doxycycline is good), chloramphenicol, gentamicin, or erythromycin may be used if the patient has penicillin hypersensitivity. The fluoroquinolone, ciprofloxacin, has also been shown to be effective in monkeys and guinea pigs and would be expected to be effective in treatment of cases of human anthrax.

Avoidance of other types of Bacillus infections is largely a matter of observing proper hygiene. Bacillus cereus and its close relatives B thuringiensis and B mycoides produce potent β-lactamases and thus are not responsive to penicillin, ampicillin, or the cephalosporins. They are mostly resistant to trimethoprim as well. These species are generally sensitive to standard empirical treatment with an aminoglycoside combined with vancomycin and to chloramphenicol, erythromycin, tetracycline, clindamycin, and sulfonamides.

Bacillus food poisoning, like all types of food poisoning, can largely be prevented by proper food handling. Food should be cooked adequately; cooked food should not be recontaminated from uncooked food (separate utensils and cutting surfaces should be used for cooked and uncooked food); and, of particular importance, cooked food should be stored under proper refrigeration.

## Chapter 18

# Miscellaneous Pathogenic Bacteria

The microorganisms discussed in this chapter are taxonomically unrelated. The human infections they cause are rare except in the case of Listeria and Propionibacterium acnes. Some of these infections are fatal; others tend to be self-limited. Recognition depends largely on the proper use of bacteriological methods, which is important not only to ensure appropriate therapy but also to exclude other possible agents.

## *LISTERIA MONOCYTOGENES*

### Clinical Manifestations

Listeriosis is a serious disease for humans, with a mortality greater than 25 per cent. There are two main clinical manifestations, sepsis and meningitis. Meningitis is often complicated by encephalitis, which is exceptional among bacterial infections. Occasionally, pyogenic infections of various organs have been found. Relapses may occur after apparent recovery.

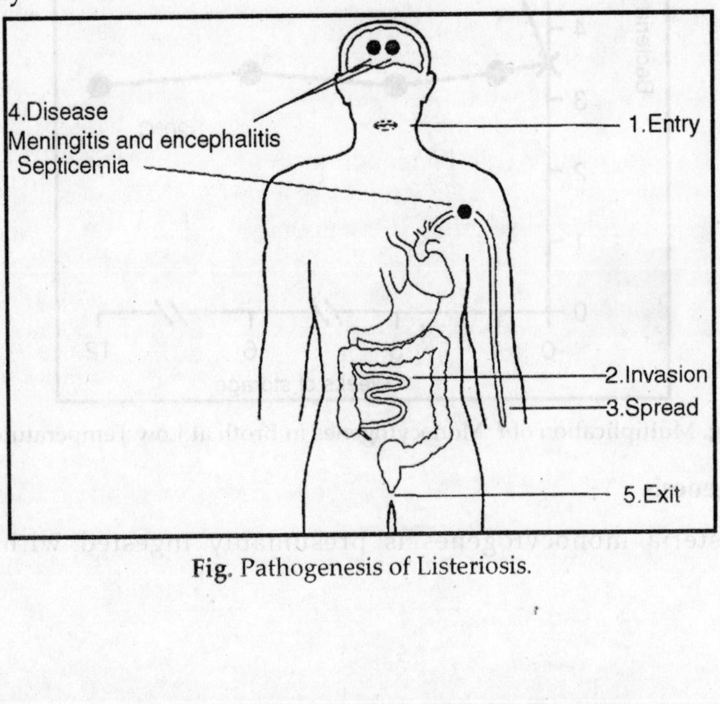


Fig. Pathogenesis of Listeriosis.

## Structure, Classification, and Antigenic Types

All Listeria species are small, Gram-positive rods, which are sometimes arranged in short chains. In direct smears they may be coccoid, so they can be mistaken for streptococci. Longer cells can be suggestive of corynebacteria. Flagella are produced at room temperature rather than at 37° C. Hemolysin production is an important marker for L monocytogenes, although it is not definitive, as L ivanovii and L seeligeri are likewise hemolytic on blood agar. Further biochemical characterization is necessary to distinguish between the different Listeria species.

It may be desirable for epidemiologic purposes to identify a particular strain by serotyping to characterize surface antigens, such as O antigens (teichoic acids) and H antigens (proteins). The serovars 1/2a and 4b are responsible for up to 90 per cent of all cases of listeriosis.

A particular property of L monocytogenes is the ability to multiply at low temperatures. Bacteria therefore can accumulate in contaminated food stored in the refrigerator.

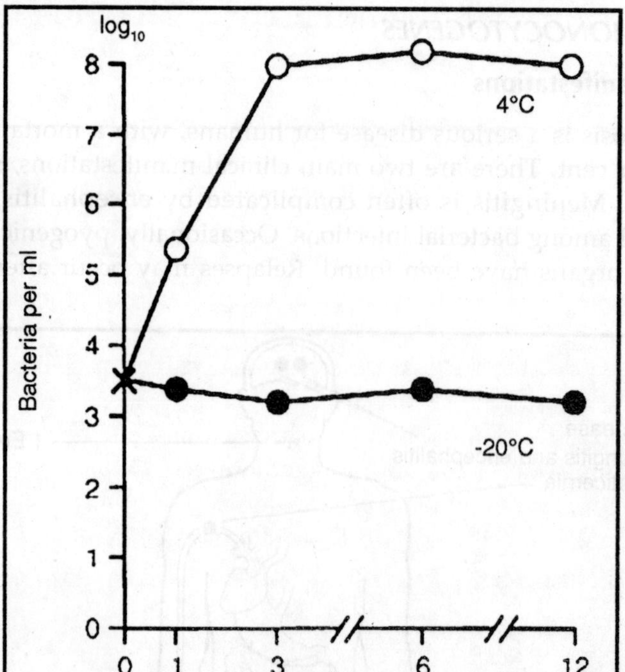


**Fig.** Multiplication of L Monocytogenes in Broth at Low Temperature.

## Pathogenesis

Listeria monocytogenes is presumably ingested with raw,

contaminated food. An invasion factor secreted by the pathogenic bacteria enables them to penetrate host cells of the epithelial lining. Since this microorganism is widely distributed, this event may occur rather often. Normally, the immune system eliminates the infection before it spreads.

Indeed, most adults who have no history of listeriosis have T lymphocytes primed specifically by Listeria antigens. If the immune system is compromised, however, systemic disease may develop. Listeria monocytogenes multiplies not only extracellularly but also intracellularly within macrophages after phagocytosis and even within parenchymal cells which are entered by induced phagocytosis. It therefore belongs to the large group of facultatively intracellular pathogens.

**Table. List of Some Facultative or Obligate Intracellular Microorganisms.**

| *Type of Microorganism* | *Agent* | *Disease* |
|---|---|---|
| Bacteria | Mycobacterium tuberculosis | Tuberculosis |
| | Salmonella typhi | Typhoid fever |
| | Yersinia postis | Plague |
| | Legionella pneumophila | Legionellosis |
| | Listeria monocytogenes | Listeriosis |
| Fungi | Histoplasma capsulatum | Histoplasmosis |
| Protozoa | Toxoplasma gondil | Kala azar |
| | Trypanosoma cruzi | Chagas disease |

Survival within the phagosomes and eventual escape into the cytoplasm are mediated by a toxin, which also acts as a hemolysin. This toxin is one of the so-called SH-activated hemolysins, which are produced by a number of different bacteria such as serogroup A streptococci, pneumococci, and Clostridium perfringens. Obviously, nature has preserved the genetic code for this bacterial product in several species, and consequently the hemolysins from these different bacteria have common biochemical, biologic, and antigenic properties. Nonhemolytic variants of L monocytogenes are completely avirulent, as are the nonhemolytic species L innocua and L welshimeri. Hemolysin is not the only Listeria virulence factor, however, since the hemolytic Listeria species besides L monocytogenes (i.e., L seeligeri and L ivanovii) possess rather limited pathogenicity. The hemolysin gene is located on the chromosome within a cluster of other virulence genes which are all regulated by a common promotor. These additional genetic determinants are necessary for further steps in the intracellular life cycle of L monocytogenes. One particular gene product promotes the polymerization of actin, a component of the host cell cytoskeleton, on the bacterial surface. In this peculiar environment within host cells, surrounded by a sheet of actin filaments, the bacteria reside and even multiply. The growing actin sheet functions as a propulsive force which drives the bacterium across the intracellular pathways until it finally reaches the surface. Then, the host cell is urged to form slim, long protrusions containing living *L. monocytogenes*. Those cellular projections are engulfed by adjacent cells, even by non-professional phagocytes such as parenchymal cells. By such a mechanism a direct cell-to-cell spread in an infected organ may occur without an extracellular stage.

## Host Defenses

Because it multiplies intracellularly, *L monocytogenes* is largely protected against humoral immune factors such as antibodies, and the effective host response is cell- mediated, involving both lymphokines (especially interferon) produced by CD4+ (T-helper) cells and direct lysis of infected cells by CD8+ (cytotoxic) T lymphocytes.

Both of these fundamental defense mechanisms are expressed in the microenvironment of the infective foci. Histologically, these foci are organized as granulomas, characterized by a central accumulation of epitheliod cells (macrophages) with irregularly shaped nuclei and large, delicately structured cytoplasm and by peripheral lymphocytes recognizable by a round nucleus and a narrow border of intensely staining cytoplasm.

## Epidemiology

Listeria species are found in living and nonliving matter. Various foodstuffs of vegetable and animal origin are sources of infection. Animal and human carriers also have been described. Most human cases of listeriosis develop in immunocompromised hosts: newborns, old people, cancer patients, and transplant recipients. Reports of sporadic cases of listeriosis are becoming more frequent as the number of persons at risk, especially because of immunosuppression by medical therapy, increases. Outbreaks of listeriosis are due mainly to a common source of contaminated food.

Listeriosis also may be transmitted congenitally across the placenta. The immunocompetent mother suffers at worst a brief, flu-like febrile illness, but the fetus, whose defense system is still immature, becomes seriously ill. Depending on the stage of gestation, the fetus is either stillborn or born with signs of congenital infection. Typically, multiple pyogenic foci are found in several organs (granulomatosis infantiseptica). The onset of listeriosis is delayed (i.e., a few days after birth) when infection is acquired during labour by bacteria colonizing the genital tract of the mother.

## Diagnosis

*Listeria monocytogenes* is implicated when monocytosis is observed in the peripheral blood as well as the cerebrospinal fluid. Early diagnosis may be obtained by finding pleocytosis with Gram-positive rods in a Gram stain of smears of the cerebrospinal fluid. Final proof is obtained by culture. Serologic tests are highly unreliable.

## Control

Hygienic food processing and storage may reduce the risk of listeriosis. Individuals in high-risk groups (i.e., immunocompromised individuals and pregnant women) should avoid uncooked food or should at least marinate salads for a long time in a vinegar-based dressing to kill adherent bacteria.

Since a cell-mediated immune response (the most potent weapon against *L monocytogenes*) is induced only by injection of living antigen, vaccination is difficult. Even an attenuated living vaccine is dangerous for persons with impaired defenses, the proper target group. Completely avirulent live bacteria do not trigger an effective, cell-mediated immune response.

Antimicrobial agents are the mainstay of treatment. Most of the common antibiotics, except cephalosporins, are active against *L monocytogenes* in vitro. In practice, ampicillin combined with an aminoglycoside has given the best results. However, because infection occurs mainly in infirm patients and because intracellular bacteria are hardly accessible to most drugs, the cure rate is low. Furthermore, Listeria cells, although inhibited, are not killed by ampicillin. High doses for prolonged periods are indicated.

# *ERYSIPELOTHRIX RHUSIOPATHIAE*

## Clinical Manifestations

The most common human infection by *E rhusiopathiae* is erysipeloid, a well-defined, violet or wine-colored inflammatory lesion of the skin of the fingers or hand. Itching is typical. Infrequently, septicemia develops, followed by various organ manifestations such as endocarditis or arthritis without fever.

## Structure and Classification

*Erysipelothrix rhusiopathiae* is a slender, Gram-positive rod similar to L monocytogenes. In general, E rhusiopathiae rods are longer, especially in rough variants. They grow on routine culture media under aerobic conditions, but preferentially in a $CO_2$ atmosphere. In contrast to *L monocytogenes,* they are nonmotile, nonhemolytic, and catalase negative. The production of H2S is highly indicative, since very few other Gram-positive bacteria have this property.

### Pathogenesis

A minor skin injury may facilitate the penetration of E rhusiopathiae after contact with infected material. After an incubation of 1 to 4 days the local lesion develops; spontaneous recovery occurs in 2 to 3 weeks. Septicemia has been observed without previous local lesions so that an oral infection is assumed. Endocarditis may develop in a few cases.

### Epidemiology

*Erysipelothrix rhusiopathiae* is found in mammals, poultry, and fish. Individuals who have occupational exposure to such animals (i.e., farmers, veterinarians, slaughterhouse workers, and fish handlers) are at risk.

### Diagnosis

The typical, nonpyogenic lesions on occupationally exposed persons suggest erysipeloid. Since there is no wound, a swab is not useful. Bacteria can be cultured from a biopsy of the progressing, inflamed edge of the lesion. Blood culture is indicated in the setting of sepsis and endocarditis.

### Control

Penicillin is the drug of choice to treat serious infections. Since local skin infection is self-limited, therapy is not essential.

## *PROPIONIBACTERIUM ACNES*

### Clinical Manifestations

The pathogenic role of *Propionibacterium* acnes is still disputed. Although it is often detected in anaerobic blood cultures, it normally colonizes the skin crypts and is transported to cultures by pure chance. Nevertheless, in compromised patients even this nonpathogenic species may induce pathologic reactions, such as endocarditis. In skin lesions P acnes is often found with other pathogenic bacteria, such as *Staphylococcus aureus* or *actinomycetes,* and is thought to support the damaging effect of those pathogens. It is doubtful whether P acnes alone is able to induce acne.

### Structure and Classification

The club-shaped, Gram-positive rods of P acnes resemble the diphtheroids but, unlike the latter, are slow-growing and anaerobic, so that their presence in blood cultures is detected after about a week.

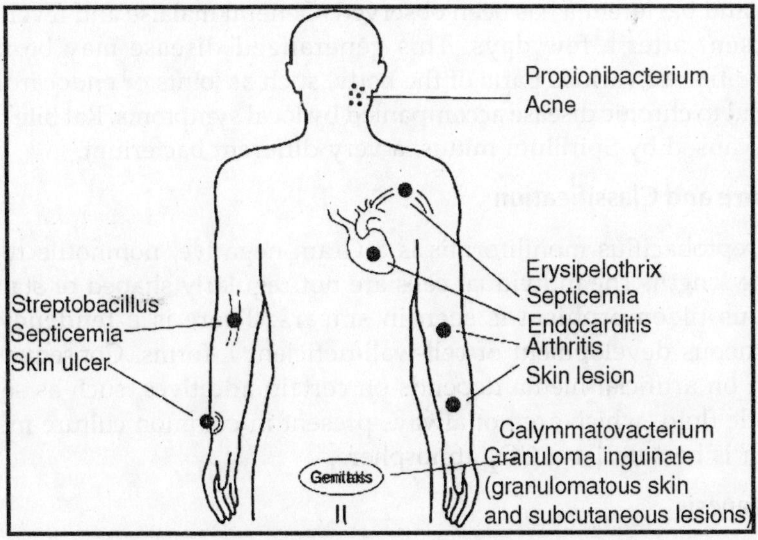


**Fig.** Disease Manifestations of *E rhusiopathiae, P Acnes, C Granulomatis,* and *S moniliformis.*

## Pathogenesis

Propionibacterium acnes produces several metabolic products, hemolysin, and various enzymes such as lipase and neuraminidase, which are excreted into the surroundings. These metabolites may clear the way for other bacteria. Furthermore, *P acnes* degrades sebaceous matter to produce fatty acids that stimulate an inflammatory reaction.

## Epidemiology

Propionibacterium acnes normally colonizes the deep crypts of the skin, where the availability of oxygen is reduced. The same applies to mucous membranes of the oroanal areas. They may be transported to other sites by chance.

## Control

Practically all common antibiotics, including penicillins, erythromycin, and tetracyclines, can be used to treat P acnes infections.

## *STREPTOBACILLUS MONILIFORMIS*

## Clinical Manifestations

*Streptobacillus moniliformis* causes the clinical disease called rat bite fever. At the site of the rodent bite, an ulcer appears; this may heal spontaneously. Occasionally, the infection spreads to the regional lymph nodes, and bacteremia has been observed. General malaise and fever may be present after a few days. This generalized disease may be fatal. Colonization of various parts of the body, such as joints or endocardium, may lead to chronic disease accompanied by local symptoms. Rat bite fever also is caused by Spirillum minus, a very different bacterium.

### Structure and Classification

Streptobacillus moniliformis is a Gram-negative, nonmotile rod of variable length. The individual cells are not regularly shaped or stained, and thus pleomorphism is seen in smears. There is a tendency for spontaneous development of cell-wall-deficient L-forms. Consequently, growth on artificial media depends on certain additives, such as serum or ascitic fluid, which are not always present in common culture media. Growth is best under a $CO_2$ atmosphere.

### Pathogenesis

Humans usually become infected with *S moniliformis* through the bite of an infected rat. Ingestion of contaminated food has rarely been incriminated as the source of infection.

### Host Defenses

The nonspecific resistance mechanisms in the skin and draining lymph nodes prevent dissemination. The low pH in the stomach normally guarantees that *S moniliformis* cannot survive gastric passage.

### Epidemiology

*S moniliformis* belongs to the common bacterial flora of the nasopharynx of rats, from which it reaches humans directly by a rat bite or indirectly via food.

### Diagnosis

The coincidence of fever after a rat bite draws attention to this infection. Positive cultures can be obtained from blood or synovial fluid. Mice are highly susceptible to S moniliformis, exhibiting a rapid lethal infection after inoculation. Because of the existence of L-forms, special media must be used for culture, since on conventional agar plates L-form colonies hardly are visible and are likely to be overlooked.

## *CALYMMATOBACTERIUM GRANULOMATIS*

### Clinical Presentation

*C granulomatis* causes granuloma inguinale. This infection typically is localized in the genital region. It spreads to adjacent areas, and the regional lymph nodes also may be inflamed. Persistent granulomatous lesions tend to ulcerate, destroying skin and subcutaneous tissue.

### Structure and Classification

*Calymmatobacterium granulomatis* is a Gram-negative, nonmotile rod. The capsule that surrounds the bacterial cell appears similar to that of Klebsiella. Addition of egg yolk and incubation in a $CO_2$ atmosphere are required for growth on artificial media.

### Pathogenesis

Calymmatobacterium granulomatis is normally present in the gut flora and may be transmitted to the genital area by autoinoculation or sexual contact. After penetrating the skin the bacteria induce an inflammatory reaction, which may lead to destruction of the infected tissue. Within the inflammatory foci C *granulomatis* is found mainly intracellularly inside tissue macrophages (Donovan bodies). This is highly typical for granuloma inguinale. Superinfection of ulcers with other pathogenic organisms is possible.

### Host Defenses

Antibodies against C *granulomatis* are produced during acute infection; their role in defense remains unclear. Cell-mediated defense mechanisms, expressed by a granulomatous reaction, are important in recovery.

### Epidemiology

Granuloma inguinale occurs most frequently in people living under poor socioeconomic conditions (e.g., in the tropics). In the United States, infection of blacks is seven times more frequent than infection of whites. Transmission by sexual contacts is most common. Other sexually transmitted diseases, such as syphilis, may be associated.

### Diagnosis

Microscopic evidence of intracellular Gram-negative encapsulated rods in ulcerative skin wounds of the genitoinguinal region is highly indicative for granuloma inguinale. Since experience with C *granulomatis* is lacking in most laboratories, cultural diagnosis probably will fail. Eventually, the yolk sac of 5-day-old chicken embryos can be inoculated directly with the infected material.

### Control

Infection can be prevented by cleanliness or by avoiding sexual contacts with infected persons. Antibiotics active against intracellular bacteria, such as tetracycline or erythromycin, are effective in treatment.

# Index


**A**

Adherence 75, 96, 116, 162, 255, 268, 289, 301
Altered Receptors 191
Anaerobes 35, 62, 63, 118, 216, 218, 240, 256
Analytic 158, 160
Anerobic 65
Anthrax 10, 146, 153, 284, 285, 286, 287, 289, 290, 291, 292, 294, 295, 296
Antibacterial 146, 177, 181, 185, 187, 203
Antibiotic 1, 2, 19, 79, 81, 86, 97, 98, 113, 130, 155, 157, 164, 176, 181, 188, 199, 201, 205, 206, 224, 236, 239, 245, 248, 251, 254, 256, 257, 258, 259, 260, 272, 274
Antibiotics 1, 2, 14, 18, 24, 30, 32, 76, 97, 99, 103, 120, 126, 127, 144, 168, 175, 183, 184, 185, 187, 190, 191, 192, 193, 203, 205, 207, 208, 219, 225, 237, 238, 239, 247, 256, 257, 259, 260, 282, 284, 291, 293, 298, 301
Antibody 35, 124, 128, 134, 143, 144, 145, 146, 158, 168, 169, 170, 174, 175, 230, 247, 249, 254, 255, 260, 268, 271, 276
Antigenic 8, 15, 83, 86, 136, 155, 158, 169, 173, 233, 243, 246, 247, 255, 260, 264, 265, 266, 274, 295
Antigens 2, 7, 17, 19, 30, 78, 86, 96, 124, 134, 139, 143, 144, 145, 146, 149, 157, 169, 170, 172, 173, 242, 243, 244, 245, 252, 260, 266, 268, 269, 274, 276, 278, 294, 295
Antimicrobial 1, 3, 34, 68, 69, 120, 127, 166, 172, 173, 177, 178, 181, 185, 189, 191, 192, 207, 208, 239, 272, 273, 274, 278
Antimicrobial drugs 3, 164, 237
Appendages 6, 8, 76, 263
Aureus 5, 12, 13, 14, 16, 22, 30, 66, 75, 99, 116, 129, 132, 155, 169, 170, 191, 192, 193, 194, 197, 200, 204, 224, 225, 226, 227, 228, 229, 230, 231, 232, 233, 234, 235, 236, 237, 238, 244, 259, 298
Autotrophy 63

**B**

Bacilli 6, 23, 116, 118, 136, 156, 168, 171, 173, 182, 289, 290, 291
Bacteria 1, 2, 3, 4, 5, 6, 7, 8, 9, 10, 11, 12, 13, 14, 15, 16, 17, 18, 19, 20, 21, 22, 25, 26, 30, 35, 36, 39, 41, 42, 43, 46, 59, 60, 61, 62, 63, 64, 65, 66, 67, 68, 79, 80, 81, 82, 86, 87, 88, 89, 90, 92, 124, 125, 126, 127, 128, 130, 131, 174, 175, 179, 180, 181, 182, 183, 268, 276, 281, 284, 288, 290, 295, 296, 297, 298, 299, 301
Bacterial cell 5, 6, 8, 9, 10, 20, 21, 29, 41, 74, 77, 85, 104, 179, 186, 190, 200, 226, 234, 301
Bacterial Infectivity 126
Bacterial resistance 190, 195, 203, 205

Bacteriophages 8, 26, 30, 68, 69, 70, 73, 74, 77, 78, 88, 89, 128, 234, 244, 257
Biochemical 7, 8, 20, 21, 22, 23, 24, 28, 29, 30, 31, 32, 33, 34, 35, 37, 41, 42, 44, 45, 47, 69, 84, 85, 88, 93, 102, 103, 173, 207, 215, 225, 241, 242, 252, 271, 279, 294, 295
Biologic 3, 17, 22, 39, 43, 51, 60, 64, 67, 83, 86, 102, 135, 136, 138, 139, 156, 158, 240, 295
Biosafety Levels 39

C

Calymmatobacterium 300, 301
Capsular 10, 20, 30, 134, 144, 145, 148, 226, 238, 243, 244, 254, 274, 276, 278
Cell wall 8, 9, 10, 11, 13, 15, 16, 19, 21, 35, 120, 146, 178, 179, 180, 181, 182, 183, 190, 192, 194, 198, 199, 204, 205, 236, 239, 242, 243, 246, 257, 259, 285, 288
Cell-mediated 147, 249, 301
Chemical hazards 38
Chromosomal 71, 74, 88, 91, 92, 93, 94, 97, 128, 169, 190, 193, 194, 201, 236, 237, 272
Clinical laboratory 22, 23, 34, 35, 37, 38, 152, 169
Coagulase 30, 174, 191, 224, 225, 226, 234, 235, 259
Colonization 8, 75, 116, 118, 124, 127, 129, 146, 230, 235, 238, 251, 257, 259, 266, 268, 276, 277
Combatting 238
Conjugation 8, 69, 86, 87, 90, 91, 92, 97, 98, 257
Conjunctival flora 120
Coryneforms 117
Curative 147, 150
Cytoplasmic 8, 19, 20, 21, 57, 60, 87, 90, 91, 111, 178, 179, 181, 183, 193, 195, 200, 244

D

Descriptive 151, 158, 159, 160, 209, 218
Detection 33, 139, 169, 170, 172, 173, 174, 175, 254, 256, 259
Diagnosis 3, 5, 35, 37, 102, 162, 164, 165, 166, 167, 172, 174, 251, 255, 260, 263, 270, 271, 277, 289, 297, 301
Diphtheroids 116, 117, 118, 120, 298
Disease-acquired 147

E

Eikenella 261, 280
Electron transport 44, 55, 57, 58, 60, 65, 187, 219
Endothelial cells 230
Endotoxin 135, 136, 137, 138, 139, 246, 261
Enzymes 10, 13, 18, 19, 25, 30, 32, 51, 57, 61, 73, 85, 93, 99, 102, 103, 108, 117, 128, 130, 131, 135, 141, 144, 156, 178, 180, 181, 187, 192, 194, 195, 197, 199, 200, 224, 237, 252, 259, 266, 267, 281, 285, 288, 299
Epidemic 22, 37, 148, 151, 152, 153, 159, 160, 161, 277
Epidemiologic 30, 36, 37, 151, 157, 158, 159, 160, 161, 162, 163, 201, 239, 243, 250, 255, 264, 294
Epidemiology 3, 151, 158, 159, 160, 162, 226
Epidermidis 12, 116, 120, 194, 197, 225, 226, 227, 229, 232, 235, 236
Epidermolytic 233
Epilogue 142
Erysipelothrix 297, 298
Exfoliative 75, 233
Exotoxins 17, 135, 139, 244
Extracellular 9, 68, 76, 77, 88, 93, 111, 128, 132, 133, 145, 224, 227, 228, 234, 239, 244, 245, 246, 255, 267, 270, 296

Phenotypes 69, 84, 85, 87, 98, 101, 266
Phenotypic 22, 25, 28, 31, 32, 68, 78, 83, 85, 86, 88, 103, 219, 226, 242
Phosphorylation 19, 44, 49, 54, 55, 57, 200, 209
Photosynthesis 43, 61
Phylogenetic 26, 28, 31, 32, 220, 223
Pili 6, 7, 8, 90, 91, 93, 133, 261, 264, 268, 275
Plasma 8, 9, 10, 12, 18, 19, 20, 127, 128, 141, 144, 145, 215, 224, 234
Plasmids 22, 28, 68, 69, 74, 75, 76, 86, 88, 90, 91, 92, 93, 97, 98, 99, 100, 101, 128, 149, 155, 157, 169, 191, 192, 193, 194, 201, 206, 226, 236, 257, 272, 287
Pneumoniae 10, 88, 89, 99, 167, 191, 193, 194, 200, 204, 239, 240, 241, 242, 243, 244, 245, 246, 247, 249, 250, 252, 254, 255, 256, 257, 258, 259, 260, 272
Polymerase 2, 102, 103, 104, 106, 107, 108, 109, 170, 179, 185, 192
Polyphasic 28
Polysaccharide 1, 16, 17, 18, 20, 135, 136, 137, 144, 145, 146, 148, 149, 179, 213, 215, 226, 230, 235, 238, 243, 244, 258, 260, 274, 278, 288
Postinfectious 247
Preventive 147
Propionibacterium 53, 117, 293, 298, 299
Protective antigens 143, 144, 146, 149
Public health 23, 30, 31, 37, 102, 124, 152, 153, 162, 290, 291
Pyogenes 78, 128, 129, 132, 134, 167, 194, 200, 239, 240, 242, 243, 244, 245, 246, 247, 249, 250, 251, 252, 254, 255, 256, 260

## R

Recombinant 2, 69, 87, 88, 89, 92, 93, 98, 100, 102, 149, 229
Recombination 86, 87, 89, 90, 91, 92, 93, 94, 95, 97
Regulons 110, 111
Relatedness 22, 24, 25, 26, 28, 29, 31, 75, 78, 220, 226
Replication 5, 41, 69, 71, 72, 73, 74, 75, 76, 79, 97, 102, 153, 169, 179, 185, 186
Resistance 19, 21, 30, 68, 75, 80, 86, 96, 97, 98, 100, 102, 123, 124, 125, 126, 130, 133, 134, 136, 137, 142, 143, 147, 155
Respiration 43, 45, 54, 55, 57, 58, 59, 61, 65, 208, 210, 212, 218, 219
Reversion 83, 85
Rhusiopathiae 297, 298
Ribosome 104, 108, 109, 110, 178, 187, 188, 189, 195, 202, 203, 237

## S

Saureus 227, 228, 231, 232
Sensitivity 31, 36, 103, 139, 155, 168, 169, 170, 252, 254, 256, 270, 271, 289, 291
Sequelae 138, 240, 246, 247, 250, 255
Serodiagnosis 174
Serologic Titers 255
Siderophores 128, 141, 142
Solid media 170, 225
Specificity 7, 8, 16, 17, 36, 78, 85, 86, 95, 116, 143, 155, 157, 168, 185, 233, 243, 254, 267, 270, 271
Specimen 22, 24, 34, 36, 37, 120, 166, 168, 169, 170, 171, 173, 174, 225, 270, 290
Spontaneous 1, 79, 82, 92, 298, 300
Staphylococci 15, 30, 40, 116, 118, 120, 174, 183, 191, 192, 195, 197, 200, 224, 225, 226, 236, 244, 254, 259
Staphylococcus 5, 13, 22, 30, 66, 75, 88, 99, 116, 129, 132, 169, 191, 192, 193, 224, 226, 229, 240, 298
Staphylokinase 232, 234
Streptobacillus 299, 300
Streptococcal 123, 151, 239, 240, 244, 245, 246, 247, 248, 250, 251, 252,

254, 255, 256, 257, 260, 272
Streptococcus 5, 9, 10, 51, 52, 78, 88, 93, 128, 129, 132, 134, 145, 155, 167, 191, 193, 224, 239, 245, 248, 272
Suppression 83, 86
Supraoptimal 26, 28
Susceptibility 2, 22, 30, 34, 35, 38, 69, 78, 85, 124, 152, 175, 246, 252, 256, 268, 288

**T**

Taxonomy 3, 23, 28, 239, 241
Teichoic acids 11, 15, 19, 294
Thermal stability 26, 27
Toxicology 205
Tract Flora 118
Transcription initiation 69, 105
Transcription termination 100, 109
Transcriptional Units 104
Transducing Proteins 110
Transduction 89
Translation 68, 69, 72, 86, 108, 109, 110
Transmission 34, 37, 147, 153, 154, 155, 156, 158, 159, 163, 250, 258, 262, 270, 271, 275
Transposons 83, 90, 95, 96, 97, 98, 128, 192, 194, 200, 201, 236, 257

**U**

Urogenital Flora 120

**V**

Vaccination 143, 144, 147, 278, 297
Vaccines 1, 2, 69, 102, 142, 147, 148, 149, 224, 240, 259, 260, 271, 278, 281, 287, 288, 291
Viridans 116, 239, 241, 242, 245, 249, 253
Virulence 2, 9, 30, 36, 38, 39, 68, 78, 96, 104, 111, 123, 124, 126, 127, 128, 130, 133, 134, 142, 154, 164, 169, 224, 227, 228, 229, 230, 231, 232, 234, 235, 236, 243, 244, 246, 249, 261, 266, 267, 274, 285, 287, 288, 295

**W**

Wall polymers 11, 16